An Introduction to Global Spectral Modeling

Second Revised and Enlarged Edition

ATMOSPHERIC AND OCEANOGRAPHIC SCIENCES LIBRARY

VOLUME 35

The titles published in this series are listed at the end of this volume.

An Introduction to Global Spectral Modeling

Second Revised and Enlarged Edition

T.N. Krishnamurti

Department of Meteorology
Florida State University
Tallahassee, FL, USA

H.S. Bedi

Delhi, India

V.M. Hardiker

Houston, TX, USA

L. Ramaswamy

Department of Meteorology
Florida State University
Tallahassee, FL, USA

Springer

T.N. Krishnamurti
Department of Meteorology
Florida State University
Tallahassee, FL 32306
USA

H.S. Bedi
c/o Vinny Bedi
84 Mausam Bhawan
Delhi 11051
India

V.M. Hardiker
Bell South
Houston, TX 75063
USA

L. Ramaswamy
Department of Meteorology
Florida State University
Tallahassee, FL 32306
USA

ISBN 978-1-4419-2137-6 e-ISBN 978-0-387-32962-8

Printed on acid-free paper.

9 8 7 6 5 4 3 2 1

springer.com

Preface

This is an introductory textbook on global spectral modeling designed for senior-level undergraduates and possibly for first-year graduate students. This text starts with an introduction to elementary finite-difference methods and moves on towards the gradual description of sophisticated dynamical and physical models in spherical coordinates. Computational aspects of the spectral transform method, the planetary boundary layer physics, the physics of precipitation processes in large-scale models, the radiative transfer including effects of diagnostic clouds and diurnal cycle, the surface energy balance over land and ocean, and the treatment of mountains are some issues that are addressed. The topic of model initialization includes the treatment of normal modes and physical processes. A concluding chapter covers the spectral energetics as a diagnostic tool for model evaluation.

This revised second edition of the text also includes three additional chapters. Chapter 11 deals with the formulation of a regional spectral model for mesoscale modeling which uses a double Fourier expansion of data and model equations for its transform. Chapter 12 deals with ensemble modeling. This is a new and important area for numerical weather and climate prediction. Finally, yet another new area that has to do with adaptive observational strategies is included as Chapter 13. It foretells where data deficiencies may reside in model from an exploratory ensemble run of experiments and the spread of such forecasts.

These classroom lectures emerged from discussions with a large number of former colleagues that include: Masao Kanamitsu, Richard Pasch, Hua-Lu Pan, Steve Cocke, Chia Bo Chang, John Molinari, Naomi Surgi, Lahouari Bounoua, Fred Carr, Simon Low-Nam, Takeo Kitade, Masato Sugi, Mukut Mathur, and Jishan Xue. Many others who were part of Krish's laboratory at the Florida State University also contributed in many ways towards the material presented here. In addition, we owe a great deal to the United States funding agencies who supported our research in this area: NSF (Pamela Stephens), ONR (Scott Sandgathe), and NOAA (Kenneth Mooney).

Contents

Chapter 1

Introduction

The spectral modeling approach to *numerical weather prediction* is being practiced in many parts of the world. Historically, the spectral approach came into atmospheric sciences from studies of *geomagnetism*, where it was introduced by Elsasser in the late nineteenth century. The first attempt of spectral representation of data sets and its use via simple vorticity conserving models came in the 1940s. The works of Neamtan (1946) from the University of Manitoba and Haurwitz (1940) and Craig (1945) from New York University were pioneering during this era. These were still close to linear problems, lacking any formalism for addressing the nonlinear advective dynamics.

It was in the late 1950s when we saw the emergence of formal proposals for the solution of the *nonlinear barotropic vorticity equation*. Pioneering work from the University Chicago (Platzman 1960; Baer 1964) explored what is sometimes called the *interaction coefficients approach* for the nonlinear problem. Around the same time at MIT we saw the elucidation of what are now called the *low-order systems*. This pioneering work of Lorenz (1960b) and Saltzman (1959) brought to us the first exposure to simple nonlinear systems and the concept of chaos. These simple three-component systems demonstrated some of the essentials of nonlinear dynamics and the growth of errors arising from initial state uncertainties.

The interaction coefficients approach to the solution of the *weather forecast problem* led to unmanageably large memory requirements that were not easily amenable to the then available, or even to the present, memory of computers. It was during the mid-1950s when the *Cooley-Tukey algorithm* (1965) emerged and provided a break though via the *fast Fourier transform*. This was exploited and demonstrated to provide accurate representations of the quadratic terms for fast computation of the nonlinear advective dynamics by contributions from Eliasen et al. (1970) and Machenhauer (1974) from the University of Copenhagen and from Orszag (1970) at MIT.

Thereafter, we saw a rapid development of global spectral models in many parts of the world, especially Australia, Canada, England, Japan, and the United States. Noteworthy contributions on the multilevel framework came from Machenhauer and Daley (1972) in Copenhagen, Bourke (1974) from the Bureau of Meteorology Research Center in Melbourne, and Robert (1966), Daley et al. (1976), and Merilees (1968) from the research Provision Numerique in Montreal. Numerous others have contributed to these developments. Currently there are as many as 30 global modeling groups that are active in different parts of the world.

There are several components to the global spectral modeling of the *weather prediction* problem. In this introductory text, we only provide an exposure to this approach. Knowledge of simple finite differences in the space and time domains are still useful ingredients in the overall construction of spectral models.

Chapters 2 and 3 provide an introduction to *finite-differencing* and *time-differencing* procedures. The definition of a spectral model is provided in Chapter 4, where we introduce the concept of the *Galerkin techniques*. Chapter 5 addresses the *Lorenz-type low-order systems* with an introduction to chaotic systems. The use of spherical harmonics as basis functions for the casting of meteorological equations in the spectral space is provided in Chapter 6. This chapter also provides the recurrence relations for the accurate computation of *associated Legendre functions* and their derivatives using the *Gaussian quadrature*. Spectral relationships for the kinematics of the atmospheric variables are also provided in this chapter. Using the above principles, the construction of simple *single-level barotropic* and *shallow-water spectral models* is also presented here. This entails the use of the semi-implicit algorithm and solutions of *Helmholtz-type equations*. The use of *Fourier-Legendre transforms* and *inverse transforms* is an integral part of these models, and this is brought out throughout the chapter.

The multilevel spectral weather prediction model is elaborated on in Chapter 7, where the use of a vertical coordinate system following the earth's surface forms the basis for the definition of variables in the vertical. The vorticity and divergence equations replace the conventional momentum equations. The closed system includes conservation laws for momentum, mass, moisture, and heat. The Fourier-Legendre transform of this basic closed system of equations leads to a coupled system of nonlinear of ordinary differential equations.

A reasonably sized weather prediction model carries 100 waves in the zonal and meridional directions over some 20 vertical layers of the atmosphere. Computationally, this amounts to close to 1.5 million coupled ordinary differential equations. These equations can be solved by various methods, such as those described in Chapter 7.

These model equations include a number of physical processes, such as the effects of cumulus convection, nonconvective precipitation, surface fluxes of heat, moisture, and momentum, the planetary boundary layer, land-surface processes, radiative transfer, cloud radiative interactions, diurnal change, surface energy balance, effects of orography, and the effects of oceans, snow cover, and ice cover. This is a rather comprehensive list. A brief treatment of the physical processes is provided in Chapter 8.

Data initialization issues are addressed in Chapter 9. Here, the emphasis is on two currently popular themes: one called *normal mode initialization* and the other called *physical initialization*. The former deals with the suppression of gravity-inertia oscillations arising from initial data imbalances in the mass and motion fields. The latter discusses the issue of improving the rain rates of the model's initial state.

Spectral energetics is a topic that deals with model output diagnostics. Kinetic and potential energy are exchanged among zonal flows and different wave components and are also mutually exchanged among the different waves. Saltzman (1957) laid the foundation for these types of inquiries in the late 1950s and carried out such studies to completion in the late 1970s. The theoretical basis for these formulations is presented in Chapter 10.

In Chapter 11, the workings of a limited area regional spectral model are provided. Chapter 12 deals with a new area in weather and climate forecasting. This includes a description of the multimodel superensemble that appears to carry higher forecast skills compared to member models of a suite. Chapter 13 addresses a new and upcoming area for forecast modeling called 'adaptive observational strategies'. This design addresses finding regions where observations are needed for improving the skill of a specific forecast.

Chapter 2

An Introduction to Finite Differencing

2.1 Introduction

This chapter on finite differencing appears oddly placed in the early part of a text on spectral modeling. Finite differences are still traditionally used for vertical differencing and for time differencing. Therefore, we feel that an introduction to finite-differencing methods is quite useful. Furthermore, the student reading this chapter has the opportunity to compare these methods with the spectral method which will be developed in later chapters.

One may use *Taylor's expansion* of a given function about a single point to approximate the derivative(s) at that point (Fig. 2.1). Derivatives in the equation involving a function are replaced by finite difference approximations. The values of the function are known at discrete points in both space and time. The resulting equation is then solved algebraically with appropriate restrictions.

Suppose u is a function of x possessing derivatives of all orders in the interval $\left(x-n\Delta x,\ x+n\Delta x\right)$. Then we can obtain the value of u at points $x \pm n\Delta x$, where n is any integer, in terms of the value of the function and its derivatives at point x, that is, $u\left(x\right)$ and its higher derivatives. For example:

$$u(x+\Delta x) = u(x) + \left.\frac{du}{dx}\right|_x \Delta x + \left.\frac{d^2u}{dx^2}\right|_x \frac{\left(\Delta x\right)^2}{2!} + \cdots$$
$$+ \left.\frac{d^n u}{dx^n}\right|_x \frac{\left(\Delta x\right)^n}{n!} + \left.\frac{d^{n+1}u}{dx^{n+1}}\right|_\theta \frac{\left(\Delta x\right)^{n+1}}{\left(n+1\right)!}, \qquad 2.1)$$

where Δx is finite increment for the value of x and $d^n u / dx^x \,|_x$ is the value of the nth derivative at point x. It should be noted that Δx may be negative. However, for convenience we let $\Delta x > 0$. Furthermore, we assume that $x < \theta < x + \Delta x$.

Similarly, we may write

$$u(x-\Delta x) = u(x) - \left.\frac{du}{dx}\right|_x \Delta x + \left.\frac{d^2u}{dx^2}\right|_x \frac{\left(\Delta x\right)^2}{2!} - \cdots \qquad (2.2)$$

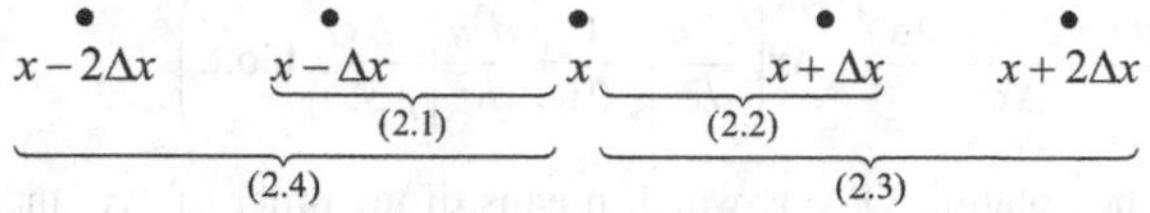

Figure 2.1. Representation of grid spacing and corresponding Taylor series expansion equations.

$$+(-1)^n \left.\frac{d^n u}{dx^n}\right|_x \frac{(\Delta x)^n}{n!} + (-1)^{n+1} \left.\frac{d^{n+1}u}{dx^{n+1}}\right|_\theta \frac{(\Delta x)^{n+1}}{(n+1)!},$$

where $x-\Delta x<\theta<x$. Likewise,

$$u(x+2\Delta x) = u(x) + \left.\frac{du}{dx}\right|_x 2\Delta x + \left.\frac{d^2u}{dx^2}\right|_\theta \frac{(2\Delta x)^2}{2!} + \cdots$$
$$+\left.\frac{d^n u}{dx^n}\right|_x \frac{(2\Delta x)^n}{n!} + \left.\frac{d^{n+1}u}{dx^{n+1}}\right|_\theta \frac{(2\Delta x)^{n+1}}{(n+1)!}, \tag{2.3}$$

where $x<\theta<x+2\Delta x$, and

$$u(x-2\Delta x) = u(x) - \left.\frac{du}{dx}\right|_x 2\Delta x + \left.\frac{d^2u}{dx^2}\right|_x \frac{(2\Delta x)^2}{2!} - \cdots \tag{2.4}$$
$$+(-1)^n \left.\frac{d^n u}{dx^n}\right|_x \frac{(2\Delta x)^n}{n!} + (-1)^{n+1} \left.\frac{d^{n+1}u}{dx^{n+1}}\right|_\theta \frac{(2\Delta x)^{n+1}}{(n+1)!},$$

where $x-2\Delta x<\theta<x$. Notice that the right hand sides of $u(x-\Delta x)$ and $u(x-2\Delta x)$ contain alternating signs.

The value of Δx is taken to be small ($\Delta x<1$), such that $(\Delta x)^n$ would be even smaller than Δx for $n\geq 2$. We can then approximate the series on the right-hand side by truncating the higher-order terms. As a result, we incorporate some error which is known as *truncation error*. In general, the more terms we keep on the right-hand side, then the better and more accurate the value of $u(x\pm n\Delta x)$ would be.

2.2 Application of Taylor's Series to Finite Differencing

If a function u is defined by an array of points in a single dimension, then by Taylor series expansion (2.1) we get

$$\left.\frac{du}{dx}\right|_x \Delta x = u(x+\Delta x) - u(x) - (\Delta x)^2 \left(\left.\frac{d^2u}{dx^2}\right|_x \frac{1}{2!} + \left.\frac{d^3u}{dx^3}\right|_x \frac{\Delta x}{3!} + \text{h.o.t.} \right),$$

where h.o.t. stands for the higher-order terms. Dividing throughout by Δx we obtain

$$\left.\frac{du}{dx}\right|_x = \frac{u(x+\Delta x)-u(x)}{\Delta x} - \Delta x\left(\left.\frac{d^2u}{dx^2}\right|_x \frac{1}{2!} + \left.\frac{d^3u}{dx^3}\right|_x \frac{\Delta x}{3!} + \text{h.o.t.}\right). \tag{2.5}$$

If we introduce the notation $O(\Delta x)$, which means of the order of Δx, then (2.5) can be written as

$$\left.\frac{du}{dx}\right|_x = \frac{u(x+\Delta x)-u(x)}{\Delta x} - O(\Delta x),$$

or

$$\left.\frac{du}{dx}\right|_x \cong \frac{u(x+\Delta x)-u(x)}{\Delta x}, \tag{2.6}$$

where

$$O(\Delta x) = \Delta x\left(\left.\frac{d^2u}{dx^2}\right|_x \frac{1}{2!} + \left.\frac{d^3u}{dx^3}\right|_x \frac{\Delta x}{3!} + \text{h.o.t.}\right).$$

For finding the value of $du/dx|_x$, we are only using the values of the function u at points x and $x+\Delta x$. All other values are neglected in this calculation. Also, the order of the truncation error involved is $O(\Delta x)$, which is not very desirable.

2.3 Forward and Backward Differencing

Given the values of a function u at discrete points in one dimension, two methods are presented to obtain the approximation for the derivative at a given point. Using (2.1), we can write

$$\left.\frac{du}{dx}\right|_x \Delta x = u(x+\Delta x) - u(x) - \left(\left.\frac{d^2u}{dx^2}\right|_x \frac{(\Delta x)^2}{2!} + \left.\frac{d^3u}{dx^3}\right|_x \frac{(\Delta x)^3}{3!} + \text{h.o.t.}\right). \tag{2.7}$$

Furthermore, using (2.2), we can write

$$\left.\frac{du}{dx}\right|_x \Delta x = u(x) - u(x-\Delta x) + \left(\left.\frac{d^2u}{dx^2}\right|_x \frac{(\Delta x)^2}{2!} - \left.\frac{d^3u}{dx^3}\right|_x \frac{(\Delta x)^3}{3!} + \text{h.o.t.}\right). \tag{2.8}$$

Dividing the above two equation by Δx, we obtain

$$\left.\frac{du}{dx}\right|_x = \frac{u(x+\Delta x)-u(x)}{\Delta x} - \left(\left.\frac{d^2u}{dx^2}\right|_x \frac{\Delta x}{2!} + \left.\frac{d^3u}{dx^3}\right|_x \frac{(\Delta x)^2}{3!} + \text{h.o.t.}\right),$$

$$\left.\frac{du}{dx}\right|_x = \frac{u(x)-u(x-\Delta x)}{\Delta x} + \left(\left.\frac{d^2u}{dx^2}\right|_x \frac{\Delta x}{2!} + \left.\frac{d^3u}{dx^3}\right|_x \frac{(\Delta x)^2}{3!} + \text{h.o.t.}\right).$$

Neglecting the terms in the parentheses, we have

$$\left.\frac{du}{dx}\right|_x \cong \frac{u(x+\Delta x)-u(x)}{\Delta x}, \tag{2.9}$$

$$\left.\frac{du}{dx}\right|_x \cong \frac{u(x)-u(x-\Delta x)}{\Delta x}. \tag{2.10}$$

Equation (2.9) is known as *forward finite differencing*, and (2.10) is known as *backward finite differencing*. These are also called first-order accurate *one-sided differences*. In both of the above, the truncation error is of the order of Δx, that is, $O(\Delta x)$, which corresponds to the largest term we neglected in approximating $du/dx|_x$.

Weather forecasters are continually making truncation errors which in turn have an impact on the weather forecast. For example, in the atmosphere advection can occur on small scales. However, the grid spacing in our finite difference representation may be on the order of 100 km, which is too coarse to resolve this advection. This introduces errors into our finite differencing equations that will compound as our forecast proceeds.

2.4 Centered Finite Differencing

In the previous section we obtained the approximation to $du/dx|_x$ with a truncation error of $O(\Delta x)$. In this section we will obtain a better approximation for $du/dx|_x$ based on the *centered finite-difference approximation*, which has a truncation error of $O(\Delta x)^2$.

By adding (2.7) and (2.8) and dividing by Δx, we obtain

$$2\left.\frac{du}{dx}\right|_x \cong \frac{u(x+\Delta x)-u(x-\Delta x)}{\Delta x} - 2(\Delta x)^2\left(\left.\frac{d^3u}{dx^3}\right|_x \frac{1}{3!} + \text{h.o.t.}\right),$$

$$\left.\frac{du}{dx}\right|_x \cong \frac{u(x+\Delta x)-u(x-\Delta x)}{2\Delta x} - O(\Delta x)^2. \tag{2.11}$$

Here the order of the truncation error is $(\Delta x)^2$, that is, $O(\Delta x)^2$, which is the largest term omitted in approximating $du/dx|_x$. Higher-order derivatives, namely $d^n u/dx^n\,|_x$ for $n \geq 2$, can also be found using finite-difference methods.

Adding (2.1) and (2.2) gives

$$u(x+\Delta x)+u(x-\Delta x) = 2u(x) + 2\left.\frac{d^2u}{dx^2}\right|_x \frac{(\Delta x)^2}{2!} + 2\left.\frac{d^4u}{dx^4}\right|_x \frac{(\Delta x)^4}{4!} + \text{h.o.t.},$$

where terms with derivatives of the order $d^n u/dx^n\,|_x$ for n=1, 3, 5, … have cancelled out. This can be written as

$$2\left.\frac{d^2u}{dx^2}\right|_x \frac{(\Delta x)^2}{2!} = u(x+\Delta x)+u(x-\Delta x)-2u(x)-(\Delta x)^4\left(2\left.\frac{d^4u}{dx^4}\right|_x \frac{1}{4!} + \text{h.o.t.}\right).$$

Now dividing by $(\Delta x)^2$ throughout, we obtain

Table 2.1. Some finite difference formulas, along with their accuracy and order of truncation error.

Formula	O(error)	Accuracy
$\left.\frac{du}{dx}\right\|_x \cong \frac{u(x+\Delta x)-u(x)}{\Delta x}$	$O(\Delta x)$	First-order forward
$\left.\frac{du}{dx}\right\|_x \cong \frac{u(x)-u(x-\Delta x)}{\Delta x}$	$O(\Delta x)$	First-order backward
$\left.\frac{du}{dx}\right\|_x \cong \frac{u(x+\Delta x)-u(x-\Delta x)}{2\Delta x}$	$O(\Delta x)^2$	Second-order centered
$\left.\frac{d^2u}{dx^2}\right\|_x \cong \frac{u(x+\Delta x)+u(x-\Delta x)-2u(x)}{(\Delta x)^2}$	$O(\Delta x)^2$	Second-order centered

$$\left.\frac{d^2u}{dx^2}\right|_x = \frac{u(x+\Delta x)+u(x-\Delta x)-2u(x)}{(\Delta x)^2} - (\Delta x)^2\left(2\left.\frac{d^4u}{dx^4}\right|_x \frac{1}{4!} + \text{h.o.t.}\right).$$

This can be expressed as

$$\left.\frac{d^2u}{dx^2}\right|_x \cong \frac{u(x+\Delta x)+u(x-\Delta x)-2u(x)}{(\Delta x)^2} - O(\Delta x)^2, \qquad (2.12)$$

which is the second-order accurate, second derivative formula. Here the truncation error in the leading term is of the order of $(\Delta x)^2$, that is, $O(\Delta x)^2$.

In summary, we may write the first-order forward and backward finite difference formulas and the second-order centered difference formulas as in Table 2.1. We next derive the fourth-order accurate formulas for $du/dx|_x$ and $d^2u/dx^2|_x$, respectively.

2.5 Fourth-Order Accurate Formulas

2.5.1 *First Derivative*

The fourth-order finite-differencing schemes can be obtained by an appropriate linear combination of (2.1) to (2.4) such that the terms of order $(\Delta x)^2$, $(\Delta x)^3$, and $(\Delta x)^4$ are eliminated in the sum and the leading error term is $O(\Delta x)^4$. That is,

$$Au(x) + B\left[u(x+\Delta x) - u(x-\Delta x)\right] + C\left[u(x+2\Delta x) - u(x-2\Delta x)\right]$$
$$= \left.\frac{du}{dx}\right|_x \Delta x + O(\Delta x)^5. \qquad (2.13)$$

To determine the coefficients A, B, and C, consider the following Taylor series representations:

$$u(x+\Delta x)-u(x-\Delta x)=2\frac{du}{dx}\bigg|_x \Delta x+\frac{d^3u}{dx^3}\bigg|_x \frac{(\Delta x)^3}{3}+\frac{d^5u}{dx^5}\bigg|_x \frac{(\Delta x)^5}{60} \tag{2.14}$$

and

$$u(x+2\Delta x)-u(x-2\Delta x)=2\frac{du}{dx}\bigg|_x 2\Delta x+\frac{d^3u}{dx^3}\bigg|_x \frac{(2\Delta x)^3}{3}+\frac{d^5u}{dx^5}\bigg|_x \frac{(2\Delta x)^5}{60}. \tag{2.15}$$

Using equation (2.13) and substituting in the right hand sides of (2.14) and (2.15) we obtain

$$Au(x)+B\left[2\frac{du}{dx}\bigg|_x \Delta x+\frac{d^3u}{dx^3}\bigg|_x \frac{(\Delta x)^3}{3}+\mathrm{O}(\Delta x)^5\right]+$$

$$C\left[2\frac{du}{dx}\bigg|_x 2\Delta x+\frac{d^3u}{dx^3}\bigg|_x \frac{(2\Delta x)^3}{3}+\mathrm{O}(\Delta x)^5\right]\cong\frac{du}{dx}\bigg|_x \Delta x+\mathrm{O}(\Delta x)^5,$$

$$Au(x)+B\left[2\frac{du}{dx}\bigg|_x \Delta x\right]+C\left[4\frac{du}{dx}\bigg|_x \Delta x\right]+B\left[\frac{d^3u}{dx^3}\bigg|_x \frac{(\Delta x)^3}{3}+\right.$$

$$\left.C\left[8\frac{d^3u}{dx^3}\bigg|_x \frac{(\Delta x)^3}{3}\right]+\mathrm{O}(\Delta x)^5\cong\frac{du}{dx}\bigg|_x \Delta x,$$

$$Au(x)+(2B+4C)\frac{du}{dx}\bigg|_x \Delta x+(B+8C)\frac{d^3u}{dx^3}\bigg|_x \frac{(\Delta x)^3}{3}+O(\Delta x)^5\cong\frac{du}{dx}\bigg|_x \Delta x.$$

Equating coefficients of $u(x)$, $du/dx|_x\,\Delta x$, and $[d^3u/dx^3|_x\,(\Delta x)^3]/3$, we obtain

$$A=0, \qquad 2B+4C=1, \qquad B+8C=0$$

or

$$A=0, \qquad B=2/3, \qquad C=\text{-}1/12.$$

Using (2.13) and the values of constants *A*, *B*, and *C*, we obtain

$$\frac{du}{dx}\bigg|_x=\frac{2}{3}\frac{u(x+\Delta x)-u(x-\Delta x)}{\Delta x}-\frac{1}{12}\frac{u(x+2\Delta x)-u(x-2\Delta x)}{\Delta x}+\mathrm{O}(\Delta x)^4,$$

or by neglecting the terms $\mathrm{O}(\Delta x)^4$, we obtain

$$\frac{du}{dx}\bigg|_x\cong\frac{4}{3}\frac{u(x+\Delta x)-u(x-\Delta x)}{2\Delta x}-\frac{1}{3}\frac{u(x+2\Delta x)-u(x-2\Delta x)}{4\Delta x}. \tag{2.16}$$

In deriving (2.16), we have neglected terms $\mathrm{O}(\Delta x)^5$ and higher in the Taylor series. Therefore this expression for du/dx is accurate to the fourth order.

2.5.2 *Second Derivative*

We want to write a general form invoking coefficients *A, B,* and *C* such that

$$Au(x)+B\left[u(x+\Delta x)+u(x-\Delta x)\right]+C\left[u(x+2\Delta x)+u(x-2\Delta x)\right]$$
$$\cong \left.\frac{d^2u}{dx^2}\right|_x (\Delta x)^2. \tag{2.17}$$

We next determine coefficients *A, B,* and *C* such that terms of order $(\Delta x)^3$, $(\Delta x)^4$, and $(\Delta x)^5$ are zero.

Consider the two equations which come about from the addition of (2.1) and (2.2) and the addition of (2.3) and (2.4), respectively,

$$u(x+\Delta x)+u(x-\Delta x)=2u(x)+\left.\frac{d^2u}{dx^2}\right|_x (\Delta x)^2+\left.\frac{d^4u}{dx^4}\right|_x \frac{(\Delta x)^4}{12}+O\left(\Delta x\right)^6, \tag{2.18}$$

$$u(x+2\Delta x)+u(x-2\Delta x)=2u(x)+4\left.\frac{d^2u}{dx^2}\right|_x (\Delta x)^2+\frac{4}{3}\left.\frac{d^4u}{dx^4}\right|_x (\Delta x)^4+O\left(\Delta x\right)^6, \tag{2.19}$$

so that (2.17) can be written as

$$Au(x)+B\left(2u(x)+\left.\frac{d^2u}{dx^2}\right|_x (\Delta x)^2+\left.\frac{d^4u}{dx^4}\right|_x \frac{(\Delta x)^4}{12}+O(\Delta x)^6\right)$$
$$+C\left(2u(x)+4\left.\frac{d^2u}{dx^2}\right|_x (\Delta x)^2+\frac{4}{3}\left.\frac{d^4u}{dx^4}\right|_x (\Delta x)^4+O(\Delta x)^6\right)\cong \left.\frac{d^2u}{dx^2}\right|_x (\Delta x)^2.$$

This equation can be rearranged as

$$(A+2B+2C)u(x)+(B+4C)\left.\frac{d^2u}{dx^2}\right|_x (\Delta x)^2$$
$$+\left(\frac{B}{12}+\frac{4C}{3}\right)\left.\frac{d^4u}{dx^4}\right|_x (\Delta x)^4+O\left(\Delta x\right)^6\cong \left.\frac{d^2u}{dx^2}\right|_x (\Delta x)^2.$$

Equating coefficients of $u(x)$, $d^2u/dx^2\,|_x(\Delta x^2)$, and $d^4u/dx^4\,|_x(\Delta x^4)$ on both sides gives $A+2B+2C=0$, $B+4C=1$, and $B/12+4C/3=0$. This gives $A=-5/2$, $B=4/3$, and $C=-1/12$, so that

$$\left.\frac{d^2u}{dx^2}\right|_x=\frac{1}{(\Delta x)^2}\left(-\frac{5}{2}u(x)+\frac{4}{3}[u(x+\Delta x)+u(x-\Delta x)]\right.$$
$$\left.-\frac{1}{12}[u(x+2\Delta x)+u(x-2\Delta x)]\right)+O\left(\Delta x\right)^4. \tag{2.20}$$

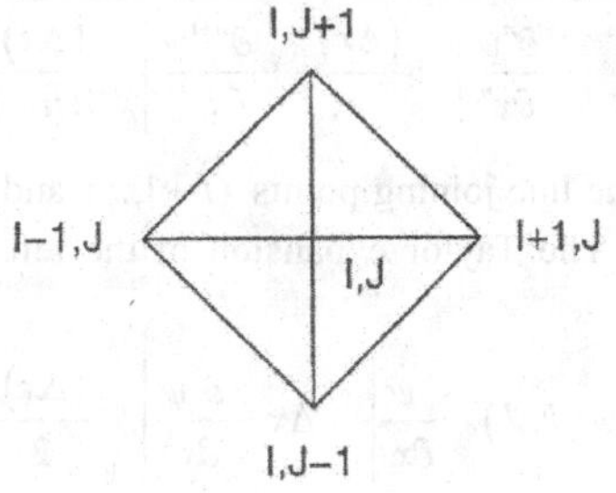

Figure 2.2. A 5-point diamond stencil.

This is a fourth-order accurate representation of the second derivative. It can be shown that fourth-order accurate finite difference representation more closely mirrors the exact derivative than second-order accurate finite difference representation.

2.6 Second-Order Accurate Laplacian

2.6.1 *5-Point Diamond Stencil*

We start with Laplace's equation,

$$\nabla^2\psi = 0, \tag{2.21}$$

where ∇^2 is the three-dimensional or two-dimensional Laplacian operator in the Cartesian coordinate system, that is,

$$\nabla^2 = \frac{\partial^2}{\partial x^2} + \frac{\partial^2}{\partial y^2} + \frac{\partial^2}{\partial z^2},$$

$$\nabla^2 = \frac{\partial^2}{\partial x^2} + \frac{\partial^2}{\partial y^2}.$$

Let $\psi = \psi(x, y)$ be any scalar function of x and y. The finite-difference analog of the second-order accurate Laplacian will be discussed with reference to the 5-point stencil shown in Fig. 2.2.

Let us suppose that we want to calculate $\nabla^2\psi$, where ψ is the streamfunction. Using Taylor's expansion (see Appendix A) for a function of two variables about point (I, J), we can write the Taylor expansion of the right point with respect to the central point, that is,

$$\psi(I+1, J) = \psi(I, J) + \left.\frac{\partial\psi}{\partial x}\right|_{I,J} \Delta x + \left.\frac{\partial^2\psi}{\partial x^2}\right|_{I,J} \frac{(\Delta x)^2}{2!} + \ldots$$

$$+\left.\frac{\partial^n\psi}{\partial x^n}\right|_{I,J}\frac{(\Delta x)^n}{n!}+\left.\frac{\partial^{n+1}\psi}{\partial x^{n+1}}\right|_{\theta,J}\frac{(\Delta x)^{n+1}}{(n+1)!}, \tag{2.22}$$

where (θ,J) is a point on the line joining points $(I+1,J)$ and (I,J) and Δx is the grid spacing in the x-direction. The Taylor expansion of the left point with respect to the central point is given by

$$\psi(I-1,J)=\psi(I,J)-\left.\frac{\partial\psi}{\partial x}\right|_{I,J}\Delta x+\left.\frac{\partial^2\psi}{\partial x^2}\right|_{I,J}\frac{(\Delta x)^2}{2!}+\ldots \tag{2.23}$$
$$+(-1)^n\left.\frac{\partial^n\psi}{\partial x^n}\right|_{I,J}\frac{(\Delta x)^n}{n!}+(-1)^{n+1}\left.\frac{\partial^{n+1}\psi}{\partial x^{n+1}}\right|_{\theta,J}\frac{(\Delta x)^{n+1}}{(n+1)!},$$

where (θ,J) is a point on the line joining points $(I-1,J)$ and (I,J). Similarly,

$$\psi(I,J+1)=\psi(I,J)+\left.\frac{\partial\psi}{\partial y}\right|_{I,J}\Delta y+\left.\frac{\partial^2\psi}{\partial y^2}\right|_{I,J}\frac{(\Delta y)^2}{2!}+\ldots$$
$$+\left.\frac{\partial^n\psi}{\partial y^n}\right|_{I,J}\frac{(\Delta y)^n}{n!}+\left.\frac{\partial^{n+1}\psi}{\partial y^{n+1}}\right|_{I,\theta}\frac{(\Delta y)^{n+1}}{(n+1)!}, \tag{2.24}$$

where (I,θ) is a point on the line joining points $(I,J+1)$ and (I,J) and Δy is the grid spacing in the y-direction. Also,

$$\psi(I,J-1)=\psi(I,J)-\left.\frac{\partial\psi}{\partial y}\right|_{I,J}\Delta y+\left.\frac{\partial^2\psi}{\partial y^2}\right|_{I,J}\frac{(\Delta y)^2}{2!}-\ldots \tag{2.25}$$
$$+(-1)^n\left.\frac{\partial^n\psi}{\partial y^n}\right|_{I,J}\frac{(\Delta y)^n}{n!}+(-1)^{n+1}\left.\frac{\partial^{n+1}\psi}{\partial y^{n+1}}\right|_{I,\theta}\frac{(\Delta y)^{n+1}}{(n+1)!},$$

where (I,θ) is a point on the line joining points $(I,J-1)$ and (I,J).

In general, we can have $\Delta x\neq\Delta y$. For simplicity, assume that we have an even mesh, that is, $\Delta x=\Delta y=\Delta$. Adding (2.22), (2.23), (2.24), and (2.25), we obtain

$$\psi(I+1,J)+\psi(I-1,J)+\psi(I,J+1)+\psi(I,J-1)-4\psi(I,J)$$
$$=\nabla^2\psi\big|_{I,J}\Delta^2+2\left.\left(\frac{\partial^4\psi}{\partial x^4}\frac{\Delta^4}{4!}+\frac{\partial^4\psi}{\partial y^4}\frac{\Delta^4}{4!}\right)\right|_{I,J}+\text{h.o.t.} \tag{2.26}$$

After dividing by Δ^2 and neglecting terms of order Δ^2 and higher, we obtain the second-order accurate Laplacian

$$\nabla^2\psi\big|_{I,J}\cong\left[\psi(I+1,J)+\psi(I-1,J)+\psi(I,J+1)+\psi(I,J-1)-4\psi(I,J)\right]/\Delta^2. \tag{2.27}$$

Weight assigned to the five points $(I+1,J)$, $(I-1,J)$, $(I,J+1)$, $(I,J-1)$, and (I,J) in this 5-point stencil are 1, 1, 1, 1 and -4, respectively.

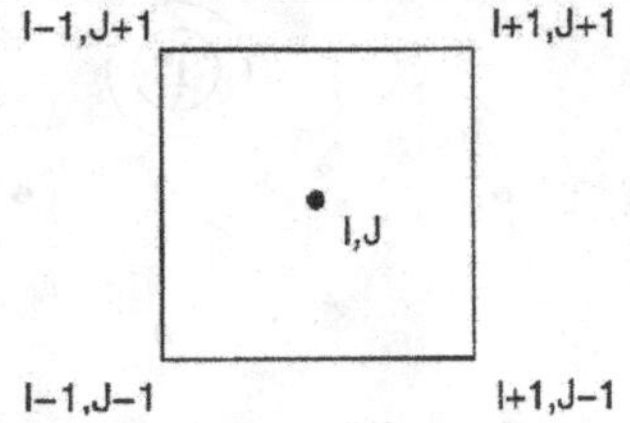

Figure 2.3. A 5-point square stencil.

2.6.2 *5-Point Square Stencil*

There is yet another 5-point stencil which also gives second-order accurate results. This uses the points shown in Fig. 2.3. To obtain the finite-difference expression for the ∇^2 operator, we use the following equations:

$$\psi(I+1,J+1)=\psi(I,J)+\Delta\left(\frac{\partial\psi}{\partial x}+\frac{\partial\psi}{\partial y}\right)\Bigg|_{I,J} +\frac{\Delta^2}{2!}\left(\frac{\partial^2\psi}{\partial x^2}+\frac{\partial^2\psi}{\partial y^2}\right)\Bigg|_{I,J}+\Delta^2\frac{\partial^2\psi}{\partial x\partial y}\Bigg|_{I,J}+\ldots, \tag{2.28}$$

$$\psi(I-1,J-1)=\psi(I,J)-\Delta\left(\frac{\partial\psi}{\partial x}+\frac{\partial\psi}{\partial y}\right)\Bigg|_{I,J} +\frac{\Delta^2}{2!}\left(\frac{\partial^2\psi}{\partial x^2}+\frac{\partial^2\psi}{\partial y^2}\right)\Bigg|_{I,J}+\Delta^2\frac{\partial^2\psi}{\partial x\partial y}\Bigg|_{I,J}+\ldots, \tag{2.29}$$

$$\psi(I-1,J+1)=\psi(I,J)-\Delta\left(\frac{\partial\psi}{\partial x}-\frac{\partial\psi}{\partial y}\right)\Bigg|_{I,J} +\frac{\Delta^2}{2!}\left(\frac{\partial^2\psi}{\partial x^2}+\frac{\partial^2\psi}{\partial y^2}\right)\Bigg|_{I,J}-\Delta^2\frac{\partial^2\psi}{\partial x\partial y}\Bigg|_{I,J}+\ldots, \tag{2.30}$$

$$\psi(I+1,J-1)=\psi(I,J)+\Delta\left(\frac{\partial\psi}{\partial x}-\frac{\partial\psi}{\partial y}\right)\Bigg|_{I,J} +\frac{\Delta^2}{2!}\left(\frac{\partial^2\psi}{\partial x^2}+\frac{\partial^2\psi}{\partial y^2}\right)\Bigg|_{I,J}-\Delta^2\frac{\partial^2\psi}{\partial x\partial y}\Bigg|_{I,J}+\ldots. \tag{2.31}$$

After adding the above four equations, we obtain

$$\psi(I+1,J+1)+\psi(I-1,J-1)+\psi(I-1,J+1)+\psi(I+1,J-1)-4\psi(I,J) =2\Delta^2\left(\frac{\partial^2\psi}{\partial x^2}+\frac{\partial^2\psi}{\partial y^2}\right)\Bigg|_{I,J}+\frac{\Delta^4}{6}\left(\frac{\partial^4\psi}{\partial x^4}+\frac{\partial^4\psi}{\partial y^4}+6\frac{\partial^4\psi}{\partial x^2\partial y^2}\right)\Bigg|_{I,J}+\text{h.o.t.} \tag{2.32}$$

By taking differences along the diagonal, (2.32) can also be written as

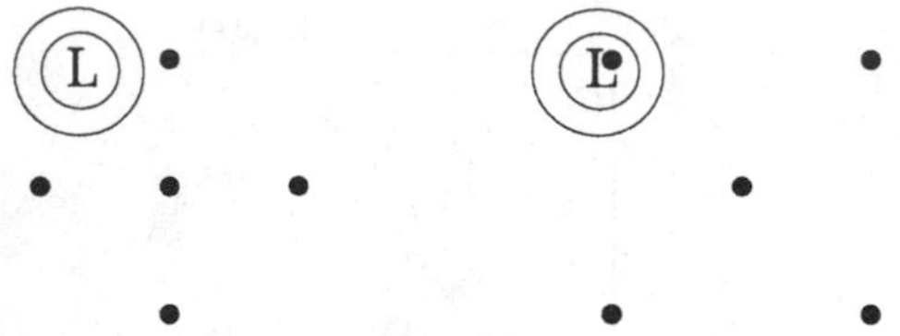

Figure 2.4. Diamond and square finite-difference representations of the Laplacian and their ability to represent an area of low-pressure.

$$\psi(I+1,J+1)+\psi(I-1,J-1)+\psi(I-1,J+1)+\psi(I+1,J-1)-4\psi(I,J)$$
$$=\nabla^2\psi\big|_{I,J}\left(2^{1/2}\Delta\right)^2+2\frac{\left(2^{1/2}\Delta\right)^4}{4!}\nabla^4\psi+\ldots$$
$$=2\nabla^2\psi\big|_{I,J}\Delta^2+\frac{1}{3}\nabla^4\psi\Delta^4+\ldots. \tag{2.33}$$

Dividing both sides by $2\Delta^2$, we can write

$$\nabla^2\psi=\frac{1}{2\Delta^2}\big[\psi(I+1,J+1)+\psi(I-1,J-1)+\psi(I-1,J+1)$$
$$+\psi(I+1,J-1)-4\psi(I,J)\big]+O(\Delta)^2. \tag{2.34}$$

This gives the finite-difference expression for the second-order Laplacian represented by the 5-point square stencil.

Which 5-point stencil is more accurate, and why? Using a 5-point stencil with diamond configuration of points is slightly more accurate than a 5-point square stencil. This is because the distance from the center point (I,J) to the other points is less in the diamond stencil (Δ) than the corresponding distance in the square stencil ($2^{1/2}\Delta$).

It is important to note that the analytical Laplacian is invariant with respect to the rotation of the coordinate system, i.e. ∇^2 has one and only one value. However, finite difference representations (diamond or square) of the Laplacian are not invariant to coordinate transformations. That is, we will get different answers for different finite difference schemes. For example, suppose we have a 5-point diamond stencil with an area of low-pressure located to the northwest of the grid points (Fig. 2.4). In this particular case, the diamond stencil will not allow us to catch this low-pressure area in our finite difference representation. However, by using a 5-point square stencil we were able to catch this area.

2.6.3 *9-Point Stencil*

There is yet another second-order accurate Laplacian which is based on a 9-point stencil. These nine points are shown in Fig. 2.5, with the corresponding weights written in

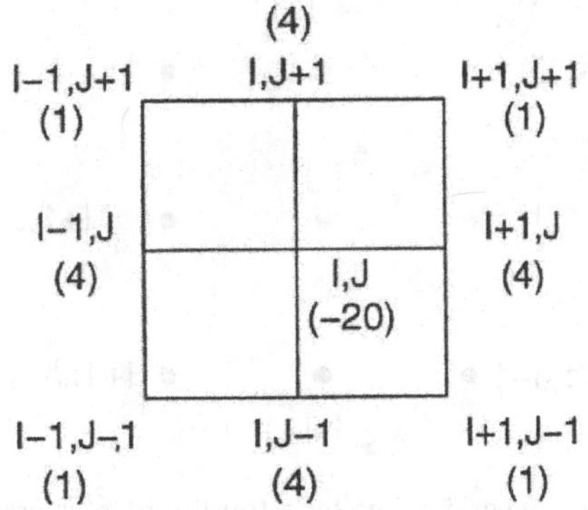

Figure 2.5. A 9-point stencil used for the second-order Laplacian with the weights assigned to each node given in parentheses.

parentheses. It should be noted that this configuration is made up of the two 5-point stencils discussed earlier.

Consider now the Taylor expansion of the eight outer points with respect to the central point (I,J). Combining (2.26) and (2.33) linearly in such a way that the terms $O(\Delta)^2$ in the two equations are neglected at the same accuracy, we obtain

$$\begin{aligned}&\psi(I+1,J+1)+\psi(I-1,J-1)+\psi(I-1,J+1)+\psi(I+1,J-1)\\&+4\left[\psi(I,J+1)+\psi(I-1,J)+\psi(I,J-1)+\psi(I+1,J)\right]-20\psi(I,J)\\&=6\Delta^2\nabla^2\psi\big|_{I,J}+\frac{2}{3}\Delta^4\nabla^4\psi\big|_{I,J}+\text{h.o.t.}\end{aligned}\tag{2.35}$$

Dividing both sides by $6\Delta^2$, we obtain

$$\begin{aligned}\nabla^2\psi\big|_{I,J}=&\frac{1}{6\Delta^2}\{\psi(I+1,J+1)+\psi(I-1,J-1)+\psi(I-1,J+1)+\psi(I+1,J-1)\\&+4\left[\psi(I,J+1)+\psi(I-1,J)+\psi(I,J-1)+\psi(I+1,J)\right]-20\psi(I,J)+O(\Delta)^2.\end{aligned}\tag{2.36}$$

This formulation uses more information around the central point. Therefore the results can be locally more accurate than the 5-point stencil given by (2.27) or (2.34).

2.7 Fourth-Order Accurate Laplacian

Let us attempt to formulate a fourth-order accurate Laplacian using the two different 5-point stencils discussed earlier. We make use of the gridpoint representation in Fig. 2.6. Using (2.26), we can write

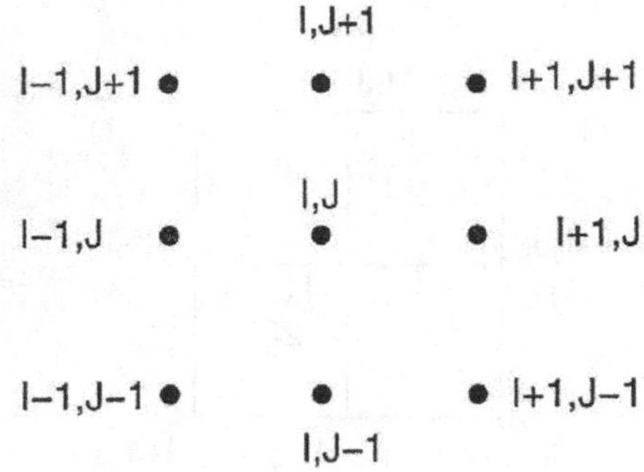

Figure 2.6. The nine grid points for a fourth-order accurate Laplacian.

$$\psi(I+1,J)+\psi(I-1,J)+\psi(I,J+1)+\psi(I,J-1)-4\psi(I,J)$$
$$=\nabla^2\psi\big|_{I,J}\Delta^2+2\frac{\Delta^4}{4!}\left(\frac{\partial^4\psi}{\partial x^4}+\frac{\partial^4\psi}{\partial y^4}\right)\Bigg|_{I,J}+2\frac{\Delta^6}{6!}\left(\frac{\partial^6\psi}{\partial x^6}+\frac{\partial^6\psi}{\partial y^6}\right)\Bigg|_{I,J}+\text{h.o.t.} \qquad (2.37)$$

Similarly, using the five grid points $(I-1,J+1)$, $(I-1,J-1)$, $(I+1,J-1)$, $(I+1,J+1)$, and (I,J), we can write

$$\psi(I-1,J+1)+\psi(I-1,J-1)+\psi(I+1,J-1)+\psi(I+1,J+1)-4\psi(I,J)$$
$$=\nabla^2\psi\big|_{I,J}\left(2^{1/2}\Delta\right)^2+2\frac{\left(2^{1/2}\Delta\right)^4}{4!}\left(\frac{\partial^4\psi}{\partial x^4}+\frac{\partial^4\psi}{\partial y^4}\right)\Bigg|_{I,J}$$
$$+2\frac{\left(2^{1/2}\Delta\right)^6}{6!}\left(\frac{\partial^6\psi}{\partial x^6}+\frac{\partial^6\psi}{\partial y^6}\right)\Bigg|_{I,J}+\text{h.o.t.} \qquad (2.38)$$

Note that here the distance between grid point (I,J) and its surrounding points is $2^{1/2}\Delta$.

To eliminate the terms $O(\Delta^4)$ from the above two equations, we multiply (2.37) by 4, from the resultant equation subtract (2.38), and divide by $2\Delta^2$ to obtain

$$\nabla^2\psi\big|_{I,J}=\frac{1}{2\Delta^2}\{4[\psi(I,J+1)+\psi(I-1,J)+\psi(I,J-1)$$
$$+\psi(I+1,J)]-[\psi(I+1,J+1)+\psi(I-1,J+1)$$
$$+\psi(I-1,J-1)+\psi(I+1,J-1)]-12\psi(I,J)\}$$
$$+\frac{\Delta^4}{180}\left(\frac{\partial^6\psi}{\partial x^6}+\frac{\partial^6\psi}{\partial y^6}\right)\Bigg|_{I,J}+\text{h.o.t.}$$

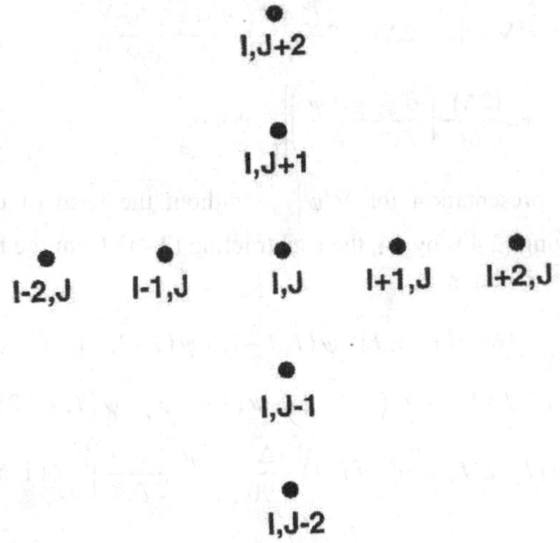

Figure 2.7. Another 9-point stencil for a fourth-order accurate Laplacian.

Hence the fourth-order accurate Laplacian obtained by using two 5-point stencils (in all we use nine grid points) is expressed as

$$\begin{aligned}\nabla^2\psi\big|_{I,J} = \frac{1}{2\Delta^2}\{4[\psi(I,J+1)+\psi(I-1,J)+\psi(I,J-1) \\ +\psi(I+1,J)]-[\psi(I+1,J+1)+\psi(I-1,J+1) \\ +\psi(I-1,J-1)+\psi(I+1,J-1)]-12\psi(I,J)\}+O(\Delta)^4.\end{aligned} \tag{2.39}$$

Continuing on the same idea, we will next formulate a fourth-order Laplacian using the nine-grid-point representation in Fig. 2.7. First, using the inner diamond grid points $(I+1,J)$, $(I-1,J)$, $(I,J+1)$, $(I,J-1)$, and (I,J), we can write

$$\begin{aligned}&\psi(I-1,J)+\psi(I,J-1)+\psi(I+1,J)+\psi(I,J+1)-4\psi(I,J) \\ &\quad=\nabla^2\psi\big|_{I,J}\Delta^2+2\frac{\Delta^4}{4!}\left(\frac{\partial^4\psi}{\partial x^4}+\frac{\partial^4\psi}{\partial y^4}\right)\Bigg|_{I,J} \\ &\quad+2\frac{\Delta^6}{6!}\left(\frac{\partial^6\psi}{\partial x^6}+\frac{\partial^6\psi}{\partial y^6}\right)\Bigg|_{I,J}+\text{h.o.t.}\end{aligned} \tag{2.40}$$

Next using the outer diamond grid points $(I+2,J)$, $(I-2,J)$, $(I,J+2)$, $(I,J-2)$, and (I,J), which have grid spacing 2Δ we can write

$$\psi(I,J+2)+\psi(I-2,J)+\psi(I,J-2)+\psi(I+2,J)-4\psi(I,J)$$

$$= \nabla^2 \psi \big|_{I,J} (2\Delta)^2 + 2 \frac{(2\Delta)^4}{4!} \left(\frac{\partial^4 \psi}{\partial x^4} + \frac{\partial^4 \psi}{\partial y^4} \right) \bigg|_{I,J}$$

$$+ 2 \frac{(2\Delta)^6}{6!} \left(\frac{\partial^6 \psi}{\partial x^6} + \frac{\partial^6 \psi}{\partial y^6} \right) \bigg|_{I,J} + \text{h.o.t.} \tag{2.41}$$

We seek a representation for $\nabla^2 \psi \big|_{I,J}$ without the term of order Δ^2. This is achieved by multiplying (2.40) by 16, then subtracting (2.41) from the resulting equation, and dividing by $12\Delta^2$ to obtain

$$\nabla^2 \psi \big|_{I,J} = \frac{1}{12\Delta^2} \Big\{ 16 \big[\psi (I-1,J) + \psi (I,J-1) + \psi (I+1,J)$$

$$+ \psi (I,J+1) \big] - \big[\psi (I,J+2) + \psi (I-2,J) + \psi (I,J-2)$$

$$+ \psi (I+2,J) \big] - 60\psi (I,J) - \frac{\Delta^4}{90} \left(\frac{\partial^6 \psi}{\partial x^6} + \frac{\partial^6 \psi}{\partial y^6} \right) \bigg|_{I,J} + \text{h.o.t.},$$

or

$$\nabla^2 \psi \big|_{I,J} = \frac{1}{12\Delta^2} \Big\{ 16 \big[\psi (I-1,J) + \psi (I,J-1) + \psi (I+1,J)$$

$$+ \psi (I,J+1) \big] - \big[\psi (I,J+2) + \psi (I-2,J) + \psi (I,J-2)$$

$$+ \psi (I+2,J) \big] - 60\psi (I,J) + O(\Delta^4). \tag{2.42}$$

Equation (2.42) represents yet another fourth-order accurate Laplacian. In general, this idea of adding and subtracting finite difference analogs can be used to develop finite-differencing schemes of various higher orders.

In review, the various finite difference representations of the Laplacian and their corresponding equations are given below in order from most accurate to least accurate representation:

Square stencil	• • •	Equation (2.39)
9-point	• • •	
Fourth-order accurate	• • •	
	•	
Interspersed diamonds	•	Equation (2.42)
9-point	• • • • •	
Fourth-order accurate	•	
	•	
Square stencil	• • •	Equation (2.36)
9-point	• • •	
Second-order accurate	• • •	

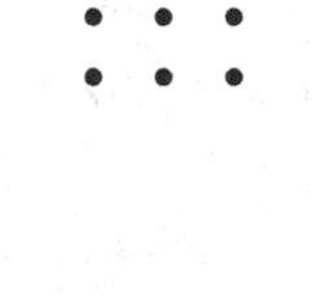

Diamond stencil
5-point
Second-order accurate

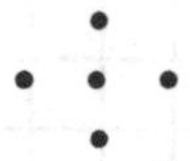

Equation (2.27)

Square stencil
5-point
Second-order accurate

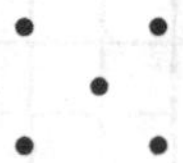

Equation (2.34)

2.8 Elliptic Partial Differential Equation in Meteorology

The most commonly occurring elliptic partial differential equations in atmospheric modeling are of the type

$$\nabla^2\phi = G \qquad \text{Poisson's equation,} \tag{2.43}$$

$$\nabla^2\phi + H\phi = G \qquad \text{Helmholtz equation.} \tag{2.44}$$

These equations occur in the relationship between vorticity and streamfunction, that is, $\nabla^2\psi = \zeta$, where ψ is the streamfunction and ζ is the relative vorticity. A similar type of relationship occurs between divergence and velocity potential, or when going from ϕ to ψ and from ψ to ϕ in a balance equation, or when obtaining time tendencies in the quasigeostrophic prediction models or in the solution of the omega (ω) equation (Holton 1992).

There are two ways in which we can solve these equations. One is based on the direct solution of simultaneous equations, which works efficiently for small domains but is computationally heavy for large domains. The other is called the *relaxation method*, which is more efficient for large domains.

2.9 Direct Method

For solving the three-dimensional elliptic boundary value problem of the type (2.43) and (2.44), one uses a *similarity transform method* involving matrices. A general outline of the method is to first write the given equation in finite difference form for each of the N vertical levels. This gives rise to a set of N linear algebraic equations. This set of N linear algebraic equations is then written in matrix form and solved.

Consider the even-mesh domain shown in Fig. 2.8 on a horizontal plane with $\Delta x = \Delta y = \Delta$. Here i and j are nodes in x and y, respectively. $i = 1(1)K$ and $j = 1(1)L$, that is, i and j vary from 1 to K and L, respectively, with increments of 1.

Assume that ϕ and G are functions of x, y, and z. Then (2.44) can be written as

$$\nabla_2^2\phi + \frac{\partial^2\phi}{\partial z^2} + H\phi = G. \tag{2.45}$$

The finite-difference representation for $\partial^2\phi/\partial z^2$ can be written as

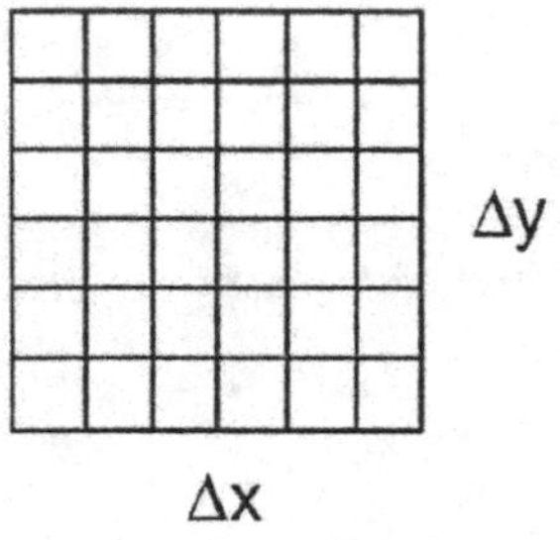

Figure 2.8. A horizontal mesh with $\Delta x = \Delta y = \Delta$.

$$\frac{\partial^2 \phi}{\partial z^2} = \frac{\phi_{k+1} - 2\phi_k + \phi_{k-1}}{(\Delta z)^2}, \tag{2.46}$$

where Δz is the distance between levels k and $k+1$. The finite-difference form of (2.45) can be written as

$$\frac{\nabla_2^2 \phi_{k+1} - 2\nabla_2^2 \phi_k + \nabla_2^2 \phi_{k-1}}{\Delta^2} + \frac{\phi_{k+1} - 2\phi_k + \phi_{k-1}}{(\Delta z)^2} + H_k \phi_k = G_k \tag{2.47}$$

for $k + 2(1)(N-1)$ where $\Delta = \Delta x = \Delta y$. Thus the three-dimensional elliptic partial differential equation is transformed into a two-dimensional equation.

For the top layer $(k = 1)$ we have

$$\frac{\nabla_2^2 \phi_2 - 2\nabla_2^2 \phi_1}{\Delta^2} + \frac{\phi_2 - 2\phi_1}{(\Delta z)^2} + H_1 \phi_1 = G_1 . \tag{2.48}$$

For the bottom layer $(k = N)$ we have

$$\frac{-2\nabla_2^2 \phi_N + \nabla_2^2 \phi_{N-1}}{\Delta^2} + \frac{\phi_{N-1} - 2\phi_N}{(\Delta z)^2} + H_N \phi_N = G_N , \tag{2.49}$$

where we assume $\phi_0 = \phi_{N+1} = 0$. This is equivalent to saying that the geopotential anomaly vanishes at the top and bottom.

Rewriting (2.47), (2.48), and (2.49) in matrix form, we obtain

$$\nabla_2^2 \left(A\Phi \right) + B\Phi = G^* , \tag{2.50}$$

where A and B are $N \times N$ coefficient matrices while Φ and G^* are column vectors, that is,

$$\Phi = \begin{bmatrix} \phi_1 \\ \phi_2 \\ \phi_3 \\ \vdots \\ \phi_N \end{bmatrix}$$

and

$$G^* = \begin{bmatrix} G_1 \\ G_2 \\ G_3 \\ \vdots \\ G_N \end{bmatrix}.$$

Here ϕ_k and G_k are the values of ϕ and G at a particular level k and a fixed grid point in the horizontal.

After premultiplying (2.50) by the matrix B^{-1}, we obtain

$$B^{-1}\nabla_2^2(A\Phi) + B^{-1}B\Phi = B^{-1}G^*. \tag{2.51}$$

Also, using the distributive property of the ∇_2^2 operator, we obtain

$$B^{-1}\nabla_2^2(A\Phi) = \nabla_2^2(B^{-1}A\Phi) - A\Phi\nabla_2^2 B^{-1} \tag{2.52}$$

and $\nabla_2^2 B^{-1} = 0$, since B^{-1} is a matrix of numbers which are constant. Noting also that $B^{-1}B = 1$, we then have

$$\nabla_2^2(B^{-1}A\Phi) + \Phi = B^{-1}G^*. \tag{2.53}$$

Let $B^{-1}A = C$, a new matrix of order $N \times N$. Then (2.53) can be written as

$$\nabla_2^2(C\Phi) + \Phi = B^{-1}G^*. \tag{2.54}$$

One can now diagonalize matrix C using a similarity transformation. This calls for two matrices U and U^{-1} such that

$$UCU^{-1} = D, \tag{2.55}$$

where D is a diagonal matrix. After rearranging, we obtain

$$C = U^{-1}DU, \tag{2.56}$$

where D is a diagonal matrix shown in Fig. 2.9. Each d_k, $k = 1(1)N$ is an eigenvalue of matrix C, that is,

$$CX_k = d_k X_k \tag{2.57}$$

for some column vector X_k. Substituting (2.56) into (2.54), we obtain

$$D = \begin{pmatrix} d_{1,1} & & & 0 \\ & d_{2,2} & & \\ & & \ddots & \\ 0 & & & d_{n,n} \end{pmatrix}$$

Figure 2.9. The elements of the diagonal matrix D.

$$\nabla_2^2\left(U^{-1}DU\Phi\right)+\Phi = B^{-1}G^*,$$

or premultiplying by *U*, we can write

$$\nabla_2^2\left(DU\Phi\right)+U\Phi = UB^{-1}G^*. \tag{2.58}$$

Let us define $U\Phi = V$ (a column vector) and $UB^{-1} = F$ (an $N \times N$ matrix). Then (2.58) can be rewritten as

$$V+\nabla_2^2\left(DV\right) = FG^*. \tag{2.59}$$

Equation (2.59) is the equivalent matrix representation for a set of *N* linearly independent Helmholtz equations. Solutions of (2.59) will provide us with the values of *V*, that is,

$$V = \begin{bmatrix} V_1 \\ V_2 \\ V_3 \\ \vdots \\ V_N \end{bmatrix}.$$

Since we have $U\Phi = V$, we can write

$$\Phi = U^{-1}V. \tag{2.60}$$

Thus if the value of *V* is known, then by premultiplying *V* by U^{-1} we can easily determine Φ. Now Φ is known, that is,

$$\Phi = \begin{bmatrix} \Phi_1 \\ \Phi_2 \\ \Phi_3 \\ \vdots \\ \Phi_N \end{bmatrix}.$$

Next we obtain matrix *U*. Let $d_k X_k$, $k = 1(1)N$ be the eigenpairs of matrix *C*; in other words, d_k is the eigenvalue and X_k is the corresponding eigenvector. It should be noted that there is the possibility that we have more than one eigenvector corresponding to a single eigenvalue, but to simplify matters we assume that there are *N* distinct eigenpairs $d_k X_k$, $k = 1(1)N$.

Since $d_k X_k$ is an eigenpair of *C*, we have $CX_k = d_k X_k$ such that

$$CX = XD, \tag{2.61}$$

where *X* is the matrix consisting of the eigenvectors of *C*. Postmultiplying (2.61) by X^{-1} gives

$$C = XDX^{-1}. \tag{2.62}$$

From (2.56) and (2.62), we obtain

$$U^{-1} = X. \tag{2.63}$$

Hence *U* is the inverse matrix of *X*.

2.10 Relaxation Method

This method is commonly used for solving Poisson- and Helmholtz-type equations. Given the 5-point stencil, the Laplacian of any function ϕ is calculated as

$$\nabla^2 \phi = \frac{\phi_{i+1,j} + \phi_{i-1,j} + \phi_{i,j+1} + \phi_{i,j-1} - 4\phi_{i,j}}{\Delta^2}, \tag{2.64}$$

where $\phi_{i,j}$ is the value of ϕ at the node point (i, j). Also assume that $\Delta x = \Delta y = \Delta$. The Helmholtz equation

$$\nabla^2 \phi - \mu\phi = F, \tag{2.65}$$

can then be written as

$$\phi_{i+1,j} + \phi_{i-1,j} + \phi_{i,j+1} + \phi_{i,j-1} - \left[4 + \mu(\Delta x)^2\right]\phi_{i,j} = F_{i,j}(\Delta x)^2. \tag{2.66}$$

In the relaxation method, the values of *F* are specified at each of the interior points of the domain. It is desired to find the values of $\phi_{i,j}$ at every point which satisfy (2.66) and the given boundary conditions. There are three types of *boundary conditions* that are commonly used. One is where the value of ϕ is prescribed at the boundaries (called the *Dirichlet* boundary condition). Another is where the value of the normal derivative $\partial\phi/\partial n$ is specified at the boundaries (called the *Neumann* boundary condition). The last one is the *mixed* boundary condition, that is, $\partial\phi/\partial n + \beta\phi$ is specified at the boundaries. These are three possible boundary conditions. Usually any other prescription defines an over specified system.

Next we describe the procedure to obtain the solution using relaxation techniques. To begin, we assume a first guess field for ϕ, say ϕ_0, over the domain and ask, does ϕ_0

satisfy the Helmholtz equation and the boundary conditions? If yes, then we have a solution. Otherwise, find the difference $\nabla^2\phi_0 - \mu\phi_0 - F$. This is the *residual* between the actual solution and the first guess, that is,

$$\nabla^2\phi_0 - \mu\phi_0 = F + R, \tag{2.67}$$

where R is residual. Now we aim to change ϕ_0 iteratively such that R is minimized to an acceptable degree of tolerance for every point in the domain.

There are two types of *relaxation schemes*, the simultaneous relaxation scheme and the sequential relaxation scheme. The *simultaneous relaxation* scheme uses the original values of $\phi_{i,j}$ from the previous iteration to calculate values at the next iteration. The *sequential relaxation* scheme uses new values of $\phi_{i,j}$ for calculating values at the next iteration. As will be shown below, the sequential relaxation scheme turns out to be faster than the simultaneous relaxation. The analysis presented here follows a procedure described by Thompson (1961).

We now address the simultaneous relaxation scheme, assuming that we are at level m of iteration at grid point (i,j). Then

$$\phi_{i+1,j}^m + \phi_{i-1,j}^m + \phi_{i,j+1}^m + \phi_{i,j-1}^m - (4+H)\phi_{i,j}^m = d^2F_{i,j} + R_{i,j}^m, \tag{2.68}$$

where $H = \mu d^2$ and $d^2 = (\Delta x)^2 = (\Delta y)^2$. $R_{i,j}^m$ is the residual at grid point (i,j) at the mth iteration. Note that $R_{i,j}^m = Rd^2$ and has the dimensions of ϕ. We change the value of $\phi_{i,j}$ such that the residual vanishes locally in the above equation. If $\phi_{i,j}^m$ is changed to $\phi_{i,j}^{m+1}$ for iteration $m+1$, then we obtain

$$\phi_{i+1,j}^m + \phi_{i-1,j}^m + \phi_{i,j+1}^m + \phi_{i,j-1}^m - (4+H)\phi_{i,j}^{m+1} = d^2F_{i,j}. \tag{2.69}$$

If we subtract (2.69) from (2.68), we obtain

$$\delta\phi_{i,j}^m = \phi_{i,j}^{m+1} - \phi_{i,j}^m = \frac{R_{i,j}^m}{4+H}, \tag{2.70}$$

where $\delta\phi_{i,j}^m$ denotes the change in $\phi_{i,j}$ necessary to make the residual vanish from (2.68).

We next address the convergence of the above interation procedure. Let us assume that $\phi_{i,j}$ is the solution. Then

$$\phi_{i+1,j} + \phi_{i-1,j} + \phi_{i,j+1} + \phi_{i,j-1} - (4+H)\phi_{i,j} = d^2F_{i,j}. \tag{2.71}$$

It should be noted that there is no residual term $R_{i,j}$ in the above equation since $\phi_{i,j}$ is the exact solution. Subtracting (2.71) from (2.68), the error equation at the mth iteration is given by

$$\epsilon_{i+1,j}^m + \epsilon_{i-1,j}^m + \epsilon_{i,j+1}^m + \epsilon_{i,j-1}^m - (4+H)\epsilon_{i,j}^m = R_{i,j}^m, \tag{2.72}$$

where $\epsilon_{i,j}^m$ is the error in $\phi_{i,j}$ at the mth iteration.

From the governing Helmholtz equation we have

$$\hat{\nabla}^2 \in_{i,j}^m - \mu d^2 \in_{i,j}^m = R_{i,j}^m, \tag{2.73}$$

where

$$\hat{\nabla}^2 \in_{i,j} = \in_{i+1,j} + \in_{i-1,j} + \in_{i,j+1} + \in_{i,j-1} - 4 \in_{i,j}.$$

Furthermore, when $\phi_{i,j}^m$ is changed to $\phi_{i,j}^{m+1}$ so as to make the residual term vanish from the above equation, we obtain

$$\in_{i,j}^{m+1} - \in_{i,j}^m = \frac{R_{i,j}^m}{4+H}. \tag{2.74}$$

Using (2.73) and (2.74), we obtain

$$\in_{i,j}^{m+1} = \in_{i,j}^m + \frac{1}{4+H}\left(\hat{\nabla}^2 \in_{i,j}^m - \mu d^2 \in_{i,j}^m\right) \tag{2.75}$$

and the ratio

$$\frac{\in_{i,j}^{m+1}}{\in_{i,j}^m} = \frac{\hat{\nabla}^2 \in_{i,j}^m + 4 \in_{i,j}^m}{(4+H)\in_{i,j}^m}. \tag{2.76}$$

If $(\in_{i,j}^{m+1} / \in_{i,j}^m) < 1$, then the error decreases as we increase the number of iterations, and the scheme converges. Now we ask ourselves, what is the condition that will guarantee convergence?

To examine this we express the error function at grid point (k,l) at iteration m by a *Fourier expansion*, that is,

$$\in_{k,l}^m = \sum_p \sum_q A_{p,q}^m e^{i(pk\Delta x)} e^{i(ql\Delta y)},$$

where p and q denote the east-west and north-south wavenumber, respectively, and m is the number of iterations. We use here the grid-point index (k,l) instead of (i,j) to avoid confusion with $i=(-1)^{1/2}$.

Substituting the above into (2.75), we obtain

$$\sum_p \sum_q A_{p,q}^{m+1} e^{i(pk\Delta x)} e^{i(ql\Delta y)} = \sum_p \sum_q A_{p,q}^m e^{i(pk\Delta x)} e^{i(ql\Delta y)} + \frac{1}{4+H}\left(\hat{\nabla}^2 \in_{i,j}^m - \mu d^2 \sum_p \sum_q A_{p,q}^m e^{i(pk\Delta x)} e^{i(ql\Delta y)}\right). \tag{2.77}$$

Furthermore, the expression for $\hat{\nabla}^2 \in_{k,l}^m$ equal to

$$\hat{\nabla}^2 \in_{k,l}^m = \sum_p \sum_q A_{p,q}^m \left(e^{i[p(k+1)\Delta x]} e^{i(ql\Delta y)} + e^{i[p(k-1)\Delta x]} e^{i(ql\Delta y)} + e^{i(pk\Delta x)} e^{i[q(l+1)\Delta y]} + e^{i(pk\Delta x)} e^{i[q(l-1)\Delta y]} - 4 e^{i(pk\Delta x)} e^{i(ql\Delta y)}\right). \tag{2.78}$$

For the sake of simplicity, we have again assumed that $\Delta x = \Delta y$.

Substituting the above expression for $\hat{\nabla}^2 \in_{k,l}^m$ into (2.77) and equating the coefficients of $e^{i(pk\Delta x)}\, e^{i(ql\Delta y)}$, we can write

$$A_{p,q}^{m+1} = A_{p,q}^m \left(1 + \frac{1}{4+H} \times \left(e^{ip\Delta x} + e^{-ip\Delta x} + e^{iq\Delta y} - 4 - \mu d^2\right)\right). \tag{2.79}$$

Using the Euler relation

$$\cos x = \frac{e^{ix} + e^{-ix}}{2}, \tag{2.80}$$

we obtain

$$A_{p,q}^{m+1} = A_{p,q}^m \left(1 + \frac{1}{4+H}\left(2\cos p\,\Delta x + 2\cos q\,\Delta y - 4\mu d^2\right)\right).$$

Furthermore, using $\cos 2x = 1 - 2\sin^2 x$ and $H = \mu d^2$, we obtain

$$A_{p,q}^{m+1} = A_{p,q}^m \left[1 + \frac{1}{4+H}\left(2 - 4\sin^2\frac{p\,\Delta x}{2} + 2 - 4\sin^2\frac{q\,\Delta y}{2} - 4 - H\right)\right],$$

or

$$A_{p,q}^{m+1} = A_{p,q}^m \left[1 - \frac{1}{4+H}\left(4\sin^2\frac{p\,\Delta x}{2} + 4\sin^2\frac{q\,\Delta y}{2} + H\right)\right], \tag{2.81}$$

so that

$$\frac{A_{p,q}^{m+1}}{A_{p,q}^m} = 1 - \frac{1}{4+H}\left(4\sin^2\frac{p\,\Delta x}{2} + 4\sin^2\frac{q\,\Delta y}{2} + H\right). \tag{2.82}$$

Let v be defined by

$$v^2 = \sin^2\frac{p\,\Delta x}{2} + \sin^2\frac{p\,\Delta y}{2} \le 2. \tag{2.83}$$

Hence we obtain

$$\frac{A_{p,q}^{m+1}}{A_{p,q}^m} = 1 - \frac{1}{4+H}\left(4v^2 + H\right). \tag{2.84}$$

The ratio on the left-hand side of the above equation can satisfy one of the following possible scenarios. If

$$\frac{A_{p,q}^{m+1}}{A_{p,q}^m} < 1,$$

then we have strong convergence. If

$$\frac{A_{p,q}^{m+1}}{A_{p,q}^m} = 1,$$

then this is the neutral or nonamplifying case. If

$$\frac{A_{p,q}^{m+1}}{A_{p,q}^{m}} > 1,$$

then the scheme will not converge. We can increase the convergence role of a relaxation scheme by the use of an appropriate relaxation factor. The change in $\phi_{i,j}$ from iteration m to $m+1$ is

$$\phi_{i,j}^{m+1} - \phi_{i,j}^{m} = \delta\phi_{i,j}^{m} = \frac{R_{i,j}^{m}}{4+H}. \tag{2.85}$$

In the case of over relaxation or underrelaxation, we change $\phi_{i,j}^{m}$ by $\hat{\delta}\phi_{i,j}^{m}$, where

$$\hat{\delta}\phi_{i,j}^{m} = \alpha\delta\phi_{i,j}^{m} = \alpha\frac{R_{i,j}^{m}}{4+H}. \tag{2.86}$$

Here α is a *relaxation factor.*

Next we show that this relaxation factor will result in faster convergence. If $\alpha = 1$, we are back to $\delta\phi_{i,j}^{m} = R_{i,j}^{m}/(4+H)$. For $\alpha \neq 1$ and for larger scales, $p\Delta x$ and $q\Delta y$ are small. The term v^2 is very small if we choose $\alpha > 1$. Then (2.84) becomes

$$\left|\frac{A_{p.q}^{m+1}}{A_{p,q}^{m}}\right|_{\alpha} = 1 - \frac{\alpha}{4+H}(4v^2 + H). \tag{2.87}$$

We can select the values of the relaxation factor α such that

$$\left|\frac{A_{p.q}^{m+1}}{A_{p,q}^{m}}\right|_{\alpha} < \left|\frac{A_{p.q}^{m+1}}{A_{p,q}^{m}}\right|, \tag{2.88}$$

where $\left|A_{p,q}^{m+1} / A_{p,q}^{m}\right|_{\alpha}$ denotes that we are using the relaxation factor α which is not equal to 1.

For large scales (i.e., for small p and q) in which

$$\frac{4v^2 + H}{4+H} << 1,$$

an overrelaxation factor $(\alpha > 1)$ would give rise to a faster rate of convergence. However, if scales are small (i.e., p and q are larger), then overrelaxation does not work. In that case, an underrelaxation coefficient $(\alpha < 1)$ is better suited. An overrelaxation coefficient with values around 1.2 to 1.7, depending up the domain, is generally helpful for fast convergence. The scale of the wave depends on the domain. The larger the domain, then generally the larger the wavenumbers p and q will be.

2.11 Sequential Relaxation Versus Simultaneous Relaxation

Here we answer the question, is sequential relaxation more efficient than simultaneous relaxation? We consider a Helmholtz equation in one dimension only, i.e.,

$$\frac{\partial^2 \phi}{\partial x^2} - \mu\phi = F\,. \tag{2.89}$$

To make the problem simple, we consider the domain consisting of points $i-2$, $i-1$, $i+1$, and $i+2$, where $i-2$ and $i+2$ are boundary grid points with known prescribed values of ϕ_{i-2} and ϕ_{i+2}.

We can express $\partial^2\phi/\partial x^2\,|_i$ in its finite-difference form as

$$\left.\frac{\partial^2 \phi}{\partial x^2}\right|_i \cong \frac{\phi_{i+1}+\phi_{i-1}-2\phi_i}{(\Delta x)^2}, \tag{2.90}$$

where Δx is the distance between two successive grid points and ϕ_i is the value of ϕ at the ith grid point. Substituting this into (2.89), we obtain

$$\phi_{i+1}+\phi_{i-1}-2\phi_i-\mu(\Delta x)^2\phi_i = F_i(\Delta x)^2. \tag{2.91}$$

If $H=\mu(\Delta x)^2$ and $(\Delta x)^2 = d^2$, then we obtain

$$\phi_{i-1}-(2+H)\phi_i+\phi_{i+1} = F_i d^2. \tag{2.92}$$

First consider simultaneous relaxation. At the mth level of iteration we have

$$\phi_{i-1}^m-(2+H)\phi_i^m+\phi_{i+1}^m = F_i d^2 + R_i^m. \tag{2.93}$$

At the $(m+1)$th iteration, ϕ_i^m is changed to ϕ_i^{m+1} so that R_i^m vanishes. This gives

$$\phi_{i-1}^m-(2+H)\phi_i^{m+1}+\phi_{i+1}^m = F_i d^2. \tag{2.94}$$

The corresponding error equations are

$$\epsilon_{i-1}^m-(2+H)\epsilon_i^{m+1}+\epsilon_{i+1}^m=0 \text{ at grid point } i, \tag{2.95}$$

$$0-(2+H)\epsilon_{i-1}^{m+1}+\epsilon_i^m=0 \text{ at grid point } i-1, \tag{2.96}$$

$$\epsilon_i^m-(2+H)\epsilon_{i+1}^{m+1}=0 \text{ at grid point } i+1. \tag{2.97}$$

It should be noted that there are no terms containing ϵ_{i-2}^m and ϵ_{i+2}^m, since ϕ_{i-2} and ϕ_{i+2} are known boundary values and only ϕ_{i-1}, ϕ_i, and ϕ_{i+1} are to be determined.

One additional iteration at point i results in

$$\epsilon_{i-1}^{m+1}-(2+H)\epsilon_i^{m+2}+\epsilon_{i+1}^{m+1}=0.$$

After eliminating ϵ_{i-1}^{m+1} and ϵ_{i+1}^{m+1} with the help of the error equations (2.96) and (2.97) at grid points $(i-1)$ and $(i+1)$, we obtain

$$\frac{\epsilon_i^m}{2+H}-(2+H)\epsilon_i^{m+2}+\frac{\epsilon_i^m}{2+H}=0, \tag{2.98}$$

$$2\epsilon_i^m-(2+H)^2\epsilon_i^{m+2}=0, \tag{2.99}$$

$$\in_i^{m+2} = \frac{2}{(2+H)^2} \in_i^m . \tag{2.100}$$

Next consider the sequential relaxation procedure. Here we use the new values of the function at the previously corrected points during an iteration. The corresponding error equation is

$$\in_{i-1}^{m+1} - (2+H) \in_i^{m+1} + \in_{i+1}^m = 0 \quad \text{at grid point } i. \tag{2.101}$$

We have used the changed value of ϕ at the $(i$ -1)th grid point. Furthermore,

$$0 - (2+H) \in_{i-1}^{m+1} + \in_i^m = 0 \quad \text{at grid point } i - 1. \tag{2.102}$$

Similarly, we obtain

$$\in_i^{m+1} - (2+H) \in_{i+1}^{m+1} + 0 = 0 \quad \text{at grid point } i + 1. \tag{2.103}$$

The error at the $(i$ -1)th grid point is related to the error at the ith grid point by

$$\in_{i-1}^{m+1} = \frac{\in_i^m}{2+H} . \tag{2.104}$$

Substituting the above value of $\in_{i-1}^{m+1}$ into (2.101) gives

$$\frac{\in_i^m}{2+H} - (2+H) \in_i^{m+1} + \in_{i+1}^m = 0 ,$$

or

$$\in_i^m - (2+H)^2 \in_i^{m+1} + (2+H) \in_{i+1}^m = 0 . \tag{2.105}$$

From the error equation at the $(i+1)$th grid point at the mth iteration, we can get $\in_{i+1}^m = \in_i^m / (2+H)$. Substituting into (2.105), we obtain

$$\in_i^m - (2+H)^2 \in_i^{m+1} + \in_i^m = 0 ,$$

or

$$\in_i^{m+1} = \frac{2}{(2+H)^2} \in_i^m . \tag{2.106}$$

Thus, the convergence given by (2.106) is twice as fast as that given by (2.100). In conclusion, *sequential relaxation converges faster than simultaneous relaxation.*

2.12 Advective Nonlinear Dynamics

Advective nonlinear dynamics is one of the most difficult processes to resolve. It creates a lot of noise in a forecast. Arakawa was the first to show how to correctly use finite differencing for nonlinear dynamics. He explained that to understand the numerics of nonlinear advective dynamics you must first be able to compute the finite difference of the Jacobian, which is simple nonlinear advective dynamics. The barotropic vorticity equation allows us to understand the properties of the Jacobian from which we can then ensure that the numerics meet these requirements.

2.12.1 *Barotropic Vorticity Equation*

Consider the horizontal momentum equations and the nondivergent continuity equation in the (x, y, p) coordinate system which make up the primitive form of the barotropic vorticity equation:

$$\frac{\partial u}{\partial t}+u\frac{\partial u}{\partial x}+v\frac{\partial u}{\partial y}=fv-g\frac{\partial z}{\partial x}, \quad (2.107)$$

$$\frac{\partial v}{\partial t}+u\frac{\partial v}{\partial x}+v\frac{\partial v}{\partial y}=-fu-g\frac{\partial z}{\partial x}, \quad (2.108)$$

$$\frac{\partial u}{\partial x}+\frac{\partial v}{\partial y}=0. \quad (2.109)$$

Here we have three equations with three unknowns: u, v, and z. For nondivergent flow we let $u=-\partial\psi/\partial y$, $v=\partial\psi/\partial x$, and $\nabla^2\psi=\partial v/\partial x-\partial u/\partial y$. By differentiating (2.107) with respect to y and (2.108) with respect to x and subtracting the two resulting equations, we get the barotropic vorticity equation:

$$\frac{\partial}{\partial t}\nabla^2\psi=-J\left(\psi,\nabla^2\psi\right)-\beta\frac{\partial\psi}{\partial x}, \quad (2.110)$$

which consists of the time tendency of vorticity, the horizontal advection of vorticity represented by the Jacobian function, and the beta term, respectively, where $\beta=\partial f/\partial y$. In this equation, the Jacobian is a nonlinear term and all other terms are linear. Also differentiating (2.107) with respect to x and (2.108) with respect to y and adding, we obtain the nonlinear balanced equation:

$$\nabla^2\phi=\nabla\cdot f\nabla\psi+2J\left(\frac{\partial\psi}{\partial x},\frac{\partial\psi}{\partial y}\right), \quad (2.111)$$

where we have neglected the time-derivative terms. The *Jacobian function J* is defined as

$$J(A,B)=\frac{\partial A}{\partial x}\frac{\partial B}{\partial y}-\frac{\partial A}{\partial y}\frac{\partial B}{\partial x}.$$

Here we have two equations and two unknowns, ψ and ϕ.

Arakawa showed that the barotropic vorticity model (2-D) has several integral constraints. These constraints that are maintained are kinetic energy, total vorticity, and total square vorticity. That is, the kinetic energy of nondivergent flow and the absolute vorticity have the following invariant properties over a closed domain:

$$\text{domain invariant}: \overline{\overline{\frac{u^2+v^2}{2}}},\ \overline{\overline{\zeta_a}},\ \overline{\overline{\zeta_a^n}}.$$

In this case, $\zeta_a = \zeta + f$, n is a real number, and $\overline{\overline{(\ )}}$ denotes the mean over the domain, i.e.,

$$\overline{\overline{Q}} = \frac{\iint Q\, dx\, dy}{\iint dx\, dy}.$$

Conservation of these properties precludes nonlinear computational instability. Conservation of enstrophy ($\overline{\overline{\zeta_a^2}}$) and kinetic energy is called quadratic invariance. If one satisfies quadratic invariance, then $\overline{\overline{\zeta_a^n}}$ for $n > 2$ also seems to remain well bounded. The *parcel-invariant* physical quantities are ζ_a and ζ_a^n. If we set $n = 2$ in the *domain-invariant* quantity $\overline{\overline{\zeta_a^n}}$, then we obtain $\overline{\overline{\zeta_a^2}}$ as invariant. These are important invariants (both parcel and domain).

Next we prove the domain invariance for ζ_a^2. We start with barotropic vorticity equation (2.110), which can be written as

$$\frac{\partial}{\partial t}\left(\nabla^2\psi + f\right) = -J\left(\psi,\ \nabla^2\psi + f\right). \tag{2.112}$$

By multiplying (2.112) by $\nabla^2\psi + f$, we obtain

$$\frac{\partial}{\partial t}\frac{\left(\nabla^2\psi + f\right)^2}{2} = -J\left(\psi,\ \frac{\left(\nabla^2\psi + f\right)^2}{2}\right)$$

since

$$\begin{aligned} J(A,B) &= \frac{\partial A}{\partial x}\frac{\partial B}{\partial y} - \frac{\partial A}{\partial y}\frac{\partial B}{\partial x} \\ B\, J(A,B) &= \frac{\partial A}{\partial x}\frac{\partial}{\partial y}\frac{B^2}{2} - \frac{\partial A}{\partial y}\frac{\partial}{\partial x}\frac{B^2}{2} \\ &= J\left(A, B^2/2\right). \end{aligned}$$

Integrating over the closed domain D leads to

$$\iint_D \frac{\partial}{\partial t}\frac{\left(\nabla^2\psi + f\right)^2}{2}\, dx\, dy = -\iint_D J\left(\psi,\ \frac{\left(\nabla^2\psi + f\right)^2}{2}\right) dx\, dy = 0,$$

since the integral of a Jacobian over a closed domain vanishes. Hence

$$\frac{\partial}{\partial t}\overline{\overline{\frac{(\nabla^2\psi + f)^2}{2}}} = 0. \tag{2.113}$$

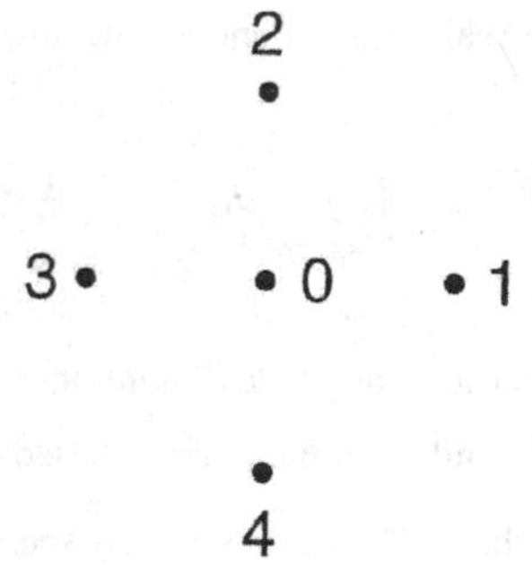

Figure 2.10. A 5-point stencil for the Jacobian using the vector identity.

Therefore

$$\overline{\overline{\frac{(\nabla^2\psi + f)^2}{2}}}$$

is invariant with time. Thus the square of absolute vorticity is conserved over a closed domain. In the same manner, it can be shown that all powers of absolute vorticity (i.e., ζ_a^n) are time-invariant.

Next we show the conservation of $\overline{\overline{k}} = \overline{\overline{(u^2+v^2)/2}}$. Multiply (2.112) by ψ to obtain

$$\psi \frac{\partial}{\partial t}\nabla^2\psi = -J\left(\frac{\psi^2}{2}, \nabla^2\psi + f\right). \tag{2.114}$$

The right-hand side vanishes on integration over a closed domain. Hence

$$\iint_D \psi \frac{\partial}{\partial t}\nabla^2\psi \, dx\, dy = 0. \tag{2.115}$$

Using the vector identity

$$\psi \frac{\partial}{\partial t}\nabla^2\psi = \nabla\cdot\left(\psi\nabla\frac{\partial\psi}{\partial t}\right) - \frac{\partial}{\partial t}\frac{\nabla\psi.\nabla\psi}{2},$$

we obtain

$$\iint_D \psi \frac{\partial}{\partial t}\nabla^2\psi \, dx\, dy = \iint_D \nabla\cdot\left(\psi\nabla\frac{\partial\psi}{\partial t}\right) dx\, dy - \iint_D \frac{\partial}{\partial t}\frac{\nabla\psi\cdot\nabla\psi}{2}\, dx\, dy. \tag{2.116}$$

As the first term on the right-hand side of (2.116) vanishes over a closed domain, we are left with

$$\iint_D \frac{\partial}{\partial t}\frac{(\nabla\psi)^2}{2}\,dx\,dy = \iint_D \psi\frac{\partial}{\partial t}\nabla^2\psi\,dx\,dy = 0,$$

or

$$\frac{\partial}{\partial t}\overline{\overline{\frac{(\nabla\psi)^2}{2}}} = \frac{\partial}{\partial t}\left[\overline{\overline{\left(\frac{\partial\psi}{\partial x}\hat{i}+\frac{\partial\psi}{\partial y}\hat{j}\right)^2}}\Big/2\right]$$

$$= \frac{\partial}{\partial t}\left[\overline{\overline{\left(\frac{\partial\psi}{\partial x}\right)^2+\left(\frac{\partial\psi}{\partial y}\right)^2}}\Big/2\right]$$

$$= \frac{\partial}{\partial t}\overline{\overline{\frac{u^2+v^2}{2}}} = 0. \tag{2.117}$$

Thus the kinetic energy over a closed domain is time-invariant. In designing any simple atmospheric model, we generally try to write the finite differencing schemes so as to satisfy conservation of kinetic energy $(\overline{\overline{k}})$, vorticity $(\overline{\overline{\zeta_a}})$, and the square of vorticity $(\overline{\overline{\zeta_a^2}})$ over a closed domain.

2.13 The 5-Point Jacobian

Let us first consider a 5-point Jacobian. Given the definition

$$J(\zeta,\psi) = \frac{\partial\zeta}{\partial x}\frac{\partial\psi}{\partial y} - \frac{\partial\zeta}{\partial y}\frac{\partial\psi}{\partial x}$$

and starting with the five grid points as shown in Fig. 2.10, the Jacobian $J(\zeta, \psi)$ can be expressed by

$$J(\zeta,\psi) \cong \frac{\zeta_1-\zeta_3}{2\Delta x}\frac{\psi_2-\psi_4}{2\Delta y} - \frac{\zeta_2-\zeta_4}{2\Delta y}\frac{\psi_1-\psi_3}{2\Delta x}.$$

If $\Delta x = \Delta y = \Delta$, then

$$J(\zeta,\psi) = \frac{1}{4\Delta^2}\left[(\zeta_1-\zeta_3)(\psi_2-\psi_4)-(\zeta_2-\zeta_4)(\psi_1-\psi_3)\right]+O(\Delta^2). \tag{2.118}$$

This is a *second-order accurate Jacobian*. It should be noted that this Jacobian does not satisfy all invariants.

2.14 Arakawa Jacobian

We discuss here the design of a Jacobian which conserves kinetic energy and enstrophy (square of the vorticity) over a closed domain. This form of the Jacobian was first designed by Arakawa (1966), and is commonly known as the Arakawa Jacobian. We consider both the second- and fourth-order accurate Arakawa Jacobians.

We can write $J(\zeta,\psi)$ in the following three forms:

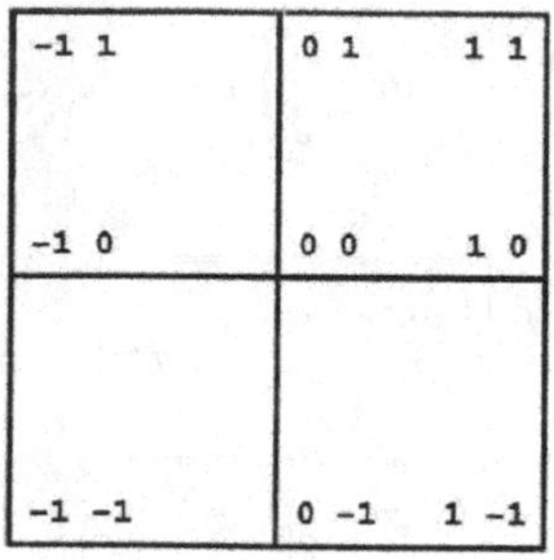

Figure 2.11. A 9-point stencil for the Jacobian.

$$J(\zeta,\psi)=\frac{\partial\zeta}{\partial x}\frac{\partial\psi}{\partial y}-\frac{\partial\zeta}{\partial y}\frac{\partial\psi}{\partial x}, \tag{2.119}$$

$$J(\zeta,\psi)=\frac{\partial}{\partial x}\left(\zeta\frac{\partial\psi}{\partial y}\right)-\frac{\partial}{\partial y}\left(\zeta\frac{\partial\psi}{\partial x}\right), \tag{2.120}$$

$$J(\zeta,\psi)=\frac{\partial}{\partial y}\left(\psi\frac{\partial\zeta}{\partial x}\right)-\frac{\partial}{\partial x}\left(\psi\frac{\partial\zeta}{\partial y}\right). \tag{2.121}$$

Analytically these three forms of $J(\zeta,\psi)$ have the same value. However, in their finite-difference form they are not identical. For this consider the 9-point stencil with its different points labeled as in Fig. 2.11 with grid distance d.

With respect to Fig. 211, the finite-difference form of the right-hand side of (2.119), (2.120), and (2.121) may be written as

$$\begin{aligned}J_{00}^{++}(\zeta,\psi)=\frac{1}{4d^2}\big[&(\zeta_{10}-\zeta_{-10})(\psi_{01}-\psi_{0-1})\\&-(\zeta_{01}-\zeta_{0-1})(\psi_{10}-\psi_{-10})\big],\end{aligned} \tag{2.122}$$

$$\begin{aligned}J_{00}^{+\times}(\zeta,\psi)=\frac{1}{4d^2}\big[&\zeta_{10}(\psi_{11}-\psi_{1-1})-\zeta_{10}(\psi_{-11}-\psi_{-1-1})\\&-\zeta_{01}(\psi_{11}-\psi_{-11})+\zeta_{0-1}(\psi_{1-1}-\psi_{-1-1})\big],\end{aligned} \tag{2.123}$$

$$\begin{aligned}J_{00}^{\times+}(\zeta,\psi)=\frac{1}{4d^2}\big[&\psi_{01}(\zeta_{11}-\zeta_{-11})-\psi_{0-1}(\zeta_{1-1}-\zeta_{1-1})\\&-\psi_{10}(\zeta_{11}-\zeta_{1-1})-\psi_{-10}(\zeta_{-11}-\zeta_{1-1})\big],\end{aligned}$$

which can also be written as

$$\begin{aligned}J_{00}^{\times+}(\zeta,\psi)=\frac{1}{4d^2}\big[&\zeta_{11}(\psi_{01}-\psi_{10})-\zeta_{1-1}(\psi_{-10}-\psi_{0-1})\\&-\zeta_{11}(\psi_{01}-\psi_{-10})-\zeta_{1-1}(\psi_{10}-\psi_{0-1})\big].\end{aligned} \tag{2.124}$$

Superscripts $^{\times}$ and $^{+}$ denote the location of ζ and ψ values (in that order) involved in the finite-difference approximation of the Jacobian. The $^{+}$ symbol indicates that values involved are located at points 10, -10, 01, and 0-1, while the ^{x} symbol indicates their location at points 11, -1-1, -11, and 1-1. The subscripts denote the point at which the Jacobian is centered.

Let us now consider the conservation of mean-square vorticity, for which the condition is

$$\overline{\zeta J(\zeta,\psi)}=0. \tag{2.125}$$

In finite-difference form this condition is satisfied if the products of ζJ at various grid points in the finite-difference from of (2.125) cancel when added over the closed domain.

Now from (2.122) we obtain nine equations, one equation for each grid point of our 9-point stencil

$$\zeta_{00}J_{00}^{++}(\zeta,\psi)=\frac{1}{4d^2}\left[\zeta_{00}\zeta_{10}(\psi_{01}-\psi_{0-1})+\ldots\right], \tag{2.126}$$

$$\zeta_{10}J_{10}^{++}(\zeta,\psi)=\frac{1}{4d^2}\left[-\zeta_{10}\zeta_{00}(\psi_{11}-\psi_{1-1})+\ldots\right], \tag{2.127}$$

and so on. Also from (2.123) we get

$$\zeta_{00}J_{00}^{+\times}(\zeta,\psi)=\frac{1}{4d^2}\left[\zeta_{00}\zeta_{10}(\psi_{11}-\psi_{1-1})+\ldots\right], \tag{2.128}$$

$$\zeta_{10}J_{10}^{+\times}(\zeta,\psi)=\frac{1}{4d^2}\left[-\zeta_{10}\zeta_{00}(\psi_{01}-\psi_{0-1})+\ldots\right], \tag{2.129}$$

and so on. Similarly from (2.124) we have

$$\zeta_{00}J_{00}^{\times+}(\zeta,\psi)=\frac{1}{4d^2}\left[\zeta_{00}\zeta_{11}(\psi_{01}-\psi_{10})+\ldots\right], \tag{2.130}$$

$$\zeta_{11}J_{11}^{\times+}(\zeta,\psi)=\frac{1}{4d^2}\left[-\zeta_{11}\zeta_{00}(\psi_{01}-\psi_{10})+\ldots\right], \tag{2.131}$$

and so on. From (2.126) and (2.127) we see that the sum of the terms involving the product $\zeta_{00}\zeta_{10}$ does not vanish. The various terms of $\zeta J^{++}(\zeta,\psi)$ therefore do not cancel when added over the whole closed domain. Therefore the domain integral of $\zeta J^{++}(\zeta,\psi)$ does not vanish.

From (2.128) and (2.129) we find that the same is true for $\zeta J^{+\times}(\zeta,\psi)$, that is, its domain integral also does not vanish. However, as the terms involving $\zeta_{00}\zeta_{10}$ in (2.126) and (2.129) are equal and opposite in sign, as are those in (2.127) and (2.128), various terms of the sum $\zeta J^{++}(\zeta,\psi)+\zeta J^{+\times}(\zeta,\psi)$ cancel, that is,

$$\overline{\zeta J^{++}(\zeta,\psi)}+\overline{\zeta J^{+x}(\zeta,\psi)}=0. \tag{2.132}$$

Also terms involving $\zeta_{00}\zeta_{10}$ in (2.130) and (2.131) are equal and opposite. Therefore various terms of $\zeta J^{\times+}(\zeta,\psi)$ cancel on addition over a closed domain. Thus

$$\overline{\zeta J^{\times+}(\zeta,\psi)} = 0\,. \tag{2.133}$$

Thus $J^{++}(\zeta,\psi)+J^{+\times}(\zeta,\psi)$ and $J^{\times+}(\zeta,\psi)$ individually conserve mean-square vorticity. Similarly it can be shown that

$$\overline{\psi J^{++}(\zeta,\psi)}+\overline{\psi J^{\times+}(\zeta,\psi)} = 0\,, \tag{2.134}$$

and

$$\overline{\psi J^{+\times}(\zeta,\psi)} = 0\,, \tag{2.135}$$

so that $J^{++}(\zeta,\psi)+J^{\times+}(\zeta,\psi)$ and $J^{+\times}(\zeta,\psi)$ individually conserve mean kinetic energy.

From these we can take the three forms of the Jacobian, take a weighted average of the three, and add them up,

$$\alpha J^{++}(\varsigma,\psi)+\beta J^{+\times}(\varsigma,\psi)+\gamma J^{+\times}(\varsigma,\psi)=0\,,$$

such that $\alpha+\beta+\gamma=1$. Arakawa searched for values of α, β, and γ that satisfied the invariants and found that $\alpha=\beta=\gamma=\frac{1}{3}$ best satisfies the invariants. From this, we conclude that the finite difference Jacobian is

$$J_1(\zeta,\psi)=\frac{1}{3}\left[J^{++}(\zeta,\psi)+J^{+\times}(\zeta,\psi)+J^{\times+}(\zeta,\psi)\right] \tag{2.136}$$

as it conserves the square of vorticity and kinetic energy over a closed domain. To investigate the accuracy of the finite-difference scheme, we expand ζ and ψ in Taylor series:

$$F_{\pm 10} = F_{00} \pm d\left(\frac{\partial F}{\partial x}\right)_{00} + \frac{d^2}{2}\left(\frac{\partial^2 F}{\partial x^2}\right)_{00} \pm \frac{d^3}{6}\left(\frac{\partial^3 F}{\partial x^3}\right)_{00} + \mathrm{O}\left(d^4\right),$$

$$F_{0\pm 1} = F_{00} \pm d\left(\frac{\partial F}{\partial y}\right)_{00} + \frac{d^2}{2}\left(\frac{\partial^2 F}{\partial y^2}\right)_{00} \pm \frac{d^3}{6}\left(\frac{\partial^3 F}{\partial y^3}\right)_{00} + \mathrm{O}\left(d^4\right),$$

$$\begin{aligned} F_{1\pm 1} = F_{00} &+ d\left(\frac{\partial F}{\partial x} \pm \frac{\partial F}{\partial y}\right)_{00} + \frac{d^2}{2}\left(\frac{\partial^2 F}{\partial x^2} \pm 2\frac{\partial^2 F}{\partial x \partial y} + \frac{\partial^2 F}{\partial y^2}\right)_{00} \\ &+ \frac{d^3}{6}\left(\frac{\partial^3 F}{\partial x^3} \pm 3\frac{\partial^3 F}{\partial x^2 \partial y} + 3\frac{\partial^3 F}{\partial x \partial y^2} \pm \frac{\partial^3 F}{\partial y^3}\right)_{00} + \mathrm{O}\left(d^4\right), \end{aligned}$$

$$\begin{aligned} F_{-1\pm 1} = F_{00} &- d\left(\frac{\partial F}{\partial x} \mp \frac{\partial F}{\partial y}\right)_{00} + \frac{d^2}{2}\left(\frac{\partial^2 F}{\partial x^2} \mp 2\frac{\partial^2 F}{\partial x \partial y} + \frac{\partial^2 F}{\partial y^2}\right)_{00} \\ &- \frac{d^3}{6}\left(\frac{\partial^3 F}{\partial x^3} \mp 3\frac{\partial^3 F}{\partial x^2 \partial y} + 3\frac{\partial^3 F}{\partial x \partial y^2} \mp \frac{\partial^3 F}{\partial y^3}\right)_{00} + \mathrm{O}\left(d^4\right). \end{aligned}$$

In these expansions, F represents either ζ or ψ. After substituting these expansions into (2.122), (2.123), and (2.124), we obtain

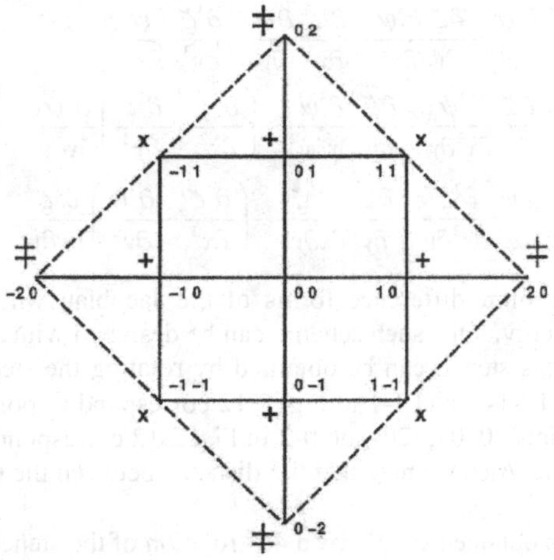

Figure 2.12. A 13-point stencil for the Jacobian.

$$J^{++}(\zeta,\psi)=J(\zeta,\psi)+\frac{d^2}{6}\left(\frac{\partial\zeta}{\partial x}\frac{\partial^3\psi}{\partial y^3}-\frac{\partial\zeta}{\partial y}\frac{\partial^3\psi}{\partial x^3}+\frac{\partial^3\zeta}{\partial x^3}\frac{\partial\psi}{\partial y}-\frac{\partial^3\zeta}{\partial y^3}\frac{\partial\psi}{\partial x}\right)+O\left(d^4\right), \tag{2.137}$$

$$J^{+\times}(\zeta,\psi)=J^{++}(\zeta,\psi)+\frac{d^2}{2}\left[\frac{\partial\zeta}{\partial x}\frac{\partial^3\psi}{\partial x^2\partial y}-\frac{\partial\zeta}{\partial y}\frac{\partial^3\psi}{\partial x\partial y^2}+\left(\frac{\partial^2\zeta}{\partial x^2}-\frac{\partial^2\zeta}{\partial y^2}\right)\frac{\partial^2\psi}{\partial x\partial y}\right]+O\left(d^4\right), \tag{2.138}$$

$$J^{\times+}(\zeta,\psi)=J^{++}(\zeta,\psi)-\frac{d^2}{2}\left[\frac{\partial\psi}{\partial x}\frac{\partial^3\zeta}{\partial x^2\partial y}-\frac{\partial\psi}{\partial y}\frac{\partial^3\zeta}{\partial x\partial y^2}+\left(\frac{\partial^2\psi}{\partial x^2}-\frac{\partial^2\psi}{\partial y^2}\right)\frac{\partial^2\zeta}{\partial x\partial y}\right]+O\left(d^4\right). \tag{2.139}$$

Note that equations (2.138) and (2.139) contain all of equation (2.137). Therefore the second-order accurate Jacobian is

$$\begin{aligned}J_1(\zeta,\psi)&=\frac{1}{3}\left[J^{++}(\zeta,\psi)+J^{+\times}(\zeta,\psi)+J^{\times+}\right]\\&=J(\zeta,\psi)+\in d^2+O\left(d^4\right),\end{aligned} \tag{2.140}$$

where

$$\begin{aligned}
6\in &= \frac{\partial\zeta}{\partial x}\frac{\partial^3\psi}{\partial y^3} - \frac{\partial\zeta}{\partial y}\frac{\partial^3\psi}{\partial x^3} + \frac{\partial^3\zeta}{\partial x^3}\frac{\partial\psi}{\partial y} - \frac{\partial^3\zeta}{\partial y^3}\frac{\partial\psi}{\partial x} \\
&+ \frac{\partial\zeta}{\partial x}\frac{\partial^3\psi}{\partial x^2\partial y} - \frac{\partial\zeta}{\partial y}\frac{\partial^3\psi}{\partial x\partial y^2} + \left(\frac{\partial^2\zeta}{\partial x^2} - \frac{\partial^2\zeta}{\partial y^2}\right)\frac{\partial^2\psi}{\partial x\partial y} \\
&- \frac{\partial\psi}{\partial x}\frac{\partial^3\zeta}{\partial x^2\partial y} + \frac{\partial\psi}{\partial y}\frac{\partial^3\zeta}{\partial x\partial y^2} - \left(\frac{\partial^2\zeta}{\partial x^2} - \frac{\partial^2\psi}{\partial y^2}\right)\frac{\partial^2\zeta}{\partial x\partial y}.
\end{aligned} \tag{2.141}$$

There are other finite-difference forms of the Jacobian which conserve square vorticity and kinetic energy. One such scheme can be designed with a 13-point stencil as shown in Fig. 2.12. This stencil can be obtained by rotating the stencil in Fig. 2.11 by 45° so that point 11, -11, -1-1, and 1-1 in Fig. 2.12 correspond to points 10, 01, -10, and 0-1 in Fig. 2.11 and points 20, 02, -20, and 0-2 in Fig. 2.12 correspond to points 1-1, 11, -11, and -1-1 in Fig. 2.11. Also we note that the distance between the points (i.e., between 00 and 11) is now $2^{1/2}d$.

As this stencil is obtained simply by a 45° rotation of the stencil in Fig. 2.11,

$$J_2(\zeta,\psi) = \frac{1}{3}\left[J^{\times\times}(\zeta,\psi) + J^{\times\ddagger}(\zeta,\psi) + J^{\ddagger\times}(\zeta,\psi)\right] \tag{2.142}$$

will also conserve kinetic energy and the square of vorticity, where

$$\begin{aligned}
J_{00}^{\times\times}(\zeta,\psi) = \frac{1}{8d^2}\big[&(\zeta_{11} - \zeta_{-1-1})(\psi_{-11} - \psi_{1-1}) \\
&-(\zeta_{-11} - \zeta_{1-1})(\psi_{11} - \psi_{-1-1})\big],
\end{aligned} \tag{2.143}$$

$$\begin{aligned}
J_{00}^{\times\ddagger}(\zeta,\psi) = \frac{1}{8d^2}\big[&\zeta_{11}(\psi_{02} - \psi_{20}) - \zeta_{1-1}(\psi_{-20} - \psi_{0-2}) \\
&-\zeta_{11}(\psi_{02} - \psi_{-20}) + \zeta_{1-1}(\psi_{20} - \psi_{0-2})\big],
\end{aligned} \tag{2.144}$$

$$\begin{aligned}
J_{00}^{\ddagger\times}(\zeta,\psi) = \frac{1}{8d^2}\big[&\zeta_{20}(\psi_{11} - \psi_{1-1}) - \zeta_{20}(\psi_{-11} - \psi_{-1-1}) \\
&-\zeta_{02}(\psi_{11} - \psi_{-11}) + \zeta_{0-2}(\psi_{1-1} - \psi_{-1-1})\big].
\end{aligned} \tag{2.145}$$

As $J_2(\zeta,\psi)$ also has centered differences, it is a second-order accurate Jacobian as well.

Using Taylor series we get an equation similar to (2.140), that is,

$$\begin{aligned}
J_2(\zeta,\psi) &= \frac{1}{3}\left[J^{\times\times}(\zeta,\psi) + J^{\times\ddagger}(\zeta,\psi) + J^{\ddagger\times}(\zeta,\psi)\right] \\
&= J(\zeta,\psi) + 2\in d^2 + O(d^4),
\end{aligned} \tag{2.146}$$

where $2\in d^2$ is from $(2^{1/2}d)^2$. From (2.140) and (2.146) we get

$$J_3(\zeta,\psi) = 2J_1(\zeta,\psi) - J_2(\zeta,\psi) = J(\zeta,\psi) + O(d^4) \tag{2.147}$$

as the fourth-order accurate Arakawa Jacobian, which conserves kinetic energy and mean-square vorticity.

Now that we have computed the fourth-order accurate Jacobian we ask what is the importance of finding it? If we were to make a barotropic forecast, which utilizes both the Laplacian and Jacobian,

$$\frac{\partial}{\partial t}\nabla^2\psi = -J\left(\psi,\nabla^2\psi\right) - \beta\frac{\partial\psi}{\partial x}$$

and used a fourth-order accurate scheme for each the Laplacian, Jacobian, and $\partial/\partial x$, we could obtain a good forecast. In other words, understanding of fourth-order accurate schemes makes for better forecasts.

2.15 Exercises

2.1. Consider the function

x	0	$\pi/6$	$\pi/4$	$\pi/2$
$u(x)$	0	1/2	$1/(2^{1/2})$	1.0

Evaluate du/dx at $x=\pi/4$ using (a) forward differencing, (b) centered differencing, (c) backward differencing, and (d) the second-order finite-difference analog. Analytically, $u(x)$ represents $\sin x$. Which of the finite-difference formulas represents the best approximation to $du/dx\big|_{x=\pi/4}$?

2.2. For the function of Exercise 2.1, evaluate $u(x)$ at $x=7\pi/12$, given that the second-order derivative of $u(x)$ at $x=\pi/2$ is exactly equal to its analytic value.

2.3. Consider Poisson's equation $\nabla^2\psi = G$, where $\psi=\psi(x,y)$ and $G=G(x,y)$. Assume an even mesh in x and y and that $0\le x\le 10$, $0\le y\le 10$. Also take $\Delta x=\Delta y=1$ and let

$$\psi(x,0)=\psi(x,10)=0 \qquad 0\le x\le 10,$$
$$\psi(0,y)=\psi(10,y)=0 \qquad 0\le y\le 10,$$
$$G(x,y)=S(x,y)=0 \qquad 0\le x\le 10 \text{ and } 0\le y\le 10.$$

Find ψ if $S=0$, $S=2(x^2+y^2)$, $S=2\exp(x+y)/e^{10}$, and $S=-2\cos x\sin y$. Plot the analytic solution as well as the solution obtained using the finite-difference method.

2.4. Let

$$J=J\left(\varphi,\psi\right)=\frac{\partial\varphi}{\partial x}\frac{\partial\psi}{\partial y}-\frac{\partial\varphi}{\partial y}\frac{\partial\psi}{\partial x}$$

represent the Jacobian. Show that $\iint_D J\,dx\,dy=0$ over the closed domain D with $a\le x\le b$ and $c\le y\le d$.

Chapter 3

Time-Differencing Schemes

3.1 Introduction

Atmospheric models generally require the solutions of partial differential equations. In spectral models, the governing partial differential equations reduce to a set of coupled ordinary nonlinear differential equations where the dependent variables contain derivatives with respect to time as well.

To march forward in time in numerical weather prediction, one needs to use a time-differencing scheme. Although much sophistication has emerged for the spatial derivatives (i.e., second- and fourth-order differencing), the time derivative has remained constructed mostly around the first- and second-order accurate schemes. Higher-order schemes in time require the specification of more than a single initial state, which has been considered to be rather cumbersome. Therefore, following the current state of the art, we focus on the first- and second-order accurate schemes. However, higher-order schemes, especially for long-term integrations such as climate modeling, deserve examination.

We start with the differential equation $dF / dt = G$, where $F = F(t)$ and $G = G(t)$. If the exact solution of the above equation can be expressed by trigonometric functions, then our problem would be to choose an appropriate time step in order to obtain a solution which behaves properly; that is, it remains bounded with time. This is illustrated in Fig. 3.1. We next show that: (1) if an improper time step is chosen, then the approximate finite difference solution may become unbounded, and (2) if a proper time step is chosen, then the finite difference solution will behave quite similar to the exact solution.

The stability or instability of a numerical scheme will be discussed for a single Fourier wave. This would also be valid for a somewhat more general case, since the total solution is a linear combination of sine and cosine functions, which are all bounded. We next define an amplification factor $|\lambda|$, the magnitude of which would determine whether a scheme is stable or not.

3.2 Amplification Factor

Consider a *linear wave equation* such as

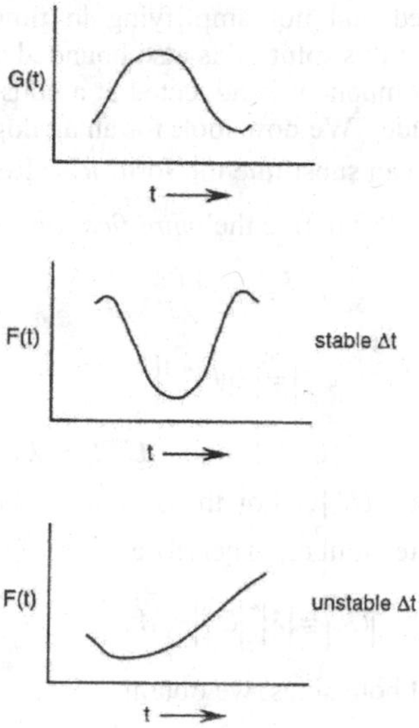

Figure 3.1. Stable and unstable time step Δt.

$$\frac{\partial u}{\partial t} + c\frac{\partial u}{\partial x} = 0 \tag{3.1}$$

which contains the wave speed, *c*, and the advection term $\partial u/\partial x$. What we learn from this equation is also applicable to the full spectral model. In other words, we can find the time-differencing scheme from this equation and then apply that scheme to the full spectral model. It happens that if the time-differencing scheme works with this equation then it seems to also work for the full model. Let the analytic solution of this equation be in the form of a single harmonic given by $u(x,t) = \text{Re}[U(t)e^{ikx}]$, where $U(t)$ is the wave amplitude and *k* is the wavenumber. Substituting this into (3.1), we obtain an equation for the *amplitude function* $U(t)$ as

$$\frac{dU}{dt} + ikcU = 0 \cdot \tag{3.2}$$

Thus we have reduced the above partial differential equation to an ordinary differential equation whose exact solution is given by

$$U(t) = U(0)e^{-ikct}. \tag{3.3}$$

Hence the desired *harmonic solution* is given by

$$u(x,t) = \text{Re}\left(U(0)e^{ik(x-ct)}\right). \tag{3.4}$$

The exact solution is bounded and not amplifying in time so any finite difference representation that is made from this solution is also bounded in time.

Therefore each wave component is advected at a constant velocity c along the x-axis with no change in amplitude. We now look for an analogous solution for the finite-difference equation, where we can substitute the form $u_j^n = \mathrm{Re}(U^n e^{ikj\Delta x})$, where U^n is the amplitude at the nth time level. We define the *amplification factor* $|\lambda|$ so that

$$U^n = \lambda U^{n-1}, \tag{3.5}$$

or

$$|U^n| = |\lambda||U^{n-1}|. \tag{3.6}$$

Substituting for $|U^{n-1}| = |\lambda||U^{n-2}|$ and $|U^{n-2}| = |\lambda||U^{n-3}|$ and following this procedure, we obtain $|U^n| = |\lambda|^n |U^0|$. For the scheme to be stable, it is required that $|U^n| < A$, where A is some finite number. Therefore

$$|U^n| = |\lambda|^n |U^0| < A. \tag{3.7}$$

Taking the natural logarithm of both sides, we obtain

$$\ln|U^n| = n\ln|\lambda| + \ln|U^0| < \ln A.$$

Simplifying, we obtain

$$n\ln|\lambda| < \ln\frac{A}{|U^0|}.$$

Now let

$$\ln\frac{A}{|U^0|} = A'.$$

Dividing the time t into n equal intervals, each of which is equal to Δt, we obtain $t = n\Delta t$. Hence

$$\ln|\lambda| < \frac{A'}{n}, \tag{3.8}$$

$$\ln|\lambda| < \frac{A'}{n}\Delta t. \tag{3.9}$$

As we require the boundedness of the solution for a finite time t, we obtain

$$\ln|\lambda| \le \mathrm{O}(\Delta t). \tag{3.10}$$

If we now define $|\lambda| \equiv 1 + \alpha$, then

$$\ln(1+\alpha) = \alpha - \frac{\alpha^2}{2} + \frac{\alpha^3}{3} - \frac{\alpha^4}{4} + \ldots \quad \text{for } -1 \le \alpha \le 1.$$

The stability condition obtained is equivalent to $\alpha \le \mathrm{O}(\Delta t)$, so that

$$|\lambda| \le 1 + \mathrm{O}(\Delta t). \tag{3.11}$$

This is the *Von Neumann necessary condition for stability*. This condition allows for an exponential growth of the solution, and it is required when the true solution grows exponentially. However, when it is known that the true solution does not grow, it is customary to replace this condition by a sufficient condition

$$|\lambda| \leq 1. \tag{3.12}$$

We apply all our schemes to the linear wave equation (3.1). If the amplification factor $|\lambda|$ is less than 1, then we obtain a stable solution. The reason why we apply our schemes to the above wave equation is that the solutions of wave equations are bounded trigonometric functions, which help us to examine the stability of the solution. The *general stability criteria* are:

$$\begin{aligned} &\text{Unstable scheme} \quad \text{if} \quad |\lambda| > 1, \\ &\text{Neutral scheme} \quad \text{if} \quad |\lambda| = 1, \\ &\text{Stable scheme} \quad \text{if} \quad |\lambda| < 1. \end{aligned}$$

3.3 Stability

In this section we discuss the computational stability of the following time-differencing schemes applied to the linear wave equation, namely the *Euler*, *backward*, *trapezoidal*, *Matsuno* and *Heun predictor-corrector*, *leap-frog*, *Adams-Bashforth*, and *implicit* schemes.

3.3.1 *Euler, Backward, and Trapezoidal Schemes*

Consider the linear wave equation (3.1) and assume

$$u(x,t) = U(t)e^{ikx}. \tag{3.13}$$

Let the value of U at time $n\Delta t$ be known. We wish to predict the value of U at time $(n+1)\Delta t$. Substituting the assumed solution, we obtain the marching equation for time-differencing (3.2)

$$\frac{dU}{dt} = f(U,t) \qquad U = U(t), \tag{3.14}$$

where $f = -ikcU = i\omega U$ and $\omega = -kc$. Integrating the above equation between time $n\Delta t$ and $(n+1)\Delta t$ where n is the current time level and $n+1$ is the future time level, we obtain

$$U^{n+1} - U^n = \int_{n\Delta t}^{(n+1)\Delta t} f(U,t)\, dt.$$

We can also write

$$U^{n+1} = U^n + \int_{n\Delta t}^{(n+1)\Delta t} f(U,t)\, dt. \tag{3.15}$$

We have two possible values for $f(U,t)$, which we assume to be constant over the time interval $[n\Delta t,\ (n+1)\Delta t]$, namely, (a) $f = f^n(U^n,\ n\Delta t)$ and (b) $f = f^{n+1}[U^{n+1},\ (n+1)\Delta t]$. In the case of (b), if $f = f^{n+1}$ (i.e., *f* depends on U^{n+1}), then the scheme is called *implicit.* If $f = f^n$ (i.e., *f* depends on U^n), then the scheme is called *explicit.*

We can also use a linear combination of f^n and f^{n+1} to define the value of $f(U,t)$ in the time interval $[n\Delta t,\ (n+1)\Delta t]$, which we assume to be constant and defined as $f(U,t) = \alpha f^n + \beta f^{n+1}$, where α and β are two real constants satisfying $\alpha + \beta = 1$. Thus (3.15) reduces to

$$U^{n+1} = U^n + \int_{n\Delta t}^{(n+1)\Delta t} \left(\alpha f^n + \beta f^{n+1}\right) dt .$$

Alternatively, we can write the time-differencing scheme as

$$U^{n+1} = U^n + \Delta t\left(\alpha f^n + \beta f^{n+1}\right). \tag{3.16}$$

By assigning different values to α and β, we obtain the different time-differencing schemes. In particular, for

Euler's forward schemes,	$\alpha = 1$ and $\beta = 0$
backward schemes,	$\alpha = 0$ and $\beta = 1$
trapezoidal schemes,	$\alpha = \beta = 0.5$.

As $f^n = i\omega U^n$, $f^{n+1} = i\omega U^{n+1}$, and $U^{n+1} = U^n + \Delta t(\alpha i\omega U^n + \beta i\omega U^{n+1})$, we can then write

$$U^{n+1} = U^n + i\alpha\omega\Delta t U^n + i\beta\omega\Delta t U^{n+1}$$
$$U^{n+1} - i\beta\omega\Delta t U^{n+1} = U^n + i\alpha\omega\Delta t U^n$$
$$U^{n+1}\left(1 - i\beta\omega\Delta t\right) = U^n\left(1 + i\alpha\omega\Delta t\right)$$
$$U^{n+1} = U^n \frac{1 + i\alpha\omega\Delta t}{1 - i\beta\omega\Delta t}. \tag{3.17}$$

Comparing the above equation with $U^{n+1} = \lambda U^n$, we observe that

$$\lambda = \frac{1 + i\alpha\omega\Delta t}{1 - i\beta\omega\Delta t}. \tag{3.18}$$

Let us define $p = \omega\Delta t$, so that $\lambda = (1 + i\alpha p)/(1 - i\beta p)$, where ω is the frequency of the wave we are studying, which we have chosen, and Δt is the time step we are assigning. After dividing and multiplying by the conjugate of $(1 - i\beta p)$, we obtain

$$\lambda = \frac{1 - \alpha\beta p^2 + ip(\alpha + \beta)}{1 + \beta^2 p^2},$$

or, since $\alpha + \beta = 1$, we can write

$$\lambda = \frac{1-\alpha\beta p^2 + ip}{1+\beta^2 p^2}. \tag{3.19}$$

Now let us discuss the stability of the above-mentioned three schemes.

Euler Forward Scheme. The Euler scheme is obtained by assigning $\alpha = 1$ and $\beta = 0$. Note that f is not centered in time in the Euler scheme. This now gives us $\lambda = 1 + ip$, so knowing that if $\lambda = a + ib$ then $|\lambda| = (a^2 + b^2)^{1/2}$ we can also write

$$|\lambda| = \left(1 + p^2\right)^{1/2}. \tag{3.20}$$

Since $p = \omega\Delta t$ is real and $p^2 > 0$, it follows that $|\lambda| > 1$; thus the Euler scheme is always unstable whatever the time step may be. It is a first-order-accurate time-differencing scheme.

Backward Scheme. The backward scheme is obtained by assigning $\alpha = 0$ and $\beta = 1$. This gives

$$\lambda = \frac{1+ip}{1+p^2}, \tag{3.21}$$

or

$$|\lambda| = \frac{(1+p^2)^{1/2}}{1+p^2} = \frac{1}{(1+p^2)^{1/2}} < 1, \text{ as } p > 0. \tag{3.22}$$

Hence the backward scheme is unconditionally stable for any time step. Furthermore, it is a damping scheme, and the amount of damping increases with p or as the frequency ω increases. During numerical integration, high-frequency modes often get excited and amplified unrealistically due to errors in the initial data. The damping property of the backward scheme is therefore desirable to reduce the amplitude of such high-frequency modes and to filter them out. It is also a first-order-accurate time-differencing scheme.

Trapezoidal Scheme. By assigning $\alpha = \beta = 0.5$, the trapezoidal scheme is obtained. In this case,

$$\lambda = \frac{4 - p^2 + i4p}{4 + p^2}.$$

Thus

$$|\lambda| = \frac{(16 + p^4 - 8p^2 + 16p^2)^{1/2}}{4 + p^2} = 1. \tag{3.23}$$

Therefore $|\lambda| = 1$ for any time interval Δt, and thus this scheme is always neutral. The amplitude of the numerical solution remains constant, just as in the exact solution. It is a second-order-accurate time-differencing scheme.

3.3.2 *Predictor-Corrector Schemes*

These types of schemes work on the following basic principle. First an attempt is made to predict the value of U at the $(n+1)$th time step, which we denote by $U^{(n+1)*}$. Euler's forward scheme is usually used to predict $U^{(n+1)*}$. This value is then used to find the value of f^{n+1} at the $(n+1)$th time step, and we denote this by $f^{(n+1)*}$. This value $f^{(n+1)*}$ is used to correct the previous $U^{(n+1)*}$ to get the final value of $U^{(n+1)}$. The step for finding $U^{(n+1)*}$ is the *predictor step*, and the step for finding $U^{(n+1)}$ using $U^{(n+1)*}$ is the *corrector step*.

The finite-difference equations for such a scheme are

$$U^{(n+1)*} = U^n + \Delta t f^n \quad (\textit{predictor step}) \tag{3.24}$$

and

$$U^{(n+1)} = U^n + \Delta t\left(\alpha f^n + \beta f^{(n+1)*}\right) \quad (\textit{corrector step}), \tag{3.25}$$

where $\alpha + \beta = 1$. As mentioned before, $f^{(n+1)*}$ and f^n are defined as

$$f^{(n+1)*} = f\left[U^{(n+1)*},\ (n+1)\Delta t\right] \tag{3.26}$$

and

$$f^n = f\left(U^n,\ n\Delta t\right). \tag{3.27}$$

To look into the stability of such a scheme, we again consider the linear wave equation (3.1). Recalling $f^n = i\omega U^n$ and $f^{n+1} = i\omega U^{n+1}$ the predictor step is described as

$$U^{(n+1)*} = U^n + i\omega\Delta t U^n \tag{3.28}$$

and the corrector step is formulated as

$$U^{(n+1)} = U^n + i\omega\Delta t\left[\alpha U^n + \beta U^{(n+1)*}\right]. \tag{3.29}$$

Substituting for $U^{(n+1)*}$ from the predictor step into the corrector step, we obtain $U^{(n+1)} = U^n + i\omega\Delta t[\alpha U^n + \beta(U^n + i\omega\Delta t U^n)]$, or we can write

$$U^{n+1} = U^n + i\omega\Delta t\alpha U^n + i\omega\Delta t\beta U^n + i^2\omega^2\Delta t^2\beta U^n$$
$$U^{(n+1)} = \left[1 - \beta\omega^2\Delta t^2 + i\omega(\alpha+\beta)\Delta t\right]U^n. \tag{3.30}$$

Thus $\lambda = U^{(n+1)}/U^n = (1-\beta\omega^2\Delta t^2) + i\omega(\alpha+\beta)\Delta t$. Since $\alpha + \beta = 1$, the amplification factor for this scheme is given by

$$|\lambda| = \left[(1-\beta\omega^2\Delta t^2)^2 + \omega^2\Delta t^2\right]^{1/2}. \tag{3.31}$$

Examples of this kind of scheme are the Matsuno scheme and Heun's scheme.

Matsuno Scheme. The Matsuno scheme is obtained by taking $\alpha = 0$ and $\beta = 1$ such that

$$|\lambda| = \left(\left(1-\omega^2\Delta t^2\right)^2 + \omega^2\Delta t^2 \right)^{1/2}$$
$$= \left(1 - 2\omega^2\Delta t^2 + \omega^4\Delta t^4 + \omega^2\Delta t^2\right)^{1/2}$$
$$= \left(1 - \omega^2\Delta t^2 + \omega^4\Delta t^4\right)^{1/2}.$$

If $\omega\Delta t \le 1$, then $\omega^4(\Delta t)^4 \le \omega^2(\Delta t)^2 \le 1$. Thus if $\omega\Delta t \le 1$, then $|\lambda| \le 1$. Hence the Matsuno scheme is a stable scheme for $\omega\Delta t \le 1$. This is also a first-order-accurate time-differencing scheme.

Heun's Scheme. Heun's scheme is obtained by assigning $\alpha = \beta = 1/2$. We can then write

$$\lambda = 1 - \frac{\omega^2(\Delta t)^2}{2} + i\omega\Delta t.$$

Thus we obtain

$$|\lambda| = \left(1 + \frac{\omega^4(\Delta t)^4}{4} - \omega^2(\Delta t)^2 + \omega^2(\Delta t)^2\right)^{1/2},$$

or

$$|\lambda| = \left(1 + \frac{\omega^4(\Delta t)^4}{4}\right)^{1/2}. \tag{3.32}$$

As $\omega^4(\Delta t)^4/4 > 0$ for any $\omega\Delta t$, we observe that $|\lambda| > 1$ for all values of $\omega\Delta t$. Therefore, Heun's scheme is an unstable scheme. We next illustrate the leap-frog time-differencing scheme.

3.3.3 *Centered or Leap-Frog Scheme*

We have discussed two-time-level differencing schemes so far. One of the most widely used time-integration schemes in numerical weather prediction is the centered, or leap-frog, time-integration scheme. This is a three-time-level scheme. The values of the function at time levels $(n-1)$ and n are known (or predicted using one forward time step to start the process), and from these we predict the value of the function at time level $(n+1)$. We make a centered evaluation of the integral as

$$U^{n+1} = U^{n-1} + \int_{(n-1)\Delta t}^{(n+1)\Delta t} f(U,t)\, dt. \tag{3.33}$$

We take f = constant in the above integral. Let this constant be denoted as f^n, the value of f at time $t = n\Delta t$, so that (3.33) reduces to

$$U^{n+1} = U^{n-1} + 2\Delta t f^n. \tag{3.34}$$

Thus we use the information at two previous time levels to generate the information at the next time level. The truncation error is of the order of $(\Delta t)^2$. This scheme is called the *centered* or the *leap-frog scheme*.

Consider the leap-frog scheme applied to the linear wave equation (3.1). Substituting $u(x,t) = \text{Re}[U(t)e^{ikx}]$ as the solution, we obtain

$$\frac{dU}{dt} - i\omega U = 0, \qquad \text{where } \omega = -kc. \tag{3.35}$$

Using a centered time-differencing scheme as in (3.34), we have

$$U^{n+1} = U^{n-1} + i2\omega\Delta t U^n, \tag{3.36}$$

where $f^n = i\omega U^n$. Furthermore, we have

$$U^{n+1} = \lambda U^n = \lambda^2 U^{n-1}, \qquad \text{as } U^n = \lambda U^{n-1} \tag{3.37}$$

and we can replace all U^n and U^{n+1} with U^{n-1} and simplify equation (3.36). Since equation (3.36) is a second-order differential equation it has a quadratic equation for λ. Substituting these into the above equations, we obtain

$$\lambda^2 U^{n-1} = U^{n-1} + i2\omega\Delta t \lambda U^{n-1}$$
$$\lambda^2 - i2p\lambda - 1 = 0, \tag{3.38}$$

where $p = \omega\Delta t$. The roots of this second-order equation are

$$\lambda_1 = \left(1 - p^2\right)^{1/2} + ip \tag{3.39}$$

and

$$\lambda_2 = -\left(1 - p^2\right)^{1/2} + ip. \tag{3.40}$$

If $U^{n+1} = \lambda U^n$ is to represent an approximation to the true solution, then we must have $\lambda \to 1$ as $\Delta t \to 0$. For the above roots, since $p = \omega\Delta t \to 0$ as $\Delta t \to 0$, we have $\lambda_1 \to 1$. However, at the same time, $\lambda_2 \to -1$. The solution associated with λ_1 is an approximation of the true solution and is called the *physical mode*. On the other hand, the solution associated with λ_2 is not the true solution and is called the *computational mode*.

The complete solution of the centered time-differencing scheme is the sum of the solutions corresponding to the physical mode and the computational mode. Because λ_2^n is positive for even values of *n* and negative for odd values of *n*, λ_2^n alternatively takes positive and negative values as the integration proceeds. This results in an oscillation of the computational mode about the true solution. This is an undesirable feature of the scheme, and methods are available to suitably suppress the computational mode during the course of integration.

3.3.3.1 *Centered in Time and Space*

One method used to suppress the computational mode is to start the integration with one forward time step and then compute 49 centered difference time steps. After these 50 time steps, we want to break the process and repeat it again for as many time steps that are needed. This is done so that the computational factor (λ_2) does not blow up.

In the discussion on the centered time-differencing scheme, we have used a centered approximation for the time differential $\partial u / \partial t$ and an analytical value for the space differential $\partial u / \partial x$ in the wave equation. We now examine the scheme using centered differencing both in time and space.

With $x = m\Delta x$ and $t = n\Delta t$, we write $u(x,t) = U(t)e^{ikx}$ as

$$u\left(m\Delta x, n\Delta t\right) = U\left(n\Delta t\right)e^{ikm\Delta x} = U^n e^{ikm\Delta x} \tag{3.41}$$

which is the finite difference representation of the trial solution. Substituting (3.41) into the linear wave equation (3.1) gives

$$\frac{U^{n+1} - U^{n-1}}{2\Delta t} = -c\frac{(U^n e^{ik\Delta x} - U^n e^{-ik\Delta x})}{2\Delta x}$$

$$U^{n+1} - U^{n-1} = -\frac{c\Delta t}{\Delta x}U^n\left(e^{ik\Delta x} - e^{ik\Delta x}\right),$$

or

$$U^{n+1} = U^{n-1} - \frac{c\Delta t}{\Delta x}U^n\left(e^{ik\Delta x} - e^{ik\Delta x}\right). \tag{3.42}$$

Futhermore, using $U^{n+1} = \lambda U^n = \lambda^2 U^{n-1}$, we obtain

$$\lambda^2 U^{n-1} = U^{n-1} - \frac{c\Delta t}{\Delta x}\lambda U^{n-1}\left(e^{ik\Delta x} - e^{-ik\Delta x}\right)$$

$$\lambda^2 + \frac{c\Delta t}{\Delta x}\lambda\left(e^{ik\Delta x} - e^{-ik\Delta x}\right) - 1 = 0\,.$$

Using the identity

$$\sin k\Delta x = \frac{e^{ik\Delta x} - e^{-ik\Delta x}}{2i}$$

we obtain

$$\lambda^2 + i2c\frac{\Delta t}{\Delta x}\sin k\Delta x \lambda - 1 = 0 \tag{3.43}$$

which is the quadratic representation for space-time differencing. The roots of (3.43) are

$$\lambda_1 = \left[1 - \left(\frac{c\Delta t}{\Delta x}\right)^2 \sin^2 k\Delta x\right]^{1/2} - ic\frac{\Delta t}{\Delta x}\sin k\Delta x,$$

$$\lambda_2 = \left[1 - \left(\frac{c\Delta t}{\Delta x}\right)^2 \sin^2 k\Delta x\right]^{1/2} - ic\frac{\Delta t}{\Delta x}\sin k\Delta x\,.$$

The absolute values of λ_1 and λ_2 are

$$|\lambda_1| = |\lambda_2| = \left[1 - \left(\frac{c\Delta t}{\Delta x}\right)^2 \sin^2 k\Delta x + \left(\frac{c\Delta t}{\Delta x}\right)^2 \sin^2 k\Delta x\right]^{1/2} = 1. \tag{3.44}$$

Thus we can write $|\lambda_1| = |\lambda_2| = 1$.

When $\Delta x \to 0$ and $\Delta t \to 0$, then the numerical solution should converge to the true solution. This turns out to be true with $\lambda = \lambda_1$, since in this case $U^{n+1} = \lambda_1 U^n = U^n$. However, if we instead take λ_2 as the solution, then $U^{n+1} = \lambda_2 U^n = -U^n$. Hence λ_2 corresponds to the oscillating computational mode of the equation and λ_1 corresponds to the physical mode.

3.3.3.2 *Alternative Method for Space-time Differencing*

Consider again (3.42). If we assume U^n to be of the form $\hat{u}e^{ivn\Delta t}$ then substituting for U^{n+1}, U^{n-1}, and U^n into this equation gives

$$\hat{u}e^{iv(n+1)\Delta t} - \hat{u}e^{iv(n-1)\Delta t} = -\frac{c\Delta t}{\Delta x}\hat{u}e^{ivn\Delta t}\left(e^{ik\Delta x} - e^{-ik\Delta x}\right)$$

$$\hat{u}e^{ivn\Delta t}e^{iv\Delta t} - \hat{u}e^{ivn\Delta t}e^{-iv\Delta t} = -\frac{c\Delta t}{\Delta x}\hat{u}e^{ivn\Delta t}\left(e^{ik\Delta x} - e^{-ik\Delta x}\right)$$

$$e^{iv\Delta t} - e^{-iv\Delta t} = -\frac{c\Delta t}{\Delta x}\left(e^{ik\Delta x} - e^{-ik\Delta x}\right),$$

or

$$\sin v\Delta t = -\frac{c\Delta t}{\Delta x}\sin k\Delta x. \tag{3.45}$$

Because k is a real wavenumber, $\sin k\Delta x \le 1$ so that (3.45) is satisfied for a real value of v if $c\Delta t/\Delta x \le 1$. However, if $c\Delta t/\Delta x > 1$, v must have an imaginary component. In that case, the finite-difference solution becomes unstable. The condition $c\Delta t/\Delta x \le 1$ is a necessary condition for the stability of an explicit time-differencing scheme. It was first discovered by Courant, Friedrichs, and Levy (1928) and is commonly known as the *CFL criterion* for the stability of a time-differencing scheme. According to this, in fine-mesh models where Δx is small, we need small time steps Δt to ensure computational stability. Also Δt should be small for fast-moving waves such as gravity waves, because c is small.

Some examples of models and their respective time steps and grid spacing are listed below:

Model	Δx	Δt
Barotropic model	100 km	1 hour
Primitive equation model	100 km	6 minutes
MM5	20 km	1 minute
Micrometeorology	10 m	Few seconds

3.3.4 *Adams-Bashforth scheme*

This is a three-time-level scheme in which the value of the space derivative is approximated from a linear extrapolation of f in the center to time levels n and $(n+1)$ from its value at time levels n and $(n-1)$. With such an approximation, the finite-difference analog of (3.1) takes the form

$$U^{n+1} = U^n + \Delta t\left(\frac{3}{2}f^n - \frac{1}{2}f^{n-1}\right) \tag{3.46}$$

which is a second-order differential equation. After applying this scheme to the linear wave equation, we obtain

$$U^{n+1} = U^n - ikc\Delta t\left(\frac{3}{2}U^n - \frac{1}{2}U^{n-1}\right). \tag{3.47}$$

Substituting $U^{n+1} = \lambda U^n$ and $U^n = \lambda U^{n-1}$, we obtain

$$\lambda^2 = \lambda - ick\Delta t\left(\frac{3}{2}\lambda - \frac{1}{2}\right),$$

or

$$\lambda^2 - \left(1 - i\frac{3}{2}\omega\Delta t\right)\lambda - i\frac{1}{2}\omega\Delta t = 0, \tag{3.48}$$

where $\omega = kc$. The roots of the above equation are

$$\lambda_1 = \frac{1}{2}\left[1 - i\frac{3}{2}\omega\Delta t + \left(1 - \frac{9}{4}\omega^2(\Delta t)^2 - i\omega\Delta t\right)^{1/2}\right], \tag{3.49}$$

$$\lambda_2 = \frac{1}{2}\left[1 - i\frac{3}{2}\omega\Delta t - \left(1 - \frac{9}{4}\omega^2(\Delta t)^2 - i\omega\Delta t\right)^{1/2}\right]. \tag{3.50}$$

We note that if $\Delta t \to 0$, then $\lambda_1 = 1$ and $\lambda_2 \to 0$. Thus λ_1 denotes the physical mode and λ_2 denotes the computational mode. As the time step Δt is reduced, the Adams-Bashforth method approaches the true solution.

On power series expansion of (3.49) and (3.50), it can be shown that $|\lambda_1| > 1$ while $|\lambda_2| < 1$. Thus the computational mode in this scheme tends to dampen, which is a beneficial property. However, the physical mode tends to amplify. The rate of amplification is small if Δt is small, and the scheme is weakly unstable. The Adams-Bashforth scheme is therefore suitable for short periods of integration with a small time step such as in a squall or flow past a mountain.

3.3.5 *Implicit Schemes*

There are two types of implicit schemes, one is the fully implicit scheme discussed below and the other is a semi-implicit scheme discussed in Sec. 3.4. In the time-differencing schemes discussed so far, the time step should satisfy the CFL condition. The implicit time-integration scheme permits longer time steps than specified by the CFL condition and is therefore more economical than explicit time-differencing schemes. In the implicit scheme, the space derivative at time level n is obtained as the mean of the space derivatives at time levels (n) and $(n+1)$. Since the future value of the function is not known explicitly, the scheme is called *implicit*.

If m is the space index and n is the time index, the fully implicit finite-difference analog of the linear wave equation takes the form

$$\frac{u_m^{n+1}-u_m^n}{\Delta t}=-\frac{c}{2}\left(\frac{u_{m+1}^{n+1}-u_{m-1}^{n+1}}{2\Delta x}+\frac{u_{m+1}^{n}-u_{m-1}^{n}}{2\Delta x}\right) \tag{3.51}$$

which is a first-order non-separable difference equation. In this equation, the unknown function u at time level $(n+1)$ appears at the three space points $(m-1)$, m, and $(m+1)$. Such equations are associated with each space grid point; this system of equations, in principle, can be solved by inverting a matrix with proper boundary conditions. However, for a large number of grid points, the procedure becomes difficult. In such cases, relaxation or spectral methods are more convenient.

Assuming a solution of the form $u_m^n = U^n e^{ikm\Delta x}$, we obtain

$$\frac{U^{n+1}e^{ikm\Delta x}-U^{n}e^{ikm\Delta x}}{\Delta t}=-\frac{c}{2}\left[\frac{U^{n+1}e^{ik(m+1)\Delta x}-U^{n+1}e^{ik(m-1)\Delta x}}{2\Delta x}+\frac{U^{n}e^{ik(m+1)\Delta x}-U^{n}e^{ik(m-1)\Delta x}}{2\Delta x}\right]$$

$$\frac{U^{n+1}e^{ikm\Delta x}-U^{n}e^{ikm\Delta x}}{\Delta t}=-\frac{c}{2}\left[\frac{U^{n+1}e^{ikm\Delta x}e^{ik\Delta x}-U^{n+1}e^{ikm\Delta x}e^{-ik\Delta x}}{2\Delta x}+\frac{U^{n}e^{ikm\Delta x}e^{ik\Delta x}-U^{n}e^{ikm\Delta x}e^{-ik\Delta x}}{2\Delta x}\right]$$

$$\frac{U^{n+1}-U^{n}}{\Delta t}=-\frac{c}{2}\left[\frac{U^{n+1}e^{ik\Delta x}-U^{n+1}e^{-ik\Delta x}}{2\Delta x}+\frac{U^{n}e^{ik\Delta x}-U^{n}e^{-ik\Delta x}}{2\Delta x}\right]$$

$$\frac{U^{n+1}-U^{n}}{\Delta t}=-\frac{c}{2}\frac{U^{n+1}\left(e^{ik\Delta x}-e^{-ik\Delta x}\right)}{2\Delta x}-\frac{c}{2}\frac{U^{n}\left(e^{ik\Delta x}-e^{-ik\Delta x}\right)}{2\Delta x}. \tag{3.52}$$

Rearranging equation (3.52) and using Euler's relation, i.e., $\sin x = (e^{ix}-e^{-ix})/2i$, we obtain

$$\frac{U^{n+1}}{\Delta t}+\frac{c}{2}\frac{U^{n+1}\left(e^{ik\Delta x}-e^{-ik\Delta x}\right)}{2\Delta x}=\frac{U^{n}}{\Delta t}-\frac{c}{2}\frac{U^{n}\left(e^{ik\Delta x}-e^{-ik\Delta x}\right)}{2\Delta x}$$

$$\frac{U^{n+1}}{\Delta t}+\frac{ic}{2}U^{n+1}\frac{\sin k\Delta x}{2\Delta x}=\frac{U^{n}}{\Delta t}-\frac{ic}{2}U^{n}\frac{\sin k\Delta x}{2\Delta x}$$

$$U^{n+1}\left(1+ic\frac{\Delta t}{2\Delta x}\sin k\Delta x\right)=U^{n}\left(1-ic\frac{\Delta t}{2\Delta x}\sin k\Delta x\right). \tag{3.53}$$

Hence

$$\frac{U^{n+1}}{U^{n}}=\frac{1-ic\frac{\Delta t}{2\Delta x}\sin k\Delta x}{1+ic\frac{\Delta t}{2\Delta x}\sin k\Delta x}. \tag{3.54}$$

This is the ratio of the amplitude at time level $(n+1)$ to that at time level n at a particular spatial point m. Furthermore, multiplying (3.54) by the complex conjugate of the denominator and taking the absolute value gives $|\lambda|=|U^{n+1}/U^{n}|=1$, which implies stability regardless of the value of $c\Delta t/\Delta x$. The future values of the function u will be of the same order as the previous values. Hence this scheme is nonamplifying.

We next show another way of demonstrating the stability of the implicit scheme, which is called the trigonometric method. Starting from the finite-difference analog of the linear wave equation, that is,

$$u_m^{n+1}-u_m^{n}=-\frac{c\Delta t}{4\Delta x}\left[u_{m+1}^{n+1}+u_{m+1}^{n}-\left(u_{m-1}^{n+1}+u_{m-1}^{n}\right)\right], \tag{3.55}$$

we can write

$$\begin{aligned}u_{m+1}^{n+1}&=\hat{u}e^{ik[(m+1)\Delta x+c(n+1)\Delta t]}\\&=\hat{u}e^{ik(m\Delta x+cn\Delta t)}e^{ik(\Delta x+c\Delta t)}=u_m^{n}e^{ik(\Delta x+c\Delta t)},\end{aligned} \tag{3.56}$$

where $u_m^n=\hat{u}e^{ik(m\Delta x+nc\Delta t)}$. Similarly,

$$\begin{aligned}u_{m+1}^{n}&=\hat{u}e^{ik[(m+1)\Delta x+cn\Delta t]}\\&=\hat{u}e^{ik(m\Delta x+cn\Delta t)}e^{ik\Delta x}=u_m^{n}e^{ik\Delta x},\\u_{m-1}^{n}&=\hat{u}e^{ik[(m-1)\Delta x+cn\Delta t]}\\&=\hat{u}e^{ik(m\Delta x+cn\Delta t)}e^{-ik\Delta x}=u_m^{n}e^{-ik\Delta x},\\u_{m-1}^{n+1}&=\hat{u}e^{ik[(m-1)\Delta x+c(n+1)\Delta t]}\\&=\hat{u}e^{ik(m\Delta x+cn\Delta t)}e^{ik(-\Delta x+c\Delta t)}=u_m^{n}e^{ik(-\Delta x+c\Delta t)},\\u_{m}^{n+1}&=\hat{u}e^{ik[m\Delta x+c(n+1)\Delta t]}\\&=\hat{u}e^{ik(m\Delta x+cn\Delta t)}e^{ikc\Delta t}=u_m^{n}e^{ikc\Delta t}.\end{aligned}$$

Then substituting these equations into (3.55), we obtain

$$u_m^n\left(e^{ikc\Delta t}-1\right)=-\frac{c\Delta t}{4\Delta x}\left(e^{ik(\Delta x+c\Delta t)}+e^{ik\Delta x}-e^{ik(-\Delta x+c\Delta t)}-e^{ik(-\Delta x)}\right)u_m^n. \tag{3.57}$$

From this we can write

$$e^{ikc\Delta t}-1=-\frac{c\Delta t}{4\Delta x}\left[e^{ik\Delta x}(e^{ikc\Delta t}+1)-e^{-ik\Delta x}(e^{ikc\Delta t}+1)\right],$$

or

$$e^{ikc\Delta t}-1=-\frac{c\Delta t}{4\Delta x}\left[(e^{ik\Delta x}-e^{-ik\Delta x})(e^{ikc\Delta t}+1)\right]. \tag{3.58}$$

We then obtain

$$\frac{e^{ikc\Delta t}-1}{e^{ikc\Delta t}+1}=\frac{-c\Delta t}{4\Delta x}\left(e^{ik\Delta x}-e^{-ik\Delta x}\right),$$

or

$$\tan\frac{kc\Delta t}{2}=-\frac{c\Delta t}{2\Delta x}\sin k\Delta x. \tag{3.59}$$

The sine of an angle is bounded between -1 and 1 while the tangent of an angle varies from $-\infty$ to $+\infty$. Hence, any value of Δt can satisfy this equation, i.e. (3.59) is satisfied for all values of $(c\Delta t)/\Delta x$. Thus the stability of this scheme does not depend on the value of $(c\Delta t)/\Delta x$ and is therefore unconditionally stable. In the following section we illustrate the application of the semi-implicit time-differencing scheme to the shallow-water model equations.

3.4 Shallow-Water Model

To demonstrate the semi-implicit time-differencing scheme the shallow water equations are used instead of the linear wave equation. The linear wave equation is a one-wave solution equation, whereas the shallow water equations are, in essence, a stripped down version of the full model equations, i.e., they have some properties of the atmosphere. In the semi-implicit scheme the nonlinear part of an equation is handled explicitly while the linear part is handled implicitly.

In the shallow-water model we have a layer of fluid with constant density. Under the hydrostatic assumption, there is no vertical variation of the pressure gradient force if density is constant. Therefore, if initially the horizontal velocities are independent of height, they will remain so during all times.

As $\partial u/\partial z=\partial/\partial z=0$, the vertical advection terms do not appear in the shallow-water equations. These equations may be written in the *Cartesian coordinate system* as the *momentum equations*:

$$\frac{\partial u}{\partial t}+u\frac{\partial u}{\partial x}+v\frac{\partial u}{\partial y}-fv+\frac{\partial\phi}{\partial x}=0, \tag{3.60}$$

$$\frac{\partial v}{\partial t}+u\frac{\partial v}{\partial x}+v\frac{\partial v}{\partial y}+fu+\frac{\partial\phi}{\partial y}=0, \tag{3.61}$$

and the *mass continuity equation* :

$$\frac{\partial u}{\partial x}+\frac{\partial v}{\partial y}+\frac{\partial w}{\partial z}=0. \tag{3.62}$$

Here u is the x-component of the wind vector, v is the y-component of the wind vector, f is the Coriolis parameter, ϕ is the geopotential, and w is the vertical component of the wind vector.

The mass continuity equation can be integrated in the vertical to obtain

$$w_{top} - w_{bottom} + \left(\frac{\partial u}{\partial x} + \frac{\partial v}{\partial y}\right) h = 0, \tag{3.63}$$

since u and v are independent of height. Let H be the mean depth of the fluid, which is invariant with time, and h' be the perturbation height, so that $h = H + h'$. Furthermore, with a flat surface at the bottom we assume $w_{bottom} = 0$. We then have

$$w_{top} = \frac{dh}{dt} = -H\left(\frac{\partial u}{\partial x} + \frac{\partial v}{\partial y}\right) - h'\left(\frac{\partial u}{\partial x} + \frac{\partial v}{\partial y}\right).$$

This equation can be written as

$$\frac{dh}{dt} = \frac{\partial h}{\partial t} + u\frac{\partial h}{\partial x} + v\frac{\partial h}{\partial y} = -H\left(\frac{\partial u}{\partial x} + \frac{\partial v}{\partial y}\right) - h'\left(\frac{\partial u}{\partial x} + \frac{\partial v}{\partial y}\right). \tag{3.64}$$

After multiplying by g and defining the geopotential (ϕ) as $\phi = gh$, we obtain the continuity equation:

$$\frac{\partial \phi'}{\partial t} + \frac{\partial}{\partial x}(u\phi') + \frac{\partial}{\partial y}(v\phi') + \bar{\phi}\left(\frac{\partial u}{\partial x} + \frac{\partial v}{\partial y}\right) = 0. \tag{3.65}$$

Here, $\bar{\phi} = gH$ and $\phi' = \phi - \bar{\phi}$. Equation (3.65) describes the variation in the height of the free surface. Equations (3.60), (3.61) and (3.65) form the shallow-water model.

A *linearized shallow-water system* has a frequency equation which is cubic. Two of the solutions are gravitational modes, and the third is a Rossby wave. To look into the nature of the gravity wave solution of the shallow-water equations, we examine their linearized form (hence, no nonlinear dynamics) on a *nonrotating frame* (i.e., $f = 0$) and assume no basic flow (i.e., $\bar{u} = 0,\ \bar{v} = 0$):

$$\frac{\partial u}{\partial t} + \frac{\partial \phi'}{\partial x} = 0, \tag{3.66}$$

$$\frac{\partial v}{\partial t} + \frac{\partial \phi'}{\partial y} = 0, \tag{3.67}$$

$$\frac{\partial \phi'}{\partial t} + \bar{\phi}\left(\frac{\partial u}{\partial x} + \frac{\partial v}{\partial y}\right) = 0. \tag{3.68}$$

Differentiating (3.68) with respect to t and making use of (3.66) and (3.67), we can write

$$\frac{\partial^2 \phi'}{\partial t^2} - \bar{\phi}\nabla^2 \phi' = 0. \tag{3.69}$$

Assuming that the perturbation is only in the x-direction, the above equation further simplifies to

$$\frac{\partial^2 \phi'}{\partial t^2} - \bar{\phi}\frac{\partial^2 \phi'}{\partial x^2} = 0. \tag{3.70}$$

Assuming a solution to be of the form $\phi' = e^{ik(x-ct)}$ and solving for phase speed c, we obtain

$$c^2 = \bar{\phi} = gH, \quad \text{or} \quad c = \pm (gH)^{1/2}.$$

This is the *gravity wave phase speed.* This disturbance is forced by gravity and moves both in the positive and negative directions. For an isothermal atmosphere, taking the mean height as $H = 9$ km and $g = 9.8$ ms^{-1}, we obtain $c \cong 300$ ms^{-1}, which is very close to the speed of sound waves.

We next look at the nature of the Rossby wave solution. Consider the horizontal motion of the fluid under the influence of the Coriolis force and assume no basic flow. Furthermore, assume the flow to be nondivergent, so that

$$u = -\frac{\partial \psi}{\partial y}, \; v = \frac{\partial \psi}{\partial x}, \text{ and } \frac{\partial u}{\partial x} + \frac{\partial v}{\partial y} = 0. \tag{3.71}$$

The linearized equations of motion are then

$$\frac{\partial u}{\partial t} - fv = -\frac{\partial \phi'}{\partial x}, \tag{3.72}$$

$$\frac{\partial v}{\partial t} + fu = -\frac{\partial \phi'}{\partial y}. \tag{3.73}$$

Differentiating (3.73) with respect to x and (3.72) with respect to y and subtracting, we get the linearized form of the vorticity equation. That is,

$$\partial \zeta / \partial t = \beta v, \tag{3.74}$$

where $\beta = \partial f / \partial y$ and $\zeta = \partial v / \partial x - \partial u / \partial y = \nabla^2 \psi$. In other words, we can write

$$\frac{\partial}{\partial t} \nabla^2 \psi = -\beta \frac{\partial \psi}{\partial x}. \tag{3.75}$$

Let us assume a solution of the form $\psi = \hat{\psi} e^{i(kx+ly-\nu t)}$. Then we have $\nu(k^2 + l^2) = -k\beta$,

$$\nu = \frac{-k\beta}{k^2 + l^2}. \tag{3.76}$$

Using this, we obtain the phase speed in the x-direction

$$c_x = \frac{\nu}{k} = \frac{-\beta}{k^2 + l^2}. \tag{3.77}$$

This is the *phase speed of Rossby waves* caused by the variation in the Coriolis force (the β effect). The phase speed is directly proportional to β and inversely proportional to the square of the wavenumber $k^2 + l^2$.

Therefore, we observe that the shallow-water equations contain both slow-moving Rossby waves and high-frequency grravity waves. Handling of high-frequency gravity waves will require shorter time steps as demanded by the CFL criterion, which is computationally expensive.

To overcome this, it is economical to apply the implicit time integration scheme, which permits relatively longer time steps. Because the nonlinear part of the equations cannot be dealt with implicitly, we calculate it explicitly. The linear part of the equations is treated implicitly. Such a scheme is known as a *semi-implicit* time-integration scheme.

Based on these considerations, we write the fully nonlinear shallow-water equations as

$$\frac{\partial u}{\partial t}+\frac{\partial \phi'}{\partial x}=-\left(u\frac{\partial u}{\partial x}+v\frac{\partial u}{\partial y}-fv\right)=N_u, \tag{3.78}$$

$$\frac{\partial v}{\partial t}+\frac{\partial \phi'}{\partial y}=-\left(u\frac{\partial v}{\partial x}+v\frac{\partial v}{\partial y}-fu\right)=N_v, \tag{3.79}$$

$$\frac{\partial \phi'}{\partial t}+\bar{\phi}\left(\frac{\partial u}{\partial x}+\frac{\partial v}{\partial y}\right)=-\left(\frac{\partial u\phi'}{\partial x}+\frac{\partial v\phi'}{\partial y}\right)=N_h, \tag{3.80}$$

where the left-hand sides of the equations are the terms that excite gravity waves, the right-hand sides of the equations are the terms for the Rossby waves, and N_u, N_v, and N_h are the nonlinear terms. We assume nonlinear dynamics do not by themselves excite gravity waves. For simplicity, the prime (') is dropped from the ϕ' term in the following discussion.

The finite-difference analog of (3.78) can be written as

$$\frac{u^{n+1}-u^{n-1}}{2\Delta t}+\frac{1}{2}\left(\frac{\partial \phi^{n+1}}{\partial x}+\frac{\partial \phi^{n-1}}{\partial x}\right)=\left(N_u\right)^n, \tag{3.81}$$

where n denotes the time level. The term $1/2(\partial\phi^{n+1}/\partial x+\partial\phi^{n-1}/\partial x)$ denotes the mean value for $\partial\phi/\partial x$ at time levels $(n+1)$ and $(n-1)$. Furthermore, the nonlinear term $\left(N_u\right)^n$ is calculated at time level n.

Thus we can write

$$u^{n+1}=u^{n-1}-\Delta t\left(\frac{\partial \phi^{n+1}}{\partial x}+\frac{\partial \phi^{n-1}}{\partial x}\right)+2\Delta t\left(N_u\right)^n, \tag{3.82}$$

$$v^{n+1}=v^{n-1}-\Delta t\left(\frac{\partial \phi^{n+1}}{\partial y}+\frac{\partial \phi^{n-1}}{\partial y}\right)+2\Delta t\left(N_v\right)^n. \tag{3.83}$$

In addition, the mass continuity equation can be written as

$$\phi^{n+1}=\phi^{n-1}-\bar{\phi}\Delta t\left[\left(\frac{\partial u^{n+1}}{\partial x}+\frac{\partial v^{n+1}}{\partial y}\right)+\left(\frac{\partial u^{n-1}}{\partial x}+\frac{\partial v^{n-1}}{\partial y}\right)\right]+2\Delta t\left(N_h\right)^n. \tag{3.84}$$

At this point there are three unknowns from equations (3.82), (3.83), and (3.84): u^{n+1}, v^{n+1}, and ϕ^{n+1}. From these equations we can obtain a single equation for the geopotential. This can be accomplished by first taking $\partial/\partial x$ of (3.82) and $\partial/\partial y$ of (3.83) and substituting into the continuity equation as

$$\frac{\partial}{\partial x}u^{n+1} = \frac{\partial}{\partial x}u^{n-1} - \Delta t \frac{\partial}{\partial x}\left(\frac{\partial}{\partial x}\phi^{n+1} + \frac{\partial}{\partial x}\phi^{n-1}\right) + 2\Delta t \frac{\partial}{\partial x}(N_u)^n$$

and

$$\frac{\partial}{\partial y}v^{n+1} = \frac{\partial}{\partial y}v^{n-1} - \Delta t \frac{\partial}{\partial y}\left(\frac{\partial}{\partial y}\phi^{n+1} + \frac{\partial}{\partial x}\phi^{n-1}\right) + 2\Delta t \frac{\partial}{\partial y}(N_v)^n .$$

Next, substituting these two equations into equation (3.84) we obtain

$$\phi^{n+1} = \phi^{n-1} - \bar{\phi}\Delta t\left[\frac{\partial}{\partial x}u^{n-1} - \Delta t \frac{\partial}{\partial x}\left(\frac{\partial}{\partial x}\phi^{n+1} + \frac{\partial}{\partial x}\phi^{n-1}\right) + 2\Delta t \frac{\partial}{\partial x}(N_u)^n\right)$$

$$+\left(\frac{\partial}{\partial y}v^{n-1} - \Delta t \frac{\partial}{\partial y}\left(\frac{\partial}{\partial y}\phi^{n+1} + \frac{\partial}{\partial y}\phi^{n-1}\right) + 2\Delta t \frac{\partial}{\partial y}(N_v)^n\right) + \frac{\partial}{\partial x}u^{n-1} + \frac{\partial}{\partial y}v^{n-1}\Bigg],$$

$$\phi^{n+1} = \phi^{n-1} - \bar{\phi}\Delta t\left[\left(\frac{\partial}{\partial x}u^{n-1} + \frac{\partial}{\partial y}v^{n-1}\right) - \Delta t\left(\frac{\partial^2}{\partial x^2}\phi^{n+1} + \frac{\partial^2}{\partial y^2}\phi^{n+1} + \frac{\partial^2}{\partial x^2}\phi^{n-1} + \frac{\partial^2}{\partial y^2}\phi^{n-1}\right)\right.$$

$$\left. +2\Delta t\left(\frac{\partial}{\partial x}(N_u)^n + \frac{\partial}{\partial y}(N_v)^n\right) + \left(\frac{\partial}{\partial x}u^{n-1} + \frac{\partial}{\partial y}v^{n-1}\right)\right] + 2\Delta t (N_h)^n ,$$

$$\phi^{n+1} = \phi^{n-1} + \bar{\phi}(\Delta t)^2 \nabla^2 \phi^{n+1} + \bar{\phi}(\Delta t)^2 \nabla^2 \phi^{n-1}$$

$$-\bar{\phi}\Delta t\left(\nabla \cdot \vec{V}^{n-1}\right) - 2\bar{\phi}(\Delta t)^2\left(\frac{\partial}{\partial x}(N_u)^n + \frac{\partial}{\partial y}(N_v)^n\right)$$

$$-\bar{\phi}\Delta t\left(\nabla \cdot \vec{V}^{n-1}\right) + 2\Delta t (N_h)^n ,$$

where $\nabla \cdot \vec{V}^{n-1} = \partial u^{n-1}/\partial x + \partial v^{n-1}/\partial y$. We place all terms at time level $(n+1)$ on the left-hand side and the rest of the terms on the right-hand side. Then we write

$$\bar{\phi}(\Delta t)^2 \nabla^2 \phi^{n+1} - \phi^{n+1} = -\left[\phi^{n-1} + \bar{\phi}(\Delta t)^2 \nabla^2 \phi^{n-1} - 2\bar{\phi}\Delta t(\nabla \cdot \vec{V}^{n-1}\right]$$

$$+2\bar{\phi}(\Delta t)^2\left[\frac{\partial}{\partial x}(N_u)^n + \frac{\partial}{\partial y}(N_v)^n\right] - 2\Delta t (N_h)^n .$$

This equation can be written as

$$\nabla^2 \phi^{n+1} - \frac{\phi^{n+1}}{\bar{\phi}(\Delta t)^2} = \frac{F^{n-1} + G^n}{\bar{\phi}(\Delta t)^2}, \tag{3.85}$$

where

$$F^{n-1} = -\left[\phi^{n-1} + \bar{\phi}(\Delta t)^2 \nabla^2 \phi^{n-1} - 2\bar{\phi}\Delta t(\nabla \cdot \vec{V}^{n-1}\right],$$

$$G^n = 2\phi(\Delta t)^2\left[\frac{\partial (N_u)^n}{\partial x} + \frac{\partial (N_v)^n}{\partial y}\right] - 2\Delta t (N_h)^n .$$

This is a *Helmholtz equation* for the variable ϕ^{n+1}, i.e. one equation and one unknown. It can be solved using relaxation techniques or casting it into a tridiagonal matrix. Once a solution ϕ^{n+1} is obtained, it can be substituted into (3.82) and (3.83) to obtain future values of the velocity components u^{n+1} and v^{n+1}. Therefore, in principle, one can march forward in time and obtain future values of ϕ, u, and v. This technique can be used to forecast an event such as the propagation of an African wave and will be good for at most one day since it does not include convection.

Chapter 4

What Is a Spectral Model?

4.1 Introduction

If one takes a closed system of the basic meteorological equations and introduces within this system a finite expansion of the dependent variables using functions such as *double Fourier* or *Fourier-Legendre functions* in space, then the use of the orthogonality properties of these spatial functions enables one to obtain a set of coupled nonlinear ordinary differential equations for the coefficients of these functions. These coefficients are functions of time and the vertical coordinate, since the horizontal spatial dependence has been removed by taking a Fourier or a Fourier-Legendre transform of the equations. The coupled nonlinear ordinary differential equations for the coefficients are usually solved by simple time-differencing and vertical finite-differencing schemes. The mapping of the solution requires the multiplication of the coefficients with the spatial functions summed over a set of chosen finite spatial basis functions. This is what defines *spectral modeling.*

Meteorological application of the spectral method was initiated by Silberman (1954), who studied the nondivergent barotropic vorticity equation in the spherical coordinate system using the spectral technique. In its earlier days, the spectral method was particularly suitable for low-resolution simple models. The equations of these simple models involved nonlinear terms evaluated at each time step. Evaluation of the nonlinear terms was performed using the interaction coefficient method and thus required large memory allocations, which was an undesirable proposition.

However, with the introduction of the transform method, developed in dependently by Eliasen et al. (1970) and Orszag (1970), the method for evaluation of these nonlinear terms changed completely. This transform method also made it feasible to include nonadiabatic effects in the model equations. For the past couple of decades, the spectral method has become an increasingly popular technique for studies of general circulation and numerical weather prediction at the operational and research centers.

4.2 The Galerkin Method

This method forms the basis for spectral modeling, and it is easy to understand if the reader has some background in linear algebra. We have a set of linearly independent functions $\theta_i(x)$, which are called the *basis functions.* The dependent variables of the problem are represented by a finite sum of these basis functions. To illustrate the application of the Galerkin method, let us consider the following problem:

$$L(u)=g(x), \qquad \alpha \le x \le \beta, \tag{4.1}$$

where L is a differential operator, u is the unknown dependent variable, g is the forcing function, and x is the independent variable. The bounds of x are indicated by the real numbers α and β.

Our objective is to solve (4.1) given appropriate boundary conditions. The first step in obtaining a solution to (4.1) is to approximate $u(x)$ by a finite sum of basis functions as

$$u(x) \cong \sum_{i=1}^{N} u_i \theta_i(x). \tag{4.2}$$

Here u_i represents the coefficient of the ith basis function $\theta_i(x)$ and N is some prescribed integer. When one approximates $u(x)$ by (4.2), the error involved in satisfying (4.1) is given by

$$(\text{ERROR})_N = L\left(\sum_{i=1}^{N} u_i \theta_i(x)\right) - g(x). \tag{4.3}$$

Our objective is to determine u_i for $i=1, 2, 3, \dots, N$. This is done by imposing the condition that the error given by (4.3) is orthogonal to each and every basis function $\theta_i(x)$ in the interval $\alpha \le x \le \beta$. Mathematically, this condition can be written as

$$\int_\alpha^\beta (\text{ERROR})_N \, \theta_i(x)\,dx = 0 \qquad \text{for } i=1, 2, 3, \dots, N. \tag{4.4}$$

Then from (4.4) we can write

$$\int_\alpha^\beta L\left(\sum_{j=1}^{N} u_j \theta_j(x)\right)\theta_i(x)\,dx - \int_\alpha^\beta g(x)\theta_i(x)\,dx = 0, \tag{4.5}$$

for $i=1, 2, 3, \dots, N$. Equation (4.5) represents N algebraic equations for N unknowns $u_1, u_2, u_3, \dots, u_N$. Therefore, in principle, one can solve for these unknowns. To illustrate the Galerkin method, let us consider the following example.

We want to solve the system given by

$$\frac{d^2u}{dx^2} = F(x) \tag{4.6}$$

where $0 \le x \le \pi$ and $du/dx|_{x=0} = du/dx|_{x=\pi} = 0$ are the boundary conditions. By inspection, the following basis functions are suitable for the above problem, since they satisfy the prescribed boundary conditions:

$$\theta_n(x) = \cos nx, \qquad n=1, 2, 3, \dots, N.$$

In this example, $L = d^2/dx^2$ and

$$L\left[u_n(x)\right]=\frac{d^2}{dx^2}u_n(x)=\frac{d^2}{dx^2}\cos nx=-n^2\cos nx\,.$$

Thus

$$L\left(\sum_{i=1}^{N}u_i\theta_i(x)\right)=\sum_{i=1}^{N}-i^2u_i\theta_i(x).$$

The error in satisfying (4.6) is given by

$$(\text{ERROR})_N=L\left(\sum_{i=1}^{N}u_i\theta_i(x)\right)-F(x),$$

which can also be written as

$$(\text{ERROR})_N=\sum_{i=1}^{N}-i^2u_i\theta_i(x)-F(x). \tag{4.7}$$

Imposing the condition that $(\text{ERROR})_N$ is orthogonal to each and every basis function for the range $0\le x\le\pi$, we obtain

$$\sum_{i=1}^{N}-i^2u_i\int_0^{\pi}\theta_i(x)\theta_j(x)dx=\int_0^{\pi}\theta_j(x)F(x)dx. \tag{4.8}$$

It should be noted that in this example,

$$\int_0^{\pi}\theta_i(x)\theta_j(x)dx=\int_0^{\pi}\cos ix\cos jx\;dx\,.$$

Furthermore, we know from the orthogonality property of the Fourier function that

$$\int_0^{\pi}\cos ix\cos jx\;dx=\frac{\pi}{2}\delta_{ij}, \tag{4.9}$$

where δ is the Kronecker delta function with the values

$$\delta_{ij}=\begin{cases}1 & \text{if } i=j\\ 0 & \text{if } i\neq j\end{cases}. \tag{4.10}$$

With this, (4.8) leads to

$$-j^2u_j\frac{\pi}{2}=\int_0^{\pi}\theta_j(x)F(x)dx,\qquad j=1,\,2,\,3,\,\ldots\,,N,$$

from which we can write

$$u_j=\frac{-2}{j^2\pi}\int_0^{\pi}\theta_j(x)F(x)dx,\qquad j=1,2,3,\ldots\,,N, \tag{4.11}$$

If $F(x)$ can also be represented using a finite sum of the basis functions $\cos nx$, then the solution would be exact. Furthermore,

$$\int_0^\pi \theta_n(x)F(x)\,dx$$

represents a kind of transform using the basis functions.

Let us consider the specific case when $F(x)=\cos x+\cos 2x+\cos 3x$. Then using (4.11) we have

$$u_1=\frac{-2}{1^2\pi}\int_0^\pi \cos x(\cos x+\cos 2x+\cos 3x)\,dx$$
$$=\frac{-2}{\pi}\int_0^\pi \cos^2 x\,dx=-1. \tag{4.12}$$

Similarly we get $u_2=-1/4$ and $u_3=-1/9$. Furthermore,

$$u_4=u_5=u_6=\ \dots\ =u_N=0.$$

Thus, $u_1=-1$, $u_2=-1/4$ and $u_3=-1/9$ are the desired spectral coefficients. Hence the solution of

$$\frac{d^2u}{dx^2}=\cos x+\cos 2x+\cos 3x, \qquad 0\le x\le\pi, \tag{4.13}$$

where $du/dx|_{x=0}=du/dx|_{x=\pi}=0$, is given by

$$u=\sum_{i=1}^{N}u_i\theta_i(x)=-\cos x-\frac{1}{4}\cos 2x-\frac{1}{9}\cos 3x. \tag{4.14}$$

In this case, we have an exact solution since $F(x)$, the forcing function, can be represented exactly as the sum of our basis functions.

Two of the most useful Galerkin methods are the *spectral method* and the *finite element method.* In the above example, if we introduce time as one of the independent variables, then the ordinary differential equation has to be recast as a partial differential equation and the spectral coefficients would depend on time.

4.3 A Meteorological Application

Suppose we want to represent the observed temperature field at 850 mb around a latitude circle using wave-like functions. We use a sum of sine and cosine functions to do this. This gives us the finite discrete Fourier series as

$$T_i=A_0+\sum_{k=1}^{n}A_k\cos\frac{ik\pi}{n}+\sum_{k=1}^{n-1}B_k\sin\frac{ik\pi}{n}, \tag{4.15}$$

where n is the total number of wave components used to describe the temperature field. The points at which the temperature is known are represented by $i=0,\ 1, 2, 3,\ \dots\ , 2n-1$.

If T is defined at 100 points, then we will have a total of 50 waves. The integer k is called the *zonal wavenumber*, which denotes the number of waves along a latitude circle. Thus $k = 10$ implies a wavelength of $L = 360°/10 = 36°$. The coefficients A_k and B_k are obtained using the following formulas, i.e., the *discrete Fourier transform*:

$$A_k = \frac{1}{n}\sum_{i=0}^{2n-1} T_i \cos\frac{ik\pi}{n}, \qquad B_k = \frac{1}{n}\sum_{i=0}^{2n-1} T_i \sin\frac{ik\pi}{n}. \tag{4.16}$$

The temperature data are used to obtain A_k and B_k. Furthermore, (4.15) can be used along with the above equation in order to reconstruct the original temperature field if we know the coefficients.

The functions used in most global atmospheric spectral models as the basis functions are the *spherical harmonics*, a combination of sine and cosine functions that represent the zonal structure and associated Legendre functions that represent the meridional structure. Moreover, we work with the spherical domain. This is unlike the grid-point finite-difference models, where we generally work with a limited area in the Cartesian coordinate system.

Using the above basis functions, a dependent variable, say $A(\lambda,\phi)$, can be represented as

$$A(\lambda,\phi) = \sum_{m=-j}^{j} \sum_{n=|m|}^{j+|m|} A_n^m e^{im\lambda} P_n^m(\sin\phi), \tag{4.17}$$

where m is the zonal wavenumber, n is the two-dimensional (total) wavenumber, j is the maximum wavenumber resolved, λ is the longitude, ϕ is the latitude, A_n^m is the spectral coefficient, and P_n^m is an associated Legendre function of the first kind. A more detailed discussion of these functions is presented in Chapter 6.

4.4 Exercises

4.1 Consider the heat equation,

$$\frac{\partial u}{\partial t} = k\frac{\partial^2 u}{\partial x^2}, \qquad 0 \le x \le L,$$

with the boundary conditions $u(0,t) = 0$ and $u(L,t) = 0$ and the initial condition $u(x,0) = f(x)$. Find the required solution $u(x,t)$ of the above problem. (*Hint*: Use separation of variables).

4.2. In Exercise 4.1 assume (*a*) $f(x) = 12 + 8\cos n\pi x/L$ and (*b*) $f(x) = -8\sin\pi x/L$. Obtain a solution for Exercise 4.1.

Chapter 5

Lower-Order Spectral Model

5.1 Introduction

This system was first developed by Lorenz (1960b). It is an elegant system that provides an introduction to the concepts of spectral modeling. It is based on the premise that the dynamic equations governing the atmosphere can be simplified to the greatest extent possible and still be realistic enough to describe certain desired features. The extent to which we can simplify the equations depends on the particular phenomena we are studying. This simplification can aid in our understanding of a certain phenomena and help to form reasonable hypotheses which we can then test through a more advanced system of equations. The system is based on the use of double Fourier series representations of the basic equations in a doubly periodic domain. That is, the equations are expanded into a series of eigenfunctions, some of whose coefficients are retained as the new dependent variables. These new dependent variables correspond to features of the largest scale. Here we examine the *barotropic vorticity equation*. We start with the equation governing the conservation of vorticity of a parcel for two-dimensional, homogeneous, incompressible, and inviscid fluid flow on an f-plane given by

$$\frac{\partial}{\partial t}\nabla^2\psi = -J\left(\psi,\nabla^2\psi\right), \quad \text{or} \quad \frac{\partial}{\partial t}\nabla^2\psi = -\vec{k}\cdot\nabla\psi\times\nabla\left(\nabla^2\psi\right) \tag{5.1}$$

where ψ is the streamfunction and J is the Jacobian. Since we are working on an f-plane, the β term does not appear in this equation.

We let ψ be doubly periodic and state the periodicity property by the relation

$$\psi(x,y,t) = \psi\left(x+\frac{2\pi}{k}, y+\frac{2\pi}{l}, t\right),$$

where k and l are the wavenumbers and are specified constants. The area is now finite but unbounded. We seek a solution to (5.1) in a closed horizontal domain. For this, we expand ψ following Lorenz (1960b), that is,

$$\psi = \sum_{m=0}^{\infty}\sum_{n=n_0}^{\infty}\frac{-1}{m^2k^2+n^2l^2}\left[A_{mn}\cos(mkx+nly)+B_{mn}\sin(mkx+nly)\right]. \tag{5.2}$$

This is the *double Fourier representation* of the function ψ. The coefficients A_{mn} and B_{mn} are functions of time, where m and n are integers representing east-west and north-south wavenumbers, respectively. Note that $A_{00} = 0$. In addition, the lower limit n_0 of n is specified as

$$n_0 = \begin{cases} -\infty & \text{if } m>0 \\ 0 & \text{if } m=0 \end{cases}.$$

The series will now be truncated and we consider only those terms for which m equals 0 and +1 and n equals -1, 0, and +1; in other words, we include only one wave in both directions. Equation (5.2) for the streamfunction ψ then reduces to

$$\begin{aligned} \psi = &-\frac{A_{10}}{k^2}\cos kx - \frac{A_{01}}{l^2}\cos ly - \frac{A_{11}}{k^2+l^2}\cos(kx+ly) \\ &-\frac{A_{1-1}}{k^2+l^2}\cos(kx-ly) - \frac{B_{10}}{k^2}\sin kx - \frac{B_{01}}{l^2}\sin ly \\ &-\frac{B_{11}}{k^2+l^2}\sin(kx+ly) - \frac{B_{1-1}}{k^2+l^2}\sin(kx-ly), \end{aligned} \tag{5.3}$$

which is the time dependent solution for (5.1) for one wave. The corresponding relative vorticity is given by

$$\begin{aligned} \nabla^2\psi = &A_{10}\cos kx + A_{01}\cos ly + A_{11}\cos(kx+ly) \\ &+A_{1,-1}\cos(kx-ly) + B_{10}\sin kx \\ &+B_{01}\sin ly + B_{11}\sin(kx+ly) + B_{1,-1}\sin(kx-ly). \end{aligned} \tag{5.4}$$

Substituting the Fourier expansion of ψ and $\nabla^2\psi$ into (5.1) and taking the Fourier transform of both sides of the resulting equation, we get the prediction equations for the amplitude of the different wave components. In all, we have eight equations providing time tendencies for each of the eight amplitudes.

5.2 Maximum Simplification

After substituting for ψ and $\nabla^2\psi$ from (5.3) and (5.4) into (5.1) and equating the coefficients of the various Fourier functions on both sides of the resulting equation, we get a set of differential equations for the coefficients A_{10}, A_{01}, A_{11}, A_{1-1}, B_{10}, B_{01}, B_{11}, and B_{1-1}. Following Lorenz (1960b), we assume: (a) If B_{10}, B_{01}, B_{11}, and B_{1-1}, vanish initially, then they will remain zero for all time since their tendencies are always equal to zero, that is,

$$\frac{dB_{10}}{dt} = \frac{dB_{01}}{dt} = \frac{dB_{11}}{dt} = \frac{dB_{1-1}}{dt} = 0.$$

We thus obtain $B_{10} = B_{01} = B_{11} = B_{1-1} = 0$. (b) If $A_{1-1} = -A_{11}$ initially, then A_{1-1} will remain equal to $-A_{11}$ for all time. Furthermore, let $A_{01} = A$, $A_{10} = F$, and $-A_{11} = G$. With this, (5.3) reduces to

$$\psi = -\frac{A}{l^2}\cos ly - \frac{F}{k^2}\cos kx - 2\frac{G}{k^2+l^2}\sin kx \sin ly . \tag{5.5}$$

A, F, and G are functions of time. The term $-(A/l^2)\cos ly$ describes the basic zonal current, that is, it has no x dependence. The term $-(F/k^2)\cos kx - [2G/(k^2+l^2)]\sin kx \sin ly$ describes the eddies. The corresponding relative vorticity is given by

$$\nabla^2\psi = A\cos ly + F\cos kx + 2G\sin kx \sin ly . \tag{5.6}$$

Next we obtain an expression for

$$J\left(\psi,\nabla^2\psi\right) = \frac{\partial\psi}{\partial x}\frac{\partial\nabla^2\psi}{\partial y} - \frac{\partial\psi}{\partial y}\frac{\partial\nabla^2\psi}{\partial x} . \tag{5.7}$$

Now,

$$\frac{\partial\psi}{\partial x} = \frac{F}{k}\sin kx - \frac{2Gk}{k^2+l^2}\cos kx \sin ly , \tag{5.8}$$

$$\frac{\partial\psi}{\partial y} = \frac{A}{l}\sin ly - \frac{2Gl}{k^2+l^2}\sin kx \cos ly , \tag{5.9}$$

$$\frac{\partial}{\partial x}\nabla^2\psi = -Fk\sin kx + 2Gk\cos kx \sin ly , \tag{5.10}$$

$$\frac{\partial}{\partial y}\nabla^2\psi = -Al\sin ly + 2Gl\sin kx \cos ly . \tag{5.11}$$

Hence

$$\begin{aligned} J\left(\psi,\nabla^2\psi\right) = {} & \left(\frac{F}{k}\sin kx - \frac{2Gk}{k^2+l^2}\sin ly \cos kx\right) \\ & \times\left(-Al\sin ly + 2Gl\sin kx \cos ly\right) \\ & -\left(\frac{A}{l}\sin ly - \frac{2Gl}{k^2+l^2}\cos ly \sin kx\right) \\ & \times\left(-Fk\sin kx + 2G\cos kx \sin ly\right) . \end{aligned}$$

After simplifying, we obtain

$$\begin{aligned} J\left(\psi,\nabla^2\psi\right) = {} & \left(\frac{1}{l^2} - \frac{1}{k^2}\right) AFkl\sin kx \sin ly \\ & + \left(\frac{1}{k^2} - \frac{1}{k^2+l^2}\right) 2FGkl\sin^2 kx \cos ly \\ & + \left(-\frac{1}{l^2} + \frac{1}{k^2+l^2}\right) 2AGkl\sin^2 ly \cos kx \end{aligned} \tag{5.12}$$

Differentiating (5.6) with respect to t, we get

$$\frac{\partial}{\partial t}\nabla^2\psi = \frac{dA}{dt}\cos ly + \frac{dF}{dt}\cos kx + 2\frac{dG}{dt}\sin kx \sin ly . \tag{5.13}$$

From the barotropic vorticity equation (5.1) along with (5.12) and (5.13), we get

$$\begin{aligned}\frac{dA}{dt}\cos ly+\frac{dF}{dt}\cos kx+2\frac{dG}{dt}\sin kx\sin ly\\ =-\Bigg[\left(\frac{1}{l^2}-\frac{1}{k^2}\right)AFkl\sin kx\sin ly\\ +\left(\frac{1}{k^2}-\frac{1}{k^2+l^2}\right)2FGkl\sin^2 kx\cos ly\\ +\left(-\frac{1}{l^2}+\frac{1}{k^2+l^2}\right)2AGkl\sin^2 ly\cos kx\Bigg].\end{aligned} \tag{5.14}$$

If we multiply (5.14) by $\cos ly$ and integrate both sides over the entire doubly periodic fundamental domain, then using the orthogonality properties of the Fourier functions we obtain

$$\begin{aligned}\frac{dA}{dt}\int_0^{2\pi}\int_0^{2\pi}\cos^2 ly\, dx\, dy=-\left(\frac{1}{k^2}-\frac{1}{k^2+l^2}\right)2klFG\\ \times\int_0^{2\pi}\int_0^{2\pi}\sin^2 kx\cos^2 ly\, dx\, dy.\end{aligned} \tag{5.15}$$

Integrating, we obtain

$$2\frac{dA}{dt}\pi^2=-2\pi^2 klFG\left(\frac{1}{k^2}-\frac{1}{k^2+l^2}\right),$$

or

$$\frac{dA}{dt}=-\left(\frac{1}{k^2}-\frac{1}{k^2+l^2}\right)klFG. \tag{5.16}$$

Similarly, if we multiply (5.14) by $\cos kx$ and $\sin kx\sin ly$ and integrate over the domain, we get

$$\frac{dF}{dt}=\left(\frac{1}{l^2}-\frac{1}{k^2+l^2}\right)klAG, \tag{5.17}$$

$$\frac{dG}{dt}=-\frac{1}{2}\left(\frac{1}{l^2}-\frac{1}{k^2}\right)klAF. \tag{5.18}$$

Equations (5.16), (5.17), and (5.18) are a system of three coupled nonlinear first-order ordinary differential equations that constitute the barotropic vorticity equation in the three unknowns A, F, and G. If their initial values are known, then their future values can be obtained using numerical integration. The above system has exact solutions which can be expressed by elliptic functions (or circular functions) in time.

5.3 Conservation of Mean-Square Vorticity and Mean Kinetic Energy

In this section we will attempt to answer the following question: Does this simplified system conserve mean-square vorticity and mean kinetic energy? Consider the expression for the *total-square vorticity*, which is written as

$$\int_0^{2\pi}\int_0^{2\pi}\left(\nabla^2\psi\right)^2 dx\,dy.$$

On substituting the Fourier expansion for $\nabla^2\psi$, we obtain

$$\int_0^{2\pi}\int_0^{2\pi}\left(\nabla^2\psi\right)^2 dx\,dy=\int_0^{2\pi}\int_0^{2\pi}\left(A\cos ly+F\cos kx+2G\sin kx\sin ly\right)^2 dx\,dy.$$

This can be written as

$$\begin{aligned}\int_0^{2\pi}\int_0^{2\pi}\left(\nabla^2\psi\right)^2 dx\,dy=\int_0^{2\pi}\int_0^{2\pi}\Big(&A^2\cos^2 ly+F^2\cos^2 kx\\&+4G^2\sin^2 kx\sin^2 ly+2AF\cos ly\cos kx\\&+4AG\sin kx\sin ly\cos ly\\&+4FG\sin kx\sin ly\cos kx\Big)dx\,dy.\end{aligned}\tag{5.19}$$

The integrals involving terms *AF*, *AG*, and *FG* turn out to be zero using the orthogonality rules, and we are left with

$$\begin{aligned}\int_0^{2\pi}\int_0^{2\pi}\left(\nabla^2\psi\right)^2 dx\,dy=\int_0^{2\pi}\int_0^{2\pi}\Big(&A^2\cos^2 ly+F^2\cos^2 kx\\&+4G^2\sin^2 kx\sin^2 ly\Big)dx\,dy.\end{aligned}\tag{5.20}$$

This gives

$$\begin{aligned}\int_0^{2\pi}\int_0^{2\pi}\left(\nabla^2\psi\right)^2 dx\,dy&=2\pi^2A^2+2\pi^2F^2+4\pi^2G^2\\&=2\pi^2\left(A^2+F^2+2G^2\right).\end{aligned}\tag{5.21}$$

Thus the total-square vorticity over the domain is

$$\int_0^{2\pi}\int_0^{2\pi}\left(\nabla^2\psi\right)^2 dx\,dy=2\pi^2\left(A^2+F^2+2G^2\right).\tag{5.22}$$

Because the domain area is given by $\int_0^{2\pi}\int_0^{2\pi}dx\,dy=4\pi^2$, the *mean-square vorticity* is

$$\overline{\left(\nabla^2\psi\right)^2}=\frac{1}{2}\left(A^2+F^2+2G^2\right).\tag{5.23}$$

Next we show that the mean-square vorticity of this system is conserved. From (5.23), using the relation $\frac{d}{dt}\left(\frac{1}{2}A^2\right)=A\frac{dA}{dt}$ the time rate of change of mean-square vorticity $\overline{(\nabla^2\psi)^2}$ is given by

$$\frac{d}{dt}\overline{\left(\nabla^2\psi\right)^2}=\frac{d}{dt}\left[\frac{1}{2}(A^2+F^2+2G^2)\right]=A\frac{dA}{dt}+F\frac{dF}{dt}+2G\frac{dG}{dt}.\tag{5.24}$$

Substituting for dA/dt, dF/dt, and dG/dt from (5.16), (5.17), and (5.18), we get

$$\frac{d}{dt}\overline{\left(\nabla^2\psi\right)^2}=AFG\left[-kl\left(\frac{1}{k^2}-\frac{1}{k^2+l^2}\right)+kl\left(\frac{1}{l^2}-\frac{1}{k^2+l^2}\right)\right.$$

$$-kl\left(\frac{1}{l^2}-\frac{1}{k^2}\right)\Bigg]=0. \tag{5.25}$$

Thus, the *mean-square vorticity of this three-component low-order system is conserved.*

We next show that the mean kinetic energy is also conserved by the low-order system. Following a similar procedure,

$$\begin{aligned}
\text{Total K.E.} &= \frac{1}{2}\int_0^{2\pi}\int_0^{2\pi}\left[\left(\frac{\partial\psi}{\partial x}\right)^2+\left(\frac{\partial\psi}{\partial y}\right)^2\right]dx\,dy \\
&= \frac{1}{2}\int_0^{2\pi}\int_0^{2\pi}\left[\left(\frac{F}{k}\sin kx-\frac{2Gk}{k^2+l^2}\cos kx\sin ly\right)^2\right. \\
&\quad \left.+\left(\frac{A}{l}\sin ly-\frac{2Gl}{k^2+l^2}\sin kx\cos ly\right)^2\right]dx\,dy \\
&= \frac{1}{2}\int_0^{2\pi}\int_0^{2\pi}\left[\left(\frac{F^2}{k^2}\sin^2 kx+\frac{4k^2G^2}{(k^2+l^2)^2}\cos^2 kx\sin^2 ly\right.\right. \\
&\quad \left.-\frac{4FG}{k^2+l^2}\sin kx\sin ly\cos kx\right) \\
&\quad +\left(\frac{A^2}{l^2}\sin^2 ly+\frac{4l^2G^2}{(k^2+l^2)^2}\cos^2 ly\sin^2 kx\right. \\
&\quad \left.\left.-\frac{4AG}{k^2+l^2}\sin kx\sin ly\cos ly\right)\right]dx\,dy \\
&= \frac{1}{2}\left[\frac{F^2}{k^2}2\pi^2+\frac{4k^2G^2}{(k^2+l^2)^2}\pi^2+\frac{A^2}{l^2}2\pi^2+\frac{4l^2G^2}{(k^2+l^2)^2}\pi^2\right].
\end{aligned} \tag{5.26}$$

Hence

$$\text{Total K.E.} = \pi^2\left(\frac{F^2}{k^2}+\frac{A^2}{l^2}+\frac{2G^2}{k^2+l^2}\right). \tag{5.27}$$

Note that the domain area equals $4\pi^2$. Thus the *mean kinetic energy* is given by

$$\overline{\text{K.E.}} = \frac{1}{4}\left(\frac{F^2}{k^2}+\frac{A^2}{l^2}+\frac{2G^2}{k^2+l^2}\right). \tag{5.28}$$

Now we show that $d/dt\ \overline{\text{K.E.}}=0$,

$$\begin{aligned}
\frac{d}{dt}\overline{\text{K.E.}} &= \frac{d}{dt}\left[\frac{1}{4}\left(\frac{F^2}{k^2}+\frac{A^2}{l^2}+\frac{2G^2}{k^2+l^2}\right)\right] \\
&= \frac{1}{2}\left[\frac{A}{l^2}\frac{dA}{dt}+\frac{F}{k^2}\frac{dF}{dt}+\frac{2G}{k^2+l^2}\frac{dG}{dt}\right]
\end{aligned}$$

$$
\begin{aligned}
&= \frac{1}{2}\left[-\frac{1}{l^2}\left(\frac{1}{k^2}-\frac{1}{k^2+l^2}\right)+\frac{1}{k^2}\left(\frac{1}{l^2}-\frac{1}{k^2+l^2}\right)\right. \\
&\quad \left.-\frac{1}{k^2+l^2}\left(\frac{1}{l^2}-\frac{1}{k^2}\right)\right] AFGkl \\
&= \frac{1}{2}\left[-\frac{1}{l^2k^2}+\frac{1}{l^2(k^2+l^2)}-\frac{1}{k^2(k^2+l^2)}\right. \\
&\quad \left.+\frac{1}{k^2l^2}-\frac{1}{l^2(k^2+l^2)}+\frac{1}{k^2(k^2+l^2)}\right] AFGkl \\
&= 0 . \qquad (5.29)
\end{aligned}
$$

Thus $d/dt\,\overline{\text{K.E.}} = 0$. Hence the mean kinetic energy is conserved with time.

5.4 Energy Transformations

In the previous section, we saw that the mean kinetic energy of the low-order system is conserved with time. However, the zonal kinetic energy and the eddy kinetic energy individually are not conserved, as we show below. Thus, continuous exchanges of kinetic energy take place between the zonal flow and the eddies. At times, the zonal may gain kinetic energy at the expense of the eddies while at other times the eddies may gain kinetic energy at the expense of the zonal flow. This occurs in such a way that the total kinetic energy is invariant.

In our low-order system, the zonal flow is represented by

$$\psi_Z = -\frac{A}{l^2}\cos ly , \qquad (5.30)$$

while the eddy flow is given by

$$\psi_E = -\frac{F}{k^2}\cos kx - \frac{2G}{k^2+l^2}\sin kx \sin ly . \qquad (5.31)$$

The *mean zonal* and *mean eddy kinetic energies* therefore are given by

$$
\begin{aligned}
K_Z &= \frac{1}{4\pi^2}\int_0^{2\pi}\int_0^{2\pi}\frac{1}{2}(\nabla\psi_Z)^2\,dx\,dy = \frac{1}{4\pi^2}\int_0^{2\pi}\int_0^{2\pi}\frac{1}{2}\frac{A^2}{l^2}\sin^2 ly\,dx\,dy \\
&= \frac{1}{4}\frac{A^2}{l^2} , \qquad (5.32)
\end{aligned}
$$

$$
\begin{aligned}
K_E &= \frac{1}{4\pi^2}\int_0^{2\pi}\int_0^{2\pi}\frac{1}{2}(\nabla\psi_E)^2\,dx\,dy \\
&= \frac{1}{4\pi^2}\int_0^{2\pi}\int_0^{2\pi}\frac{1}{2}\left[\left(\frac{F}{k}\sin kx - \frac{2Gk}{k^2+l^2}\cos kx \sin ly\right)^2\right. \\
&\quad \left.+\left(-\frac{2Gl}{k^2+l^2}\sin kx\cos ly\right)^2\right] dx\,dy
\end{aligned}
$$

$$
= \frac{1}{4\pi^2} \int_0^{2\pi} \int_0^{2\pi} \frac{1}{2} \left[\frac{F^2}{k^2} \sin^2 kx + \frac{4G^2 k^2}{\left(k^2 + l^2\right)^2} \cos^2 kx \sin^2 ly \right.
$$

$$
\left. + \frac{4G^2 l^2}{(k^2 + l^2)^2} \sin^2 kx \cos^2 ly - \frac{4GF}{k^2 + l^2} \sin kx \cos kx \sin ly \right] dx\, dy
$$

$$
= \frac{1}{4}\left(\frac{F^2}{k^2} + \frac{2G^2}{k^2 + l^2} \right). \tag{5.33}
$$

Thus we have

$$
\frac{d\mathrm{K}_Z}{dt} = \frac{1}{2}\frac{d}{dt}\frac{A^2}{l^2} = \frac{A}{2l^2}\frac{dA}{dt}. \tag{5.34}
$$

This, after substituting the value of dA/dt from (5.16), gives

$$
\frac{d\mathrm{K}_Z}{dt} = -AFG\frac{kl}{2l^2}\left(\frac{1}{k^2} - \frac{1}{k^2 + l^2}\right) = -AFG\frac{k}{2l}\left(\frac{1}{k^2} - \frac{1}{k^2 + l^2}\right). \tag{5.35}
$$

Similarly,

$$
\frac{\mathrm{dK_E}}{\mathrm{dt}} = \frac{1}{2k^2} F \frac{dF}{dt} + \frac{1}{k^2 + l^2} G \frac{dG}{dt}
$$

$$
= \frac{AFGkl}{2}\left[\frac{1}{k^2}\left(\frac{1}{l^2} - \frac{1}{k^2 + l^2}\right) - \frac{1}{k^2 + l^2}\left(\frac{1}{l^2} - \frac{1}{k^2}\right)\right]
$$

$$
= \frac{AFGk}{2l}\left(\frac{1}{k^2} - \frac{1}{k^2 + l^2}\right). \tag{5.36}
$$

We denote $\partial \mathrm{K}_Z / \partial t$ as $< \mathrm{K_E}, \mathrm{K}_Z >$, which is the change in K_Z due to energy transfer from the eddy to the zonal flow. Also, $< \mathrm{K}_Z, \mathrm{K_E} >$ denotes $\partial \mathrm{K_E} / \partial t$ and represents the change in $\mathrm{K_E}$ due to transfer from the zonal to the eddy flow. Note that $< \mathrm{K_E}, \mathrm{K}_Z >$ and $< \mathrm{K}_Z, \mathrm{K_E} >$ are equal in magnitude but opposite in sign. Thus in this low-order system, the gain of kinetic energy by the zonal flow is equal to the loss of kinetic energy by the eddy flow or vice versa, so that the total kinetic energy is invariant with time. Such barotropic energy exchange is an important property of a nondivergent flow.

5.5 Mapping the Solution

In the expression for the streamfunction of a low-order system, that is,

$$
\psi = -\frac{A}{l^2}\cos ly - \frac{F}{k^2}\cos kx - 2\frac{G}{k^2 + l^2}\sin kx \sin ly, \tag{5.37}
$$

the latitudes $\pi/(2l)$, $3\pi/(2l)$, $5\pi/(2l)$, … correspond to the zonal wind maxima and are fixed. However, the intensity of the zonal flow given by A/l may vary. Thus the variable A denotes the zonal index. The last two terms represent disturbances superimposed on the zonal flow. Both of these combined describe a wave of a single wavenumber, but with a variable shape and phase. This simple model simulates the interaction of the zonal flow with the superimposed disturbance.

This problem is easy to solve on any simple calculator or a personal computer. The student is strongly advised to try this as an exercise. The initial values of A, F, G and the wavenumbers k and l need to be assigned. For numerical integration in time, one forward time step may be followed by centered time steps. For that purpose, one uses (5.16), (5.17), and (5.18). At the end of each day of integration, the predicted values of A, F and G can be substituted in (5.37) to map the forecast.

5.6 An Example of Chaos

In order to illustrate the sensitivity of the low-order model to small changes of parameters, we examine the following *sixth-order system*:

$$\dot{A}+\sigma(1+\alpha^2)A-\frac{\sigma\alpha}{1+\alpha^2}D-\frac{1}{2}\alpha\frac{3+\alpha^2}{1+\alpha^2}BC=0, \tag{5.38}$$

$$\dot{B}+\sigma B+\frac{3}{4}\alpha AC=0, \tag{5.39}$$

$$\dot{C}+\sigma(4+\alpha^2)C-\frac{\sigma\alpha}{4+\alpha^2}F+\frac{1}{2}\frac{\alpha^3}{4+\alpha^2}AB=0, \tag{5.40}$$

$$\dot{D}+(1+\alpha^2)D-R\alpha A+\alpha AE+\frac{1}{2}\alpha BF=0, \tag{5.41}$$

$$\dot{E}+4E-\frac{1}{2}\alpha AD=0, \tag{5.42}$$

$$\dot{F}+(4+\alpha^2)F+R\alpha C-\frac{1}{2}\alpha BD=0. \tag{5.43}$$

This is a coupled nonlinear system of six equations for the six dependent variables *A, B, C, D, E,* and *F.* The independent variable here is time. Furthermore, an overdot ($\dot{}$) represents a derivative with respect to time, α is the wavenumber, and R is the Rayleigh number, which is expressed as $R=g\beta\Delta Td^3/(\kappa\nu)$. Here g is the acceleration due to gravity, β is the thermal expansion coefficient, κ is the thermal diffusivity, ν is the kinematic viscosity, ΔT is the temperature excess of the bottom boundary over the top boundary, and d is the distance between these two boundaries.

This problem is solved numerically (see Fig. 5.1) in the same manner as the simple three-component system illustrated above. In fact, with $B=C=F=0$, this six-component system reduces to a *three-component Lorenz system.* A time-differencing scheme is needed to solve this coupled system numerically. In the coding of this problem, we have used an *Adams-Bashforth* time-differencing scheme, which was described in Chapter 3.

The sensitivity of the solution to small changes in the Rayleigh number R is illustrated in Fig. 5.1. Here we show a periodic solution for R = 50.2299 and a chaotic solution for R = 50.2300. The abscissa denotes the value of the coefficient B, and the ordinate denotes the value of coefficient C. The time evolution of the solution for R =

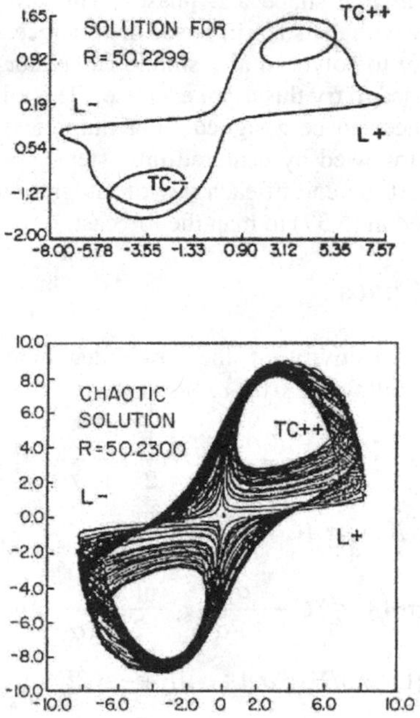

Figure 5.1. Periodic and chaotic solutions of a six-component system.

50.2299 repeats itself, whereas for $R = 50.2300$ the solution is chaotic. It should be noted that the same sort of multidimensional behavior (periodic or chaotic) is found for other pairs of variable choices as well. Hence, we see that there are huge initial uncertainties in non-linear systems.

5.7. Exercises

5.1. In the equations

$$\frac{dA}{dt} = -\left(\frac{1}{k^2} - \frac{1}{k^2 + l^2}\right) klFG,$$

$$\frac{dF}{dt} = \left(\frac{1}{l^2} - \frac{1}{k^2 + l^2}\right) klAG,$$

$$\frac{dG}{dt} = -\frac{1}{2}\left(\frac{1}{l^2} - \frac{1}{k^2}\right) klAF,$$

set $2\pi / l = 5000$ and $2\pi / k = 2500$, so that $\alpha = k / l = 2$. For initial conditions, let $A(0) = 0.06$ units, $F(0) = 0.12$ units, and $G(0) = 0$. Obtain the future values of $A(t)$, $F(t)$, and $G(t)$ using $\Delta t = 6$ hours.

5.2. For $A(t)$, $F(t)$, and $G(t)$ in Exercise 5.1, map the resulting streamfunction at time $t = 0$ and time $t = 24$ hours.

Chapter 6

Mathematical Aspects of Spectral Models

6.1 Introduction

In this chapter we provide an introduction to the topic of spherical harmonics as basis functions for a global spectral model. The *spherical harmonics* are made up of trigonometric functions along the zonal direction and associated Legendre functions in the meridional direction. A number of properties of these functions need to be understood for the formulation of a spectral model. This chapter describes some useful properties that will be used to illustrate the procedure for the representation of data sets over a sphere with spherical harmonics as basis functions. The calculations of Fourier and Legendre transforms and their inverse transforms are an important part of global spectral modeling, and these are covered in some detail in this chapter. Finally, this chapter addresses the formulation of two simple spectral models. One of these is a single-level barotropic model, and the other is a shallow-water model.

Consider the equation

$$\nabla^2 u = \frac{1}{c^2}\frac{\partial^2 u}{\partial t^2},$$

which is satisfied by the velocity potential of a compressible fluid. Here u represents the velocity potential and c is the speed of gravity waves in the fluid. If the fluid is in a steady state, this equation reduces to *Laplace's equation*, which is

$$\nabla^2 u = \frac{\partial^2 u}{\partial x^2} + \frac{\partial^2 u}{\partial y^2} + \frac{\partial^2 u}{\partial z^2} = 0. \tag{6.1}$$

A similar equation describes the steady-state diffusive process or the heat conduction process in a medium. Laplace's equation and its solutions, which are harmonic functions, are of fundamental importance in the study of fluid dynamics. The solutions of Laplace's equation in a spherical coordinate system are the spherical harmonics, and are obtained by the method of *separation of variables*. That is, the solutions can be broken up into factors, each factor being the function of a single coordinate.

The transformation between Cartesian and spherical coordinates is given by

$$x = r\cos\lambda\,\sin\theta,\qquad y = r\sin\lambda\,\cos\theta,\qquad z = r\cos\theta, \tag{6.2}$$

where r is the radius, λ is longitude, and θ is co-latitude. Substituting x, y, and z from (6.2) into (6.1), we obtain Laplace's equation on a sphere:

$$\nabla^2 u = \frac{\partial^2 u}{\partial r^2} + \frac{2}{r}\frac{\partial u}{\partial r} + \frac{1}{r^2}\frac{\partial^2 u}{\partial \theta^2} + \frac{\cot\theta}{r^2}\frac{\partial u}{\partial \theta} + \frac{1}{r^2 \sin^2\theta}\frac{\partial^2 u}{\partial \lambda^2} = 0, \tag{6.3}$$

or

$$\frac{1}{r^2}\left[\frac{\partial}{\partial r}\left(r^2 \frac{\partial u}{\partial r}\right) + \frac{1}{\sin\theta}\frac{\partial}{\partial \theta}\left(\sin\theta \frac{\partial u}{\partial \theta}\right) + \frac{1}{\sin^2\theta}\frac{\partial^2 u}{\partial \lambda^2}\right] = 0.$$

In the case of atmospheric models, latitude is usually used instead of co-latitude as one of the coordinates. Hereafter, θ denotes the latitude. Laplace's equation (6.3) in the (r, λ, θ) coordinate system (where θ is latitude and remembering sin(colat) = cos(lat)) then takes the form

$$\frac{1}{r^2}\left[\frac{\partial}{\partial r}\left(r^2 \frac{\partial u}{\partial r}\right) + \frac{1}{\cos\theta}\frac{\partial}{\partial \theta}\left(\cos\theta \frac{\partial u}{\partial \theta}\right) + \frac{1}{\cos^2\theta}\frac{\partial^2 u}{\partial \lambda^2}\right] = 0,$$

or

$$\frac{\partial}{\partial r}\left(r^2 \frac{\partial u}{\partial r}\right) + \frac{1}{\cos\theta}\frac{\partial}{\partial \theta}\left(\cos\theta \frac{\partial u}{\partial \theta}\right) + \frac{1}{\cos^2\theta}\frac{\partial^2 u}{\partial \lambda^2}\Bigg] = 0. \tag{6.4}$$

We use the method of separation of variables to solve (6.4). For this, assume a solution of the form

$$u = R(r)L(\lambda)P(\theta) \tag{6.5}$$

where R is a function of r, L is a function of λ, and P is a function of θ. Substituting (6.5) into (6.4) we obtain

$$LP\frac{d}{dr}\left(r^2 \frac{dR}{dr}\right) + \frac{RL}{\cos\theta}\frac{d}{d\theta}\left(\cos\theta \frac{dP}{d\theta}\right) + \frac{RP}{\cos^2\theta}\frac{d^2 L}{d\lambda^2} = 0,$$

or, on dividing by RLP, we get

$$\frac{1}{P\cos\theta}\frac{d}{d\theta}\left(\cos\theta \frac{dP}{d\theta}\right) + \frac{1}{L\cos^2\theta}\frac{d^2 L}{d\lambda^2} = -\frac{1}{R}\frac{d}{dr}\left(r^2 \frac{dR}{dr}\right). \tag{6.6}$$

The left-hand side of (6.6) is a function of λ and θ, while the right-hand side is a function of r only. Therefore, both sides should be equal to a constant, say k.

If we now consider the right-hand side of (6.6), we obtain

$$-\frac{1}{R}\frac{d}{dr}\left(r^2 \frac{dR}{dr}\right) = k,$$

$$k + \frac{1}{R}\frac{d}{dr}\left(r^2 \frac{dR}{dr}\right) = 0,$$

$$\frac{d}{dr}\left(r^2 \frac{dR}{dr}\right) + kR = 0. \tag{6.7}$$

Let the solution of (6.7) be of the form $R = r^n$, where n is an integer. If n is positive, then on substitution of $R = r^n$ into (6.7) we obtain

$$\frac{d}{dr}\left(r^2\frac{dr^n}{dr}\right)+r^n k=0,$$

$$\frac{d}{dr}\left(r^2 nr^{n-1}\right)+r^n k=0,$$

$$n\frac{d}{dr}r^{n+1}+r^n k=0,$$

$$n(n+1)r^n+r^n k=0,$$

$$\left[n(n+1)+k\right]R=0, \text{ or}$$

$$k=-n(n+1).$$

We see that the value of k is not altered if we replace n by $-n-1$. Therefore r^{-n-1} is also a solution of (6.7), for which $k=-n(n-1)$. Thus k is of the form $-n(n+1)$, n being a positive integer including zero. Equation (6.6) thus reduces to

$$\frac{1}{P\cos\theta}\frac{d}{d\theta}\left(\cos\theta\frac{dP}{d\theta}\right)+\frac{1}{L\cos^2\theta}\frac{d^2L}{d\lambda^2}=-n(n+1),$$

or multiplying by $\cos^2\theta$ and rearranging we obtain

$$\frac{\cos\theta}{P}\frac{d}{d\theta}\left(\cos\theta\frac{dP}{d\theta}\right)+n(n+1)\cos^2\theta=-\frac{1}{L}\frac{d^2L}{d\lambda^2}. \tag{6.8}$$

Since the left-hand side of (6.8) is a function of θ and the right-hand side is a function of λ, both sides must be equal to a constant, say m^2. Considering the right-hand side of (6.8) gives

$$-\frac{1}{L}\frac{d^2L}{d\lambda^2}=m^2,$$

$$\frac{d^2L}{d\lambda^2}+m^2L=0, \tag{6.9}$$

which has a solution of the form

$$L=e^{\pm im\lambda},\ m=0,\ 1,\ 2,\ 3,\ldots$$

Hence after multiplying (6.8) by $P/\cos^2\theta$ and remembering both sides of the equation are equal to m^2 the equation reduces to

$$\frac{1}{\cos\theta}\frac{d}{d\theta}\left(\cos\theta\frac{dP}{d\theta}\right)+n(n+1)P=\frac{m^2P}{\cos^2\theta},$$

$$\frac{1}{\cos\theta}\frac{d}{d\theta}\left(\cos\theta\frac{dP}{d\theta}\right)+\left(n(n+1)-\frac{m^2}{\cos^2\theta}\right)P=0. \tag{6.10}$$

If $\mu = \sin\theta$ and $\frac{d}{d\theta} = \frac{d}{d\mu}\frac{d\mu}{d\theta} = \frac{d}{d\mu}\frac{d\sin\theta}{d\theta} = \cos\theta\frac{d}{d\mu}$, then (6.10) can be written as

$$\frac{d}{d\mu}\left(\left(1-\mu^2\right)\frac{dP}{d\mu}\right)+\left(n(n+1)-\frac{m^2}{1-\mu^2}\right)P=0, \tag{6.11}$$

remembering $1-\sin^2\theta = \cos^2\theta$. Since θ varies from -π/2 to π/2, μ varies from -1 to 1. Equations (6.10) and (6.11) are both known as *associated Legendre equations*, as they apply to all wavenumbers, *m*.

6.2 Legendre Equation and Associated Legendre Equation

If $m = 0$ in equation (6.11), we obtain

$$\frac{d}{d\mu}\left(\left(1-\mu^2\right)\frac{dP}{d\mu}\right)+n(n+1)P=0, \tag{6.12}$$

which is called the *Legendre equation.* We next discuss the solutions of the Legendre equation and the associated Legendre equation. Furthermore, we discuss the properties of their solutions without rigorous mathematical proofs.

Solutions of the Legendre equation are known as *Legendre polynomials* and are denoted by $P_n(\mu)$. For a given *n*, $P_n(\mu)$ is a polynomial of degree *n* and is given by

$$P_n(\mu)=\sum_{r=0}^{M}(-1)^r\frac{(2n-2r)!}{2^n r!(n-r)!(n-2r)!}\mu^{n-2r}, \tag{6.13}$$

where $M = n/2$ if *n* is even and $M = (n-1)/2$ if *n* is odd.

A more convenient form of $P_n(\mu)$ is given by *Rodrigues' formula*, namely

$$P_n(\mu)=\frac{1}{2^n n!}\frac{d^n}{d\mu^n}\left(\mu^2-1\right)^n, \quad n=0,\ 1,\ 2,\ 3,\ldots,\ |\mu|\le 1. \tag{6.14}$$

In particular,

$$P_0(\mu)=1,$$
$$P_1(\mu)=\frac{1}{2}\frac{d}{d\mu}\left(\mu^2-1\right)=\mu,$$
$$P_2(\mu)=\frac{1}{8}\frac{d^2}{d\mu^2}\left(\mu^2-1\right)^2=\frac{1}{2}\left(3\mu^2-1\right),$$
$$P_3(\mu)=\frac{1}{48}\frac{d^3}{d\mu^3}\left(\mu^2-1\right)^3=\frac{1}{2}\left(5\mu^2-3\mu\right),$$
$$P_4(\mu)=\frac{1}{384}\frac{d^4}{d\mu^4}\left(\mu^2-1\right)^4=\frac{1}{8}\left(35\mu^4-30\mu^2+3\right),$$
$$P_5(\mu)=\frac{1}{3840}\frac{d^5}{d\mu^5}\left(\mu^2-1\right)^5=\frac{1}{8}\left(63\mu^5-70\mu^3+15\mu\right).$$

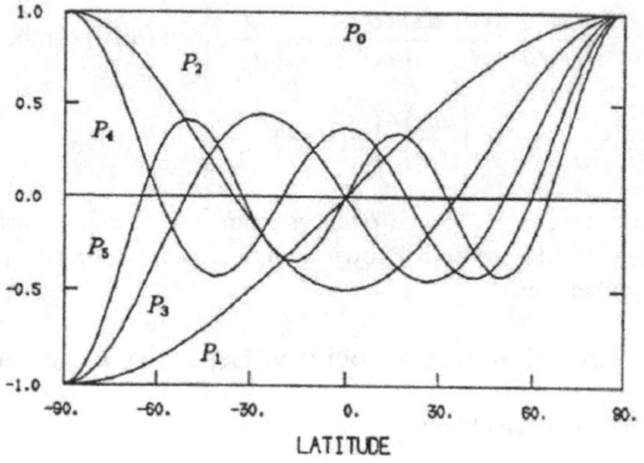

Figure 6.1. Representation of Legendre polynomials $P_0(\mu)$ to $P_5(\mu)$. The number of zero crossings is the same as the degree of the polynomial, *n*. If *n* is even, $P_n(\mu)$ is symmetric about the equator. If *n* is odd, $P_n(\mu)$ is antisymmetric about the equator.

Figure 6.1 shows the graphs of $P_0(\mu)$ to $P_5(\mu)$ for $-1 \le \mu \le 1$.

Three useful properties of $P_n(\mu)$ are as follows: (1) $P_n(\mu=1)=1$. (2) If *n* is even, $P_n(\mu)$ has only even powers of μ and is symmetric with respect to the equator ($\mu=0$). (3) If *n* is odd, $P_n(\mu)$ has only odd powers of μ and is antisymmetric with respect to the equator. In other words, for odd *n*, the graph on the negative side of the μ-axis is a mirror image of the graph on the positive side of the μ-axis, as shown in Fig. 6.1.

We now consider the associated Legendre equation (6.11). Solutions of this equation involve two parameters, *m* and *n*, and are denoted by $P_n^m(\mu)$. $P_n^m(\mu)$ are called *associated Legendre functions of the first kind of order m and degree n.* Here *m* is any integer and *n* is a non-negative integer such that $n \ge |m|$.

One can obtain $P_n^m(\mu)$ by using Rodrigues' formula,

$$P_n^m(\mu)=\frac{(1-\mu^2)^{m/2}}{2^n n!}\frac{d^{n+m}}{d\mu^{n+m}}(\mu^2-1)^n, \qquad |\mu|\le 1. \tag{6.15}$$

In particular, for $n=5$,

$$P_5^0(\mu)=\frac{1}{8}\left(63\mu^5-70\mu^3+15\mu\right),$$

$$P_5^1(\mu)=\frac{15}{8}\left(1-\mu^2\right)^{1/2}\left(21\mu^4-14\mu^2+1\right),$$

$$P_5^2(\mu)=\frac{105}{2}\left(1-\mu^2\right)\left(3\mu^3-\mu\right),$$

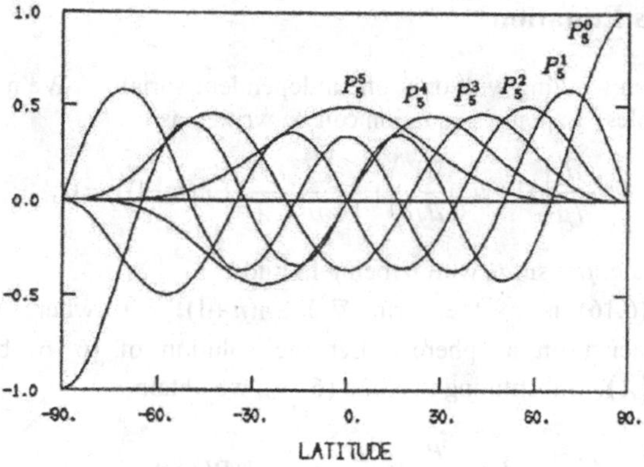

Figure 6.2. Representation of associated Legendre functions $P_5^0(\mu)$ to $P_5^5(\mu)$. Each $P_n^m(\mu)$ is normalized by multiplying by $(n-m)!/(n+m)!$. If $n-m$ is even, $P_n^m(\mu)$ is symmetric about the equator and is antisymmetric if $n-m$ is odd.

$$P_5^3(\mu)=\frac{105}{2}\left(1-\mu^2\right)^{3/2}\left(9\mu^2-1\right),$$
$$P_5^4(\mu)=945\left(1-\mu^2\right)^2\mu,$$
$$P_5^5(\mu)=945\left(1-\mu^2\right)^{5/2}.$$

Figure 6.2 shows the graphical representation of the above associated Legendre functions.

Since $P_n^m(\mu)$ is a polynomial of degree *n*, it has *n* roots given by the equation $P_n^m(\mu)=0$. It is clear from (6.15) that *m* of these roots are at the poles ($\mu=\pm 1$), while $n-m$ roots are between the poles. The $n-m$ roots between the poles are called the zeroes of the *associated Legendre function* $P_n^m(\mu)$. If $n-m$ is even, then $P_n^m(\mu)$ is symmetric with respect to the equator. If $n-m$ is odd, then $P_n^m(\mu)$ is antisymmetric with respect to the equator. Three useful properties of $P_n^m(\mu)$ are as follows:

$$P_n^m(\mu)=0 \qquad \text{if } n<m,$$
$$P_n^{-m}(\mu)=(-1)^m\frac{(n-m)!}{(n+m)!}P_n^m(\mu),$$
$$P_n^m(-\mu)=(-1)^{n-m}P_n^m(\mu).$$

6.3 Laplace's Equation

So far we have been working with only one independent variable. We next consider two independent variables. Laplace's equation can be written as

$$\frac{d}{d\mu}\left((1-\mu^2)\frac{dY}{d\mu}\right)+\frac{1}{1-\mu^2}\frac{\partial^2 Y}{\partial\lambda^2}+n(n+1)Y=0\,, \tag{6.16}$$

where λ is longitude, $\mu = \sin\theta$, with θ being latitude.

Equation (6.16) is of the form $\nabla^2 Y + n(n+1)Y = 0$, where ∇^2 is the two-dimensional Laplacian on a sphere. Let the solution of (6.16) be of the form $Y(\mu,\lambda) = P(\mu)L(\lambda)$. Substituting this into (6.16), we obtain

$$L+\frac{P}{1-\mu^2}\frac{\partial^2 L}{\partial\lambda^2}+n(n+1)PL=0\,, \tag{6.17}$$

After dividing by PL and multiplying by $1-\mu^2$, (6.17) may be written as

$$\frac{1-\mu^2}{P}\frac{d}{d\mu}\left((1-\mu^2)\frac{dP}{d\mu}\right)+n(n+1)(1-\mu^2)=\frac{-1}{L}\frac{d^2L}{d\lambda^2}\,. \tag{6.18}$$

The left-hand side of the above equation is a function of μ, while the right-hand side is a function of λ. This implies that both sides must be equal to some constant. Let this constant be m^2, so that from the right-hand side of (6.18) we obtain

$$-\frac{1}{L}\frac{d^2L}{d\lambda^2}=m^2\,,$$

$$\frac{d^2}{d\lambda^2}=-Lm^2$$

$$\frac{d^2L}{d\lambda^2}+m^2L=0\,. \tag{6.19}$$

The solution of (6.19) is given by $L = e^{\pm im\lambda}$. Furthermore, from the left-hand side of (6.18) we obtain

$$\frac{1-\mu^2}{P}\frac{d}{d\mu}\left((1-\mu^2)\frac{dP}{d\mu}\right)+n(n+1)(1-\mu^2)=m^2\,,$$

or upon multiplication by $P/(1-\mu^2)$

$$\frac{d}{d\mu}\left((1-\mu^2)\frac{dP}{d\mu}\right)+\left(n(n+1)-\frac{m^2}{1-\mu^2}\right)P=0\,, \tag{6.20}$$

which is an associated Legendre equation. As mentioned in Section 6.2, this has a solution of the form $P_n^m(\mu)$.

Thus given the solution of equation (6.16), $Y(\mu,\lambda)=P(\mu)L(\lambda)$, we found that $L=e^{\pm im\lambda}$ using equation (6.19) and that $P(\mu)$ has the form $P_n^m(\mu)$ using equation (6.20). Hence, the solution of Laplace's equation on a sphere is of the form $Y_n^m(\mu,\lambda)=P_n^m(\mu)e^{im\lambda}$. $Y_n^m(\mu,\lambda)$ is a *spherical harmonic* of order *m* and degree *n*. The factor $e^{im\lambda}$ describes the east-west variation, and the factor $P_n^m(\mu)$ describes the north-south variation of the spherical harmonic wave $Y_n^m(\mu,\lambda)$. Some useful mathematical properties of $Y_n^m(\mu,\lambda)$ are

$$Y_n^m(\mu,\lambda)=0 \qquad \text{for } n<m,$$

$$Y_n^{m*}(\mu,\lambda)=P_n^m(\mu)e^{-im\lambda},$$

$$Y_n^{-m}(\mu,\lambda)=P_n^{-m}(\mu)e^{-im\lambda}=(-1)^m\frac{(n-m)!}{(n+m)!}P_n^m(\mu)e^{-im\lambda},$$

$$\nabla^2 Y_n^m=\frac{-n(n+1)}{a^2}Y_n^m,$$

where $Y_n^{m*}(\mu,\lambda)$ is the complex conjugate of $Y_n^m(\mu,\lambda)$ and

$$\nabla^2=\frac{1}{a^2\cos\theta}\left(\frac{1}{\cos\theta}\frac{\partial^2}{\partial\lambda^2}+\frac{\partial}{\partial\theta}\cos\theta\frac{\partial}{\partial\theta}\right),$$

and *a* is the radius of the sphere.

6.4 Orthogonality Properties

The spectral equations that we work with contain a system of nonlinear differential equations that are functions of time only. In order to derive such a set of equations, one needs to remove the spatial dependence. Spherical harmonics (containing trigonometric functions in the zonal direction and associated Legendre functions in the meridional direction) describe this spatial dependence. These are removed by invoking orthogonality properties of the trigonometric and Legendre functions using what are called *Fourier* and *Legendre transforms*. In this section, we provide a theoretical background for the desired orthogonality properties.

Legendre polynomials satisfy the orthogonality condition

$$\int_{-1}^{1}P_m(\mu)P_n(\mu)d\mu=\begin{cases}0 & \text{if } m\neq n\\ \dfrac{2}{2n+1} & \text{if } m=n\end{cases}. \tag{6.21}$$

To prove this, we first evaluate the integral

$$\int_{-1}^{1}\mu^m P_n(\mu)d\mu, \qquad m\le n. \tag{6.22}$$

Using Rodrigues' formula and integrating the above equation by parts *m* times, we have

$$\int_{-1}^{1}\mu^m P_n(\mu)\,d\mu = \int_{-1}^{1}\mu^m \frac{1}{2^n n!}\frac{d^n(\mu^2-1)^n}{d\mu^n}d\mu$$
$$= \frac{1}{2^n n!}\mu^m \frac{d^{n-1}(\mu^2-1)^n}{d\mu^{n-1}}\Bigg|_{-1}^{1} - \frac{m}{2^n n!}\int_{-1}^{1}\mu^{m-1}\frac{d^{n-1}(\mu^2-1)^n}{d\mu^{n-1}}d\mu$$
$$= (-1)^m \frac{m}{2^n n!}\int_{-1}^{1}\left(\mu^2-1\right)^n d\mu \quad \text{if } m=n,$$
$$= 0 \quad \text{if } m<n. \tag{6.23}$$

Hence

$$\int_{-1}^{1}\mu^m P_n(\mu)\,d\mu = \begin{cases}(-)^n \dfrac{1}{2^n}\displaystyle\int_{-1}^{1}(\mu^2-1)^n d\mu & \text{if } m=n \\ 0 & \text{if } m<n\end{cases}. \tag{6.24}$$

Let us next evaluate $(-1)^n \int_{-1}^{1}\left(\mu^2-1\right)^n d\mu$. Writing $\mu=\sin\theta$ and after integrating by parts, we obtain

$$(-1)^n \int_{-1}^{1}\left(\mu^2-1\right)^n d\mu = \int_{-\pi/2}^{\pi/2}\cos^{2n+1}\theta d\theta$$
$$= \cos^{2n}\theta\sin\theta\Big|_{-\pi/2}^{\pi/2} + 2n\int_{-\pi/2}^{\pi/2}\cos^{2n-1}\theta\sin^2\theta d\theta$$
$$= 2n\int_{-\pi/2}^{\pi/2}\cos^{2n-1}\theta d\theta\left(1-\cos^2\theta\right)d\theta,$$

or

$$\int_{-\pi/2}^{\pi/2}\cos^{2n+1}\theta d\theta = \frac{2n}{2n+1}\int_{-\pi/2}^{\pi/2}\cos^{2n-1}\theta d\theta \tag{6.25}$$
$$= \frac{2n}{2n+1}\frac{2n-2}{2n-1}\cdots\frac{2}{3}\int_{-\pi/2}^{\pi/2}\cos\theta d\theta = \frac{2(2^n)^2(n!)^2}{(2n+1)!}.$$

Thus

$$\int_{-1}^{1}\mu^m P_n(\mu)\,d\mu = \begin{cases}\dfrac{2^{n+1}(n!)^2}{(2n+1)!} & \text{if } m=n \\ 0 & \text{if } m<n\end{cases}. \tag{6.26}$$

Now we prove the orthogonality condition given by (6.21). For this, without loss of generality, assume that $m \le n$ in (6.21). Writing $P_m(\mu)$ in polynomial form using Rodrigues' formula, we have

$$\int_{-1}^{1}P_m(\mu)P_n(\mu)\,d\mu = \int_{-1}^{1}\left(\frac{(2m)!}{2^m m!m!}\mu^m \right. \tag{6.27}$$
$$\left. -\frac{(2m-2)!\,\mu^{m-2}}{2^m 1!(m-1)!(m-2)!}+\ldots\right)P_n(\mu)\,d\mu.$$

From (6.26), if $m \neq n$ (i.e., $m < n$), then each term in the brackets, when multiplied by $P_n(\mu)$ and integrated over the interval -1 to 1, has a value of zero. If $m = n$, then only the term with the highest degree n contributes toward the integral, so that

$$\int_{-1}^{1} P_n(\mu)^2 d\mu = \frac{(2n)!}{2^n (n!)^2} \int_{-1}^{1} \mu^n P_n(\mu) d\mu$$
$$= \frac{(2n)!}{2^n (n!)^2} \frac{2^{n+1}(n!)^2}{(2n+1)!} = \frac{2}{2n+1}. \tag{6.28}$$

This proves the orthogonality conditions given by (6.21).

For associated Legendre functions, a similar orthogonality condition is

$$\int_{-1}^{1} P_{n_1}^{m_1}(\mu) P_{n_2}^{m_2}(\mu) d\mu = \begin{cases} \dfrac{(n+m)!}{(n-m)!} \dfrac{2}{2n+1} & \text{if } m_1 = m_2 = m \text{ and if } n_1 = n_2 = n \\ 0 & \text{if } m_1 \neq m_2 \text{ and/or if } n_1 \neq n_2 \end{cases}.$$

To prove this, first let $m_1 \neq m_2$ and/or $n_1 \neq n_2$ and write the associated Legendre equation with indices m_1, n_1 and m_2, n_2 as

$$\left(n_1(n_1+1) - \frac{m_1^2}{1-\mu^2} \right) P_{n_1}^{m_1}(\mu) = -\left[(1-\mu^2) P_{n_1}^{m_1\prime}(\mu) \right]' \tag{6.29}$$

and

$$\left(n_2(n_2+1) - \frac{m_2^2}{1-\mu^2} \right) P_{n_2}^{m_2}(\mu) = -\left[(1-\mu^2) P_{n_1}^{m_2\prime}(\mu) \right]', \tag{6.30}$$

where the prime denotes a derivative with respect to μ.

Multiplying the first equation by $P_{n_2}^{m_2}$, the second by $P_{n_1}^{m_1}$, and subtracting the second from the first, we obtain

$$\left(n_1(n_1+1) - n_2(n_2+1) - \frac{m_1^2 - m_2^2}{1-\mu^2} \right) P_{n_1}^{m_1}(\mu) P_{n_2}^{m_2}(\mu) \tag{6.31}$$
$$= -\left\{ \left[(1-\mu^2) P_{n_1}^{m_1\prime}(\mu) \right]' P_{n_2}^{m_2}(\mu) - \left[(1-\mu^2) P_{n_2}^{m_2\prime}(\mu) \right]' P_{n_1}^{m_1}(\mu) \right\}$$
$$= -\left\{ \left[(1-\mu^2) P_{n_1}^{m_1\prime}(\mu) P_{n_2}^{m_2}(\mu) \right]' - \left[(1-\mu^2) P_{n_2}^{m_2\prime}(\mu) P_{n_1}^{m_1}(\mu) \right]' \right\}.$$

Because the right-hand side of (6.31) vanishes on integrating between the limits -1 and 1, we have

$$\int_{-1}^{1} \left(n_1(n_1+1) - n_2(n_2+1) - \frac{m_1^2 - m_2^2}{1-\mu^2} \right) P_{n_1}^{m_1}(\mu) P_{n_2}^{m_2}(\mu) d\mu = 0. \tag{6.32}$$

As $m_1 \neq m_2$ and/or $n_1 \neq n_2$, the term within brackets does not vanish. Therefore, the integral will vanish only if $\int_{-1}^{1} P_{n_1}^{m_1}(\mu) P_{n_2}^{m_2}(\mu) d\mu = 0$ for $m_1 \neq m_2$ and/or $n_1 \neq n_2$.

We next consider the case when $m_1 = m_2 = m$ and $n_1 = n_2 = n$. Then using

$$P_n^{-m}(\mu) = (-1)^m \frac{(n-m)!}{(n+m)!} P_n^m(\mu),$$

we have

$$\begin{aligned}
\int_{-1}^{1} P_n^m(\mu)^2\, d\mu &= \int_{-1}^{1} P_n^m P_n^m d(\mu) \\
&= (-1)^m \frac{(n+m)!}{(n-m)!} \int_{-1}^{1} P_n^{-m}(\mu) P_n^m(\mu)\, d\mu \\
&= (-1)^m \frac{(n+m)!}{(n-m)!} \frac{1}{(2^n)^2 (n!)^2} \times \int_{-1}^{1} \frac{d^{n-m}(\mu^2-1)^n}{d\mu^{n-m}} \frac{d^{n+m}(\mu^2-1)^n}{d\mu^{n+m}} d\mu .
\end{aligned} \tag{6.33}$$

Let

$$K_{mn} = (-1)^m \frac{(n+m)!}{(n-m)!} \frac{1}{(2^n)^2 (n!)^2},$$

and integrating by parts we obtain

$$\begin{aligned}
\int_{-1}^{1} P_n^m(\mu)^2\, d(\mu) = K_{mn} \Bigg[& \frac{d^{n-m}(\mu^2-1)^n}{d\mu^{n-m}} \frac{d^{n+m-1}(\mu^2-1)^n}{d\mu^{n+m-1}} \Bigg|_{-1}^{1} \\
& - \int_{-1}^{1} \frac{d^{n-m+1}(\mu^2-1)^n}{d\mu^{n-m+1}} \frac{d^{n+m-1}(\mu^2-1)^n}{d\mu^{n+m-1}} d\mu \Bigg].
\end{aligned}$$

Then we can write

$$\int_{-1}^{1} P_n^m(\mu)^2\, d(\mu) = -K_{mn} \int_{-1}^{1} \frac{d^{n-m+1}(\mu^2-1)^n}{d\mu^{n-m+1}} \frac{d^{n+m-1}(\mu^2-1)^n}{d\mu^{n+m-1}} d\mu .$$

Continuing the process m times, we obtain

$$\begin{aligned}
\int_{-1}^{1} P_n^m(\mu)^2\, d(\mu) &= (-1)^m K_{mn} \int_{-1}^{1} \frac{d^n(\mu^2-1)^n}{d\mu^n} \frac{d^n(\mu^2-1)^n}{d\mu^n} d\mu \\
&= \frac{(n+m)!}{(n-m)!} \int_{-1}^{1} \left(\frac{1}{2^n n!} \frac{d^n(\mu^2-1)^n}{d\mu^n} \right)^2 d\mu \\
&= \frac{(n+m)!}{(n-m)!} \int_{-1}^{1} P_n(\mu)^2\, d\mu = \frac{(n+m)!}{(n-m)!} \frac{2}{2n+1}.
\end{aligned} \tag{6.34}$$

In practice, for spectral modeling it is more convenient to use normalized Legendre polynomials and associated Legendre functions. We may normalize a Legendre polynomial $P_n(\mu)$ by multiplying it by $\left[(2n+1)/2\right]^{1/2}$. Denoting a normalized polynomial by $\tilde{P}_n(\mu)$, we then write the orthogonality relation as

$$\int_{-1}^{1} \tilde{P}_m(\mu)\, \tilde{P}_n(\mu) = \begin{cases} 0 \text{ if } m \neq n \\ 1 \text{ if } m = n \end{cases}. \tag{6.35}$$

Likewise, a normalized associated Legendre function $\tilde{P}_n^m(\mu)$ is given by

$$\tilde{P}_n^m(\mu) = \left(\frac{(n-m)!}{(n+m)!}\right)^{1/2} \left(\frac{(2n+1)}{2}\right)^{1/2} P_n^m(\mu),$$

and the corresponding orthogonality relation becomes

$$\int_{-1}^{1} \tilde{P}_{n_1}^{m_1}(\mu)\tilde{P}_{n_2}^{m_2} d(\mu) = \begin{cases} 0 \text{ if } m_1 \neq m_2 \text{ and/or } n_1 \neq n_2 \\ 1 \text{ if } m_1 = m_2 = m \text{ and } n_1 = n_2 = n \end{cases}. \tag{6.36}$$

Note that with this normalization, $\tilde{P}_n^{-m}(\mu) = (-1)^m \tilde{P}_n^m$.

Throughout the remainder of the book, we deal with normalized Legendre polynomials and normalized associated Legendre functions. We will, however, drop the tilde (˜) so that from now on $P_n(\mu)$ and $P_n^m(\mu)$ represent normalized Legendre polynomial and normalized associated Legendre function, respectively, unless otherwise stated.

6.5 Recurrence Relations

Taking a derivative in the east-west direction of a spherical harmonic given by $Y_n^m = P_n^m e^{im\lambda}$ we obtain

$$\frac{\partial}{\partial\lambda} Y_n^m = \frac{\partial}{\partial\lambda} P_n^m e^{im\lambda}$$

$$\frac{\partial}{\partial\lambda} Y_n^m = imP_n^m e^{im\lambda}.$$

However, taking a derivative in the north-south direction of the same harmonic is more complicated. For this we need recurrence relations. There are a number of relations relating the associated Legendre functions of different orders and degrees as well as their derivatives. These are useful for calculating associated Legendre functions and their derivatives needed for spectral modeling. For our purpose, the following *four recurrence relations* for the normalized associated Legendre functions are most useful:

(1) $\mu P_n^m(\mu) = \epsilon_{n+1}^m P_{n+1}^m(\mu) + \epsilon_n^m P_{n-1}^m(\mu)$, or

$$\sin\theta P_n^m(\sin\theta) = \epsilon_{n+1}^m P_{n+1}^m(\sin\theta) + \epsilon_n^m P_{n-1}^m(\sin\theta), \tag{6.37}$$

where

$$\epsilon_n^m = \left(\frac{n^2 - m^2}{4n^2 - 1}\right)^{1/2}.$$

(2) $\left(1-\mu^2\right) dP_n^m(\mu)/d\mu = -n\,\epsilon_{n+1}^m P_{n+1}^m(\mu) + (n+1)\,\epsilon_n^m P_{n-1}^m(\mu)$, or

$$\cos\theta \frac{dP_n^m(\sin\theta)}{d\theta} = -n\,\epsilon_{n+1}^m P_{n+1}^m(\sin\theta) + (n+1)\,\epsilon_n^m P_{n-1}^m(\sin\theta). \tag{6.38}$$

Eliminating P_{n+1}^m between (6.37) and (6.38), we obtain the following relations:

(3) $\left(1-\mu^2\right)dP_n^m(\mu)/d\mu=(2n+1)\in_n^m P_{n-1}^m(\mu)-n\mu P_n^m(\mu)$, or

$$\cos\theta\frac{dP_n^m(\sin\theta)}{d\theta}=(2n+1)\in_n^m P_{n-1}^m(\sin\theta)-n\sin\theta P_n^m(\sin\theta). \tag{6.39}$$

(4) $\left(1-\mu^2\right)^{1/2} P_n^m(\mu)=g_n^m P_{n+1}^{m+1}(\mu)-h_n^m P_{n-1}^{m+1}(\mu)$, or

$$\cos\theta P_n^m(\sin\theta)=g_n^m P_{n+1}^{m+1}(\sin\theta)-h_n^m P_{n-1}^{m+1}(\sin\theta), \tag{6.40}$$

where

$$g_n^m=\left(\frac{(n+m+1)(n+m+2)}{(2n+1)(2n+3)}\right)^{1/2}$$

and

$$h_n^m=\left(\frac{(n-m-1)(n-m)}{(2n+1)(2n-1)}\right)^{1/2}.$$

Using (6.37) to (6.40), we can calculate $P_n^m(\mu)$ and their derivatives for any given *m* and *n*. Starting with the value of $P_0^0(\mu)$, the recurrence relations (1) and (4) given by (6.37) and (6.40) can generate the values of associated Legendre functions of any given order *m* and degree *n*. The value of normalized $P_0^0(\mu)$ for the global domain (μ= -1 to μ= 1) is $1/(2^{1/2})$, and for the hemispheric domain (μ= -1 to μ= 0 or μ= 0 to μ = 1) it is 1.

With $m = 0$ and $n = 0$, recurrence relation (4) becomes

$$g_0^0 P_1^1(\mu)=\left(1-\mu^2\right)^{1/2} P_0^0(\mu). \tag{6.41}$$

From the above equality, if we know $P_0^0(\mu)$, then we can determine $P_1^1(\mu)$. If we know $P_1^1(\mu)$, then the same recurrence relation can be used to obtain $P_2^2(\mu)$. Thus proceeding recursively, we can find the value of $P_m^m(\mu)$ for any given *m*. Likewise if we know $P_m^m(\mu)$, then recurrence relation (1) determines the value of $P_{m+1}^m(\mu)$. From $P_m^m(\mu)$, and $P_{m+1}^m(\mu)$, the same relation gives value of $P_{m+2}^m(\mu)$. Thus proceeding recursively, we can calculate the value of $P_n^m(\mu)$ for any given degree *n*.

Recurrence relation (2) can be used to calculate the differential of any $P_m^m(\mu)$ from the values of $P_{n-1}^m(\mu)$ and $P_{n+1}^m(\mu)$. Recurrence relation (3) serves a similar purpose, but we need $P_{n-1}^m(\mu)$ and $P_n^m(\mu)$ to calculate differentials of $P_n^m(\mu)$. To calculate the derivative of $P_n^m(\mu)$ of the highest degree *n* using recurrence relation (2), we need the value of $P_{n+1}^m(\mu)$, which is beyond the usual truncation of the series. Recurrence relation (3) uses the values of associated Legendre functions within the truncation limit only. For calculation of differentials of $P_n^m(\mu)$, it is therefore preferable to use recurrence relation (3) rather relation (2).

6.6 Gaussian Quadrature

To obtain the coefficients of the truncated spherical harmonics in the spectral model we have to perform a Fourier transform as well as evaluate the Legendre transform. This requires the evaluation of numerical integrals. The *Gaussian quadrature* is used for the Legendre transform of data in the north-south direction. We show that the Gaussian quadrature is an efficient numerical quadrature for this purpose. Furthermore, we describe the procedure to design it.

Consider the integral $\int_a^b f(x)dx$, where $f(x)$ is an integrable function on $a \le x \le b$. We can write this as

$$\int_a^b f(x)dx = \sum_{i=1}^{n} w_i f(x_i), \tag{6.42}$$

where the expression on the right-hand side is the numerical equivalent of the integral on the left-hand side. The expression on the right-hand side is known as a *numerical integral quadrature.* Our aim is to select w_i and x_i such that the summation on the right-hand side is exactly equal to the integral on the left-hand side. If we choose x_i equally spaced within the interval of integration, then we shall have n values of w_i as a function of the location of these points.

It is possible to make the quadrature $\sum_{i=1}^{n} w_i f(x_i)$ exact for $f(x)$ of degree $\le n$ - 1 by suitable selection of w_i. However, we show that if we can choose both w_i and x_i suitably, it is possible to make the numerical quadrature $\sum_{i=1}^{n} w_i f(x_i)$ exactly equal to the integral on the left-hand side for $f(x)$ of degree $\le 2n$ - 1, which is the highest possible accuracy attainable from $2n$ degrees of freedom (n for x_i and n for w_i). We call w_i the *Gaussian weights* and x_i the *Gaussian ordinates*, and the numerical quadrature is called the *Gaussian quadrature.*

For convenience, we transform the interval of integration from (a, b) to $(-1, 1)$. This can be done by defining a new variable, that is,

$$z = \frac{2x-(a+b)}{b-a}; \qquad x = \frac{(b-a)z+a+b}{2}, \tag{6.43}$$

so that

$$\int_a^b f(x)dx = \frac{b-a}{2}\int_{-1}^{1} f\left(\frac{(b-a)z+a+b}{2}\right) = \int_{-1}^{1} F(z)dz, \tag{6.44}$$

where

$$F(x) = \frac{b-a}{2} f\left(\frac{(b-a)z+a+b}{2}\right).$$

For Gaussian quadrature, one needs to find n values of x_i and w_i such that $\int_{-1}^{1} F(x)dx = \sum_{i=1}^{n} w_i F(x_i)$, which is exact for $F(x)$ of degree $\le 2n$ -1.

Consider first the case when x_i is arbitrarily predefined, and also suppose that they are equidistant with x_1 = -1 and x_n = 1. In addition, $F(x_1), F(x_2), \ldots, F(x_i), \ldots, F(x_n)$ represent the values of $F(x)$ at $x_1, x_2, \ldots, x_i, \ldots, x_n$, respectively. Then using Lagrange's interpolation formula, we can find a unique polynomial $G_{n-1}(x)$ of n-1 degree passing through these n values. This is given by

$$G_{n-1}(x)=\sum_{i=1}^{n}L(x-x_i)F(x_i), \tag{6.45}$$

where

$$L(x-x_i)=\frac{(x-x_1)(x-x_2)...(x-x_{i-1})(x-x_{i+1})...(x-x_n)}{(x_i-x_1)(x_i-x_2)...(x_i-x_{i-1})(x_i-x_{i+1})...(x_i-x_n)}.$$

We can write this as

$$L(x-x_i)=\frac{\pi^n(x-x_i)}{(x-x_i)\pi^{n!}(x_i)}, \tag{6.46}$$

where

$$\pi^n(x-x_i)=(x-x_1)(x-x_2)\dots(x-x_n)$$

and

$$\pi^{n'}(x_i-x_i)=(x_i-x_1)(x_i-x_2)\dots(x_i-x_{i-1})(x_i-x_{i+1})\dots(x_i-x_n).$$

Because

$$L(x-x_i)=\begin{cases}1 & \text{if } x=x_i\\ 0 & \text{if } x\neq x_i\end{cases},$$

$G_{n-1}(x)$ has values $F(x_i)$ at x_i, $i = 1, 2, 3, \dots, n$. Thus $G_{n-1}(x)$ is the desired polynomial passing through the given n values. Integrating (6.45) over the limit (-1, 1), we obtain

$$\int_{-1}^{1}G_{n-1}(x)dx=\int_{-1}^{1}\sum_{i=1}^{n}L(x-x_i)F(x_i)dx. \tag{6.47}$$

Interchanging the order of summation and integration operations on the right-hand side, we obtain

$$\int_{-1}^{1}G_{n-1}(x)dx=\int_{-1}^{1}\sum_{i=1}^{n}L(x-x_i)F(x_i)dx=\sum_{i=1}^{n}w_iF(x_i), \tag{6.48}$$

where

$$w_i=\int_{-1}^{1}L(x-x_i)dx. \tag{6.49}$$

Thus for such values of w_i, the summation $\sum_{i=1}^{n}w_iF(x_i)$ is exactly equal to the integral $\int_{-1}^{1}G_{n-1}(x)dx$, where $G_{n-1}(x)$ is a curve of degree $\leq n$ -1 passing through the n ordinates $F(x_1)$, $F(x_2)$,..., $F(x_n)$.

Suppose next that we have $F(x)$ as an arbitrary polynomial of degree 2n-1 having values at each x_i, $i = 1, 2, \dots, n$. By Lagrange's interpolation formula, we can again find a polynomial $G_{n-1}(x)$ passing through $F(x)$ at the given points, so that

$$G_{n-1}(x_i)=F(x_i). \tag{6.50}$$

We write this arbitrary polynomial $F(x)$ of degree 2n - 1 in the form

$$F(x) = G_{n-1}(x) + \phi_{n-1}(x)\prod_{i=1}^{n}(x - x_i), \tag{6.51}$$

where $G_{n-1}(x)$ is a polynomial of degree n -1 passing through the n ordinates $F(x_i)$, $\phi_{n-1}(x)$ is an arbitrary polynomial of degree n - 1, and $\prod_{i=1}^{n}(x - x_i)$ is a polynomial of degree n. Then the polynomial on the right-hand side of (6.51) is equal to $F(x)$ and is of degree $2n$ - 1.

If we integrate (6.51) from (-1, 1), we obtain

$$\int_{-1}^{1} F(x)dx = \int_{-1}^{1} G_{n-1}(x)dx + \int_{-1}^{1} \phi_{n-1}(x)\prod_{i=1}^{n}(x - x_i)dx,$$

or

$$\int_{-1}^{1} F(x)dx = \sum_{i=1}^{n} w_i F(x_i) + \int_{-1}^{1} \phi_{n-1}(x)\prod_{i=1}^{n}(x - x_i)dx. \tag{6.52}$$

If $\sum_{i=1}^{n} w_i F(x_i)$ is to be exactly equal to $\int_{-1}^{1} F(x)dx$ for $F(x)$ of degree $2n$ - 1, then the second term on the right-hand side of (6.52) must vanish. This can be achieved by suitable positioning of points $x_1, x_2, x_3, \ldots, x_n$ so that

$$\int_{-1}^{1} \phi_{n-1}(x)\prod_{i=1}^{n}(x - x_i)dx = 0. \tag{6.53}$$

Because $\phi_{n-1}(x)$ is an arbitrary polynomial, the integral on the left-hand side of (6.53) must vanish for every polynomial $\phi_{n-1}(x)\prod_{i=1}^{n}(x - x_i)$. This is possible if $\phi_{n-1}(x)\prod_{i=1}^{n}(x - x_i) = 0$. Since $\phi_{n-1}(x)$ is an arbitrary polynomial (which may not necessarily equal zero), we have

$$\prod_{i=1}^{n}(x - x_i) = 0. \tag{6.54}$$

Thus x_i are the roots of $\prod_{i=1}^{n}(x - x_i) = 0$.

We may represent $\phi_{n-1}(x)$ and $\prod_{i=1}^{n}(x - x_i)$ in terms of Legendre polynomials as

$$\phi_{n-1}(x) = \sum_{k=0}^{n-1} a_k P_k(x) \text{ and } \prod_{i=1}^{n}(x - x_i) = \sum_{j=0}^{n} b_j P_j(x). \tag{6.55}$$

Note that because $\phi_{n-1}(x)$ and $\prod_{i=1}^{n}(x - x_i)$ are polynomials of degree n-1 and n, respectively, the coefficients of P_{n-1} in the summation for ϕ_{n-1} and for P_n in the summation for $\prod_{i=1}^{n}(x - x_i)$ must not vanish; in other words, $a_{n-1} \neq 0$ and $b_n \neq 0$. Substituting (6.55) into (6.53) and using the orthogonality property of Legendre polynomials, we obtain

$$\int_{-1}^{1} \phi_{n-1}(x)\prod_{i=1}^{n}(x-x_i)dx = \int_{-1}^{1}\sum_{k=0}^{n-1} a_k P_k(x)\sum_{j=0}^{n} b_j P_j(x)dx$$
$$= \int_{-1}^{1}\sum_{k=0}^{n-1} a_k b_k P_k^2(x)dx. \qquad (6.56)$$

Since $\phi_{n-1}(x)$ is an arbitrary polynomial, a_k for $k = 0, 1, 2, 3, \ldots, n-1$ in general do not vanish. Therefore, the above integral will vanish only if $b_k = 0$ for k = 0, 1, 2, 3, ... , n - 1.

Thus the integral (6.53) vanishes for all values of $\phi_{n-1}(x)$ only if

$$\prod_{i=1}^{n}(x-x_i) = b_n P_n(x). \qquad (6.57)$$

From (6.54) and (6.57) we note that x_i are the roots of $P_n(x)$ (i.e., of the Legendre polynomial of degree n). Also,

$$\pi^{n'}(x_i) = \frac{d}{dx}\prod_{i=1}^{n}(x-x_i)\bigg|_{x=x_i} = \frac{d}{dx}b_n P_n(x)\bigg|_{x=x_i} = b_n P'_n(x_i). \qquad (6.58)$$

The weights w_i in (6.49) may now be expressed as

$$w_i = \int_{-1}^{1} L(x-x_i)dx = \int_{-1}^{1}\frac{\pi^n(x-x_i)}{(x-x_i)\pi^{n'}(x_i)}dx$$
$$= \int_{-1}^{1}\frac{b_n P_n(x)}{(x-x_i)b_n P'_n(x_i)}dx,$$

or

$$w_i = \frac{1}{P'_n(x_i)}\int_{-1}^{1}\frac{P_n(x)}{x-x_i}dx. \qquad (6.59)$$

To evaluate $\int_{-1}^{1} P_n(x)/(x-x_i)dx$, we make use of recurrence relation (1) given by (6.37), that is,

$$\varepsilon_{l+1}^0 P_{l+1}(x) = xP_l(x) - \varepsilon_l^0 P_{l-1}(x). \qquad (6.60)$$

For $x = x_i$, (6.60) may be written as

$$\varepsilon_{l+1}^0 P_{l+1}(x_i) = x_i P_l(x_i) - \varepsilon_l^0 P_{l-1}(x_i). \qquad (6.61)$$

If we multiply (6.60) by $P_l(x_i)$ and (6.61) by $P_l(x)$ and then subtract the resulting equations, we obtain

$$\varepsilon_{l+1}^0\left[P_{l+1}(x)P_l(x_i) - P_{l+1}(x_i)P_l(x)\right]$$
$$= P_l(x)P_l(x_i)(x-x_i) - \varepsilon_l^0\left[P_{l-1}(x)P_l(x_i) - P_{l-1}(x_i)P_l(x)\right]. \qquad (6.62)$$

After summation of the two sides of (6.62) over the range $l = 1$ to $l = n$, we obtain

$$\sum_{l=0}^{n} \varepsilon_{l+1}^{0}[P_{l+1}(x)P_l(x_i) - P_{l+1}(x_i)P_l(x)]$$
$$= \sum_{l=0}^{n} \{P_l(x)P_l(x_i)(x - x_i) - \varepsilon_l^0 [P_{l-1}(x)P_l(x_i) - P_{l-1}(x_i)P_l(x)]\}.$$

After some simplification we obtain

$$\sum_{l=0}^{n} P_l(x)P_l(x_i)(x - x_i) = \varepsilon_{n+1}^{0}[P_{n+1}(x)P_n(x_i) - P_{n+1}(x_i)P_n(x)]$$
$$+\varepsilon_1^0 [P_0(x)P_1(x_i) - P_0(x_i)P_{n1}(x)]. \qquad (6.63)$$

Noting that for the spherical domain,

$$P_0(x) = \frac{1}{2^{1/2}}, \quad P_0(x_i) = \frac{1}{2^{1/2}},$$
$$P_1(x) = \left(\frac{3}{2}\right)^{1/2} x, \quad P_1(x_i) = \left(\frac{3}{2}\right)^{1/2} x_i, \ \varepsilon_1^0 = \frac{1}{3^{1/2}},$$

and $P_n(x_i) = 0$, since x_i are the roots of the Legendre polynomial $P_n(x)$. We finally obtain

$$\sum_{l=1}^{n} P_l(x)P_l(x_i)(x - x_i) = -\varepsilon_{n+1}^{0} P_{n+1}(x_i)P_n(x) - \frac{x - x_i}{2} \qquad (6.64)$$
$$= -\varepsilon_{n+1}^{0} P_{n+1}(x_i)P_n(x) - P_0(x_i)P_0(x)(x - x_i).$$

Transferring the second term from the right-hand side to the left-hand side, we obtain

$$\sum_{l=0}^{n} P_l(x)P_l(x_i)(x - x_i) = -\varepsilon_{n+1}^{0} P_{n+1}(x_i)P_n(x),$$

or

$$-\frac{\epsilon_{n+1}^{0} P_{n+1}(x_i)P_n(x)}{x - x_i} = \sum_{l=0}^{n} P_l(x)P_l(x_i). \qquad (6.65)$$

Integrating (6.65) from -1 to 1, we obtain

$$-\epsilon_{n+1}^{0} P_{n+1}(x_i) \int_{-1}^{1} \frac{P_n(x)}{x - x_i} dx = \int_{-1}^{1} \sum_{l=0}^{n} P_l(x) P_l(x_i) dx,$$

or

$$-\epsilon_{n+1}^{0} P_{n+1}(x_i) \int_{-1}^{1} \frac{P_n(x)}{x - x_i} dx = \sum_{l=0}^{n} P_l(x_i) \int_{-1}^{1} P_l(x) dx. \qquad (6.66)$$

Because $\int_{-1}^{1} P_l(x)\, dx = 0$ for all l except $l = 0$, (6.66) reduces to

$$-\epsilon_{n+1}^{0} P_{n+1}(x_i) \int_{-1}^{1} \frac{P_n(x)}{x - x_i} dx = P_0(x_i) \int_{-1}^{1} P_0(x) dx = 1. \qquad (6.67)$$

Thus

$$-\int_{-1}^{1} \frac{P_n(x)}{x - x_i} dx = \frac{-1}{\epsilon_{n+1}^0 \, P_{n+1}(x_i)}. \tag{6.68}$$

Substituting into (6.59), we obtain

$$w_i = \frac{-1}{\epsilon_{n+1}^0 \, P_{n+1}(x_i) P_n'(x_i)}. \tag{6.69}$$

These are the corresponding weights assigned to the various x_i. The expression for w_i given by (6.69), though proper, is not very convenient, because it needs the values of $P_{n+1}(x_i)$ and the derivatives $P'(x_i)$. We express w_i in a more convenient form below.

Using recurrence relation (2) given by (6.39) at $x = x_i$, we obtain

$$\left(1 - x_i^2\right) P_n'(x_i) = (2n+1) \epsilon_n^0 \, P_{n-1}(x_i) - 0.$$

$P_n(x_i) = 0$ since x_i are the roots of $P_n(x)$, or

$$P_n'(x_i) = \frac{(2n+1)\epsilon_n^0}{1 - x_i^2} P_{n-1}(x_i). \tag{6.70}$$

Also, using recurrence relation (1) given by (6.37) at $x = x_i$ and noting that $P_n(x_i) = 0$, we obtain

$$\epsilon_{n+1}^0 \, P_{n+1}(x_i) = -\epsilon_n^0 \, P_{n-1}(x_i),$$

or

$$P_{n+1}(x_i) = -\frac{\epsilon_n^0}{\epsilon_{n+1}^0} P_{n-1}(x_i). \tag{6.71}$$

Substituting (6.70) and (6.71) into (6.69), we obtain

$$w_i = \frac{1 - x_i^2}{(2n+1)(\epsilon_n^0)^2 P_{n-1}(x_i)^2},$$

or

$$w_i = \frac{2(1 - x_i^2)(n - \frac{1}{2})}{n^2 P_{n-1}(x_i)^2}, \tag{6.72}$$

which is a more convenient expression for the Gaussian weights w_i.

Thus if a polynomial $F(x)$ of degree $2n-1$ has values $F(x_i)$ available at n points x_i, $i = 1, 2, 3, \ldots, n$, as the roots (zeroes) of the Legendre polynomial P_n, then the integral $\int_{-1}^{1} F(x)dx$ is evaluated exactly by the Gaussian quadrature $\sum_{i=1}^{n} w_i F(x_i)$ where x_i are the zeroes of the Legendre polynomial of degree n and w_i are the weights given by (6.72).

The efficiency of the Gaussian quadrature lies in the fact that it can evaluate the integral of a function of degree $2n-1$ from its values given at n points. For designing a Gaussian quadrature, we need to take the following steps:

1. If N is the highest degree of the polynomial to be integrated, then the minimum number of points n needed for the Gaussian quadrature is given by $2n-1=N$, or $n=(N+1)/2$.
2. The Gaussian points x_i, $i=1, 2, \ldots, n$, are determined as the zeroes of the Legendre polynomials of degree n, that is, as the roots of $P_n(x)=0$. The roots of $P_n(x)$ are not equally spaced, but are nearly so. They can be determined iteratively using the Newton-Raphson method starting with the first guess of x_i as n equally spaced points between -1 and 1.
3. Once x_i are found, the weights w_i can be obtained from (6.72).

In the case of a Legendre polynomial of even degree, its zeroes are located symmetrically, in other words, at $\pm x_i$. In the case of a Legendre polynomial of odd degree, $x=0$ is one of the zeroes and the rest of the zeroes would be located symmetrically at $\pm x_i$. The weight w_i has the same value for $\pm x_i$. It is therefore sufficient to calculate the zeroes of the Legendre polynomials and the corresponding weights for $x \geq 0$ to complete the Gaussian quadrature. In a spectral model, the use of a suitable fast Fourier transform (FFT) for the Fourier transform and Gaussian quadrature for the Legendre transform achieves the optimally fast and accurate calculations.

6.7 Spectral Representation of Physical Fields

Any smooth function over a sphere can be expressed as a sum of spherical harmonics. However, the convergence of term-by-term derivatives of the function is assured for much less liberal conditions, that is, the absolute convergence of the series. With most meteorological quantities (scalars), this is generally not a problem. However, sharp discontinuities across clouds, rain areas, fronts, and so on, are not easy to represent. There is also a problem with the horizontal wind components u and v, which have a singular behavior at the poles. While u and v can be expressed to any desired accuracy by a series of spherical harmonics, there is no guarantee that the derivative of such a series will converge properly (this is the pole problem). The cross-polar flow cannot be expressed by a series. At one side, the flow may appear to be northerly, while at the other side of the pole the flow might appear southerly. This poses a singularity at the pole.

In the past u and v were set equal to zero at the poles for all times. However, due to this singularity at the poles, the velocity components are carried in terms of the following pseudoscalar fields (called *Robert functions*):

$$U=\frac{u\cos\theta}{a} \qquad \text{and} \qquad V=\frac{v\cos\theta}{a}, \tag{6.73}$$

where θ is the latitude and a is the radius of the earth. At the poles, $\theta=\pm\pi/2$, so $U=V=0$, but we can find U and V very close to 90°N and 90°S. The Robert functions U and V are therefore well defined at the poles. It is these functions rather than u and v that we work with.

One can obtain a simple kinematic relationship between the flow-field variables in the spectral domain using the formula discussed in earlier sections. We deal with variables such as ψ, χ, ζ, D, U and V, where ψ is the stream function, χ is the velocity potential, ζ is the relative vorticity, D is the divergence, and U and V are the Robert functions defined in (6.73). The relationships between U, V and ψ, χ (or ζ, D) are given by

$$U = \frac{1}{a^2}\left(\frac{\partial \chi}{\partial \lambda} - \cos\theta \frac{\partial \psi}{\partial \theta}\right), \tag{6.74}$$

$$V = \frac{1}{a^2}\left(\frac{\partial \psi}{\partial \lambda} + \cos\theta \frac{\partial \chi}{\partial \theta}\right), \tag{6.75}$$

$$\zeta = \frac{1}{\cos^2\theta}\left(\frac{\partial V}{\partial \lambda} - \cos\theta \frac{\partial U}{\partial \theta}\right), \tag{6.76}$$

$$D = \frac{1}{\cos^2\theta}\left(\frac{\partial U}{\partial \lambda} + \cos\theta \frac{\partial V}{\partial \theta}\right), \tag{6.77}$$

where θ is the latitude, λ is the longitude, and a is the radius of the earth.

If ψ and χ are both well-behaved, which generally is the case, then for a given truncation we can expand ψ and χ by the relations

$$\psi = a^2 \sum_m \sum_n \psi_n^m Y_n^m(\lambda, \mu) \tag{6.78}$$

and

$$\chi = a^2 \sum_m \sum_n \chi_n^m Y_n^m(\lambda, \mu). \tag{6.79}$$

For the same truncation, remembering the mathematical property $\nabla^2 Y_n^m = -\frac{n(n+1)}{a^2} Y_n^m$ given in Sec. 6.3, we can express the vorticity and divergence as

$$\zeta = \nabla^2 \psi = \sum_m \sum_n -n(n+1)\psi_n^m Y_n^m \tag{6.80}$$

and

$$D = \nabla^2 \chi = \sum_m \sum_n -n(n+1)\chi_n^m Y_n^m. \tag{6.81}$$

Using the spectral expansions $\zeta = \sum_m \sum_n \zeta_n^m Y_n^m$ and $D = \sum_m \sum_n D_n^m Y_n^m$ we get the spectral form of ζ and D for one harmonic so that

$$\zeta_n^m = -n(n+1)\psi_n^m \tag{6.82}$$

and

$$D_n^m = -n(n+1)\chi_n^m. \tag{6.83}$$

By substituting the spectral representations of ψ (6.78) and χ (6.79) into (6.74) and (6.75) and making use of the zonal meridional differentials of $Y_n^m(\lambda,\mu)$ given by

$$
\begin{aligned}
Y_n^m(\lambda,\mu) &= P_n^m(\mu)e^{im\lambda},\\
\frac{\partial Y_n^m(\lambda,\mu)}{\partial\lambda} &= imP_n^m(\mu)e^{im\lambda},\\
\frac{\partial Y_n^m(\lambda,\mu)}{\partial\lambda} &= imY_n^m(\lambda,\mu)
\end{aligned}
\tag{6.84}
$$

and using the recurrence relation (6.38) where $e^{im\lambda}$ has been multiplied through

$$
\begin{aligned}
\left(1-\mu^2\right)\frac{\partial Y_n^m(\lambda,\mu)}{\partial\mu} &= -n\in_{n+1}^m Y_{n+1}^m(\lambda,\mu)\\
&\quad +(n+1)\in_n^m Y_{n-1}^m(\lambda,\mu),
\end{aligned}
\tag{6.85}
$$

we obtain the spectral expansions of both U and V. That is,

$$U=\sum_m\sum_n U_n^m Y_n^m=\frac{1}{a^2}\left(a^2\sum_m\sum_n\frac{\partial}{\partial\lambda}\chi_n^m Y_n^m-\cos\theta a^2\sum_m\sum_n\frac{\partial}{\partial\theta}\psi_n^m Y_n^m\right).$$

Remembering $Y_n^m=P_n^m e^{im\lambda}$, $\frac{\partial}{\partial\theta}=\cos\theta\frac{\partial}{\partial\mu}$, and $\mu=\sin\theta$ and using (6.84), we obtain

$$U=\sum_m\sum_n im\chi_n^m Y_n^m-\cos\theta\sum_m\sum_n\cos\theta\frac{\partial}{\partial\mu}\psi_n^m Y_n^m.$$

Using the identity $\cos^2\theta=1-\sin^2\theta=1-\mu^2$, we obtain

$$U=\sum_m\sum_n im\chi_n^m Y_n^m-\left(1-\mu^2\right)\sum_m\sum_n\psi_n^m\frac{\partial Y_n^m}{\partial\mu}.$$

Finally, using (6.85) we obtain

$$
\begin{aligned}
U=\sum_m\sum_n U_n^m Y_n^m &= \sum_m\sum_n im\chi_n^m Y_n^m\\
&-\sum_m\sum_n\psi_n^m\left[-n\in_{n+1}^m Y_{n+1}^m+(n+1)\in_n^m Y_{n-1}^m\right].
\end{aligned}
\tag{6.86}
$$

Similarly for V we obtain

$$
\begin{aligned}
V=\sum_m\sum_n V_n^m Y_n^m &= \sum_m\sum_n im\psi_n^m Y_n^m\\
&+\sum_m\sum_n\chi_n^m\left[-n\in_{n+1}^m Y_{n+1}^m+(n+1)\in_n^m Y_{n-1}^m\right].
\end{aligned}
\tag{6.87}
$$

Equating the coefficients of Y_n^m on both sides of the above equations, we obtain

$$U_n^m = im\chi_n^m + (n-1)\in_n^m \psi_{n-1}^m - (n+2)\in_{n+1}^m \psi_{n+1}^m, \tag{6.88}$$

$$V_n^m = im\psi_n^m - (n-1)\in_n^m \chi_{n-1}^m + (n+2)\in_{n+1}^m \chi_{n+1}^m. \tag{6.89}$$

Since $\zeta_n^m = -n(n+1)\psi_n^m$ and $D_n^m - (n+1)\chi_n^m$, we may also write

$$n(n+1)U_n^m = -imD_n^m - (n+1)\in_n^m \zeta_{n-1}^m + n\in_{n+1}^m \zeta_{n+1}^m, \tag{6.90}$$

$$n(n+1)V_n^m = -im\zeta_n^m + (n+1)\in_n^m D_{n-1}^m - n\in_{n+1}^m D_{n+1}^m. \tag{6.91}$$

For $m = 0$ and $n = 0$ and remembering $Y_n^m(\lambda,\mu) = 0$ for $n < m$,

$$U_0^0 = -2\in_1^0 \psi_1^0 = \in_1^0 \zeta_1^0, \tag{6.92}$$

$$V_0^0 = -2\in_1^0 \chi_1^0 = \in_1^0 D_1^0. \tag{6.93}$$

At this point, let us ask the following questions: How does kinematics work, given ψ_n^m and χ_n^m? If we are given the spectral coefficients of the stream function and the velocity potential (i.e., given ψ_n^m and χ_n^m), can we obtain ζ_n^m, D_n^m, U_n^m, and V_n^m? Since we know ψ_n^m and χ_n^m for all desired values of n and m, we can use (6.88) and (6.89) to obtain U_n^m and V_n^m, respectively.

Using (6.82) and (6.83), one can obtain ζ_n^m and D_n^m, respectively. Furthermore, one can obtain the true velocity components u and v using the inverse Robert functions. That is, equations (6.82), (6.83), and (6.88)-(6.91) are a series of equations that enable one to solve for ψ, χ, ζ, D, U and V given two spectral coefficients. This can be called the triangle problem. Without worrying about details, the orthogonality properties of Legendre polynomials and associated Legendre functions are used for determining the expansion coefficients. This leads to

$$\int_0^{2\pi}\int_{-1}^{1} AY_n^{m*}\,d\mu\,d\lambda = \sum_m\sum_n A_n^m \int_0^{2\pi}\int_{-1}^{1} Y_n^m Y_n^{m*}\,d\mu\,d\lambda, \tag{6.94}$$

where Y_n^{m*} is the complex conjugate of Y_n^m. We have used the orthogonality of the spherical harmonics, that is,

$$\frac{1}{2\pi}\int_0^{2\pi}\int_{-1}^{1} Y_n^m Y_j^{k*}\,d\mu\,d\lambda = \delta_{mk}\delta_{nj} = \begin{cases} 0 & \text{if } m \neq j \text{ and/or } n \neq j \\ 1 & \text{if } m = k \text{ and } n = j \end{cases}.$$

Transformation from grid space to spectral space and vice versa is important in spectral modeling. Grid space is represented by $A(\lambda,\mu)$ and is the representation on a weather map. Spectral space is represented by A_n^m and are the amplitudes of the spherical harmonics of $A(\lambda,\mu)$. It is important to learn to switch between grid ($A(\lambda,\mu)$) and spectral space (A_n^m), i.e. to get weather maps from the spectral coefficients. To do this we must utilize the Fourier transform, Legendre transform, inverse Fourier transform, and inverse Legendre transform. We can write a function A as

$$A = \sum_m \sum_n A_n^m Y_n^m,$$

where $A = A(\lambda, \mu)$ and

$$A_n^m = \frac{1}{2\pi} \int_0^{2\pi} \int_{-1}^{1} A Y_n^{m*} \, d\mu \, d\lambda. \tag{6.95}$$

It is necessary to truncate the series at some wavenumber. There are two common ways of truncating the series: (1) *Rhomboidal truncation*, which has the form

$$A(\lambda, \mu) = \sum_{m=-N}^{N} \sum_{n=|m|}^{|m|+N} A_n^m Y_n^m (\lambda, \mu), \tag{6.96}$$

and (2) *triangular truncation*, which has the form

$$A(\lambda, \mu) = \sum_{m=-N}^{N} \sum_{n=|m|}^{N} A_n^m Y_n^m (\lambda, \mu). \tag{6.97}$$

We discuss these truncations in detail in Chapter 7.

If the set of spectral coefficients A_n^m is known, then by using either (6.96) or (6.97) the function $A(\lambda, \mu)$ can be defined everywhere over the globe. In practice, it is necessary to evaluate the function at a finite set of grid points. Likewise, the evaluation of the integrals in the Fourier-Legendre transform is based on data at only a finite number of points.

6.7.1 *Grid to Spectral Space*

In practice, the evaluation of A_n^m from $A(\lambda, \mu)$ is carried out in two steps.

Step 1: Perform the Fourier transform of the space field along latitudes,

$$A^m(\mu) = \frac{1}{2\pi} \int_0^{2\pi} A(\lambda, \mu) e^{-im\lambda} d\lambda. \tag{6.98}$$

Step 2: Perform the Legendre transform of the Fourier components,

$$A_n^m = \int_{-1}^{1} A^m(\mu) P_n^m(\mu) \, d\mu. \tag{6.99}$$

The Fourier transform is evaluate using the trapezoidal quadrature formula, that is,

$$A^m(\mu_k) = \frac{1}{2M} \sum_{j=0}^{2M-1} A(\lambda_j, \mu_k) e^{-im\lambda_j}, \tag{6.100}$$

where $\lambda_j = [(2\pi)/(2M)]j = (\pi / M)j$. This integration is exact for any function which may be represented by a truncated Fourier series with wavenumbers $\leq 2M-1$. This calculation is very efficiently done by the FFT. The Legendre transform is evaluated using the Gaussian quadrature formula, that is,

$$A_n^m = \sum_{k=1}^{K} W(\mu_k) A^m(\mu_k) P_n^m(\mu_k), \tag{6.101}$$

where μ_k are the *Gaussian latitudes* and $W(\mu_k)$ are the *Gaussian weights*, as discussed in Section 6.6. This formula is exact for any polynomial [here $A^m(\mu)P_n^m(\mu)$] of degree $\leq 2K-1$.

6.7.2 *Spectral to Grid Space*

The transformation from spectral to grid space is also achieved similarly in two steps.

Step 1: Perform the reverse Legendre transform,

$$A^m(\mu_k) = \sum_n \phi A_n^m P_n^m(\mu_k). \tag{6.102}$$

Step 2: Perform the reverse Fourier transform,

$$A(\lambda_j, \mu_k) = A_n^m(\mu_k) e^{im\lambda_j}. \tag{6.103}$$

Thus one proceeds from spherical harmonic components A_n^m to Fourier components A^m and then to grid-point values $A(\lambda_j, \mu_k)$. It should be noted that the use of Gaussian latitudes and weights enables one to calculate the Legendre transform exactly.

6.8 Barotropic Spectral Model on a Sphere

In Chapter 2 we discussed the finite-difference barotropic model where we were concerned with the proper formulation of the space-differencing schemes, the Jacobian, the Laplacian, and the solution of Poisson's equation. In this section, we consider a spectral model on a sphere, the integration of which will be through a coupled system of nonlinear ordinary differential equations. This system of equations is obtained by transforming the model equations from the space to the spectral domain. The basis functions for this transformation will be surface spherical harmonics, some important properties of which have already been discussed.

If λ represents the longitude and θ the latitude, then the barotropic vorticity equation on a sphere is given by

$$\frac{\partial \zeta}{\partial t} = \frac{1}{a^2}\left(\frac{\partial \psi}{\partial \mu}\frac{\partial}{\partial \lambda}(\zeta + f) - \frac{\partial \psi}{\partial \lambda}\frac{\partial}{\partial \mu}(\zeta + f)\right), \tag{6.104}$$

where $\mu = \sin\theta$. Noting that $\partial f / \partial \lambda = 0$ and $\partial f / \partial \mu = 2\Omega$, we obtain

$$\frac{\partial \zeta}{\partial t} = \frac{1}{a^2}\left(\frac{\partial \psi}{\partial \mu}\frac{\partial \zeta}{\partial \lambda} - \frac{\partial \psi}{\partial \lambda}\frac{\partial \zeta}{\partial \mu}\right) - \frac{2\Omega}{a^2}\frac{\partial \psi}{\partial \lambda},$$

or

$$\frac{\partial \zeta}{\partial t} = F(\lambda, \mu) - \frac{2\Omega}{a^2}\frac{\partial \psi}{\partial \lambda}. \tag{6.105}$$

In the above equation, the nonlinear advective term

$$\frac{1}{a^2}\left(\frac{\partial \psi}{\partial \mu}\frac{\partial \zeta}{\partial \lambda}-\frac{\partial \psi}{\partial \lambda}\frac{\partial \zeta}{\partial \mu}\right)$$

is written as $F(\lambda,\mu)$. The second term on the right-hand side of (6.105) is the earth's rotation term and it is linear.

To transform (6.105) into its spectral form, we take the Fourier-Legendre expansion of variables ζ, ψ, and F

$$\zeta(\lambda,\mu,t)=\sum_m\sum_n \zeta_n^m(t)Y_n^m(\lambda,\mu), \tag{6.106}$$

$$\psi(\lambda,\mu,t)=\sum_m\sum_n \psi_n^m(t)Y_n^m(\lambda,\mu), \tag{6.107}$$

$$F(\lambda,\mu,t)=\sum_m\sum_n F_n^m(t)Y_n^m(\lambda,\mu). \tag{6.108}$$

Substituting into (6.105) we obtain

$$\sum_m\sum_n \frac{d\zeta_n^m(t)}{dt}Y_n^m(\lambda,\mu)=\sum_m\sum_n\Big(F_n^m(t)Y_n^m(\lambda,\mu) \\ -\frac{2\Omega}{a^2}\psi_n^m(t)\frac{\partial}{\partial\lambda}Y_n^m(\lambda,\mu)\Big). \tag{6.109}$$

Using $\zeta=\nabla^2\psi$ and looking at the left-hand side of equation (6.105) only, we get

$$\frac{d\zeta_n^m(t)}{dt}Y_n^m(\lambda,\mu)=-\frac{n(n+1)}{a^2}\frac{d\psi_n^m(t)}{dt}Y_n^m(\lambda,\mu), \tag{6.110}$$

remembering $\nabla^2 Y_n^m=\frac{-n(n+1)}{a^2}Y_n^m$. Also,

$$\frac{\partial}{\partial\lambda}Y_n^m(\lambda,\mu)=imY_n^m(\lambda,\mu). \tag{6.111}$$

Substituting (6.110) and (6.111) into (6.109) and then equating the coefficients of $Y_n^m(\lambda,\mu)$ on both sides of the above equation, we obtain

$$\frac{-n(n+1)}{a^2}\frac{d\psi_n^m(t)}{dt}Y_n^m(\lambda,\mu)=F_n^m(t)Y_n^m(\lambda,\mu)-\frac{2\Omega}{a^2}\psi_n^m(t)imY_n^m(\lambda,\mu)$$

$$\frac{d\psi_n^m}{dt}=\frac{2\Omega im}{n(n+1)}\psi_n^m-\frac{a^2}{n(n+1)}F_n^m \tag{6.112}$$

as the spectral form of the barotropic vorticity equation (6.105). The first term on the right-hand side of (6.112) is the beta term (the acceleration due to the advection of the earth's vorticity), while the second term is the advection of relative vorticity.

The spectral form of the barotropic vorticity equation consists of a set of ordinary nonlinear differential equations in spectral space. It should be noted that F_n^m is a nonlinear term. We use the *transform method* to evaluate such nonlinear terms.

The transform method involves calculating the terms $\partial\psi/\partial\lambda$, $\partial\psi/\partial\mu$, $\partial\zeta/\partial\lambda$, and $\partial\zeta/\partial\mu$ on the grid points by projecting the spectral coefficients onto the space domain. These are then multiplied to get values of the nonlinear terms $F(\lambda,\mu)$ on the grid points. The spectral analysis of $F(\lambda,\mu)$ then involves the Fourier analysis of the $F(\lambda,\mu)$ along latitude circles, followed by a Legendre transform of the resulting Fourier coefficients to obtain the spectrum F_n^m. This procedure is called the *Fourier-Legendre transform*.

We can write

$$F(\lambda,\mu)=\frac{1}{a^2}\left(\frac{\partial\psi}{\partial\mu}\frac{\partial\zeta}{\partial\lambda}-\frac{\partial\psi}{\partial\lambda}\frac{\partial\zeta}{\partial\mu}\right) \tag{6.113}$$

$$=\frac{1}{a^2(1-\mu^2)}\left((1-\mu^2)\frac{\partial\psi}{\partial\mu}\frac{\partial\zeta}{\partial\lambda}-\frac{\partial\psi}{\partial\lambda}(1-\mu^2)\frac{\partial\zeta}{\partial\mu}\right).$$

Let

$$\psi(\lambda,\mu)=\sum_m\sum_n\psi_n^m(t)Y_n^m(\lambda,\mu), \tag{6.114}$$

so that upon taking the derivative of (6.114) with respect to μ, multiplying by $(1-\mu^2)$, and remembering that $Y_n^m(\lambda,\mu)=P_n^m(\mu)e^{im\lambda}$ we obtain the north-south derivative of ψ

$$(1-\mu^2)\frac{\partial\psi}{\partial\mu}=\sum_m\sum_n\psi_n^m(t)e^{im\lambda}(1-\mu^2)\frac{\partial}{\partial\mu}P_n^m(\mu), \tag{6.115}$$

where by the recurrence relation (6.38) we have

$$(1-\mu^2)\frac{d}{d\mu}P_n^m(\mu)=-n\in_{n+1}^m P_{n+1}^m(\mu)+(n+1)\in_n^m P_{n-1}^m(\mu)$$

and

$$\in_n^m=\left(\frac{n^2-m^2}{4n^2-1}\right)^{1/2}.$$

Hence using the recurrence relation in equation (6.115) we obtain

$$(1-\mu^2)\frac{\partial\psi}{\partial\mu}=\sum_m\sum_n\psi_n^m\left[-n\in_{n+1}^m P_{n+1}^m(\mu)\right.$$
$$\left.+(n+1)\in_n^m P_{n-1}^m(\mu)\right]e^{im\lambda} \tag{6.116}$$

where the term in brackets is always known.

If we differentiate (6.114) with respect to λ, we obtain the east-west derivative of ψ

$$\frac{\partial \psi}{\partial \lambda} = \sum_m \sum_n im\psi_n^m(t) P_n^m(\mu) e^{im\lambda}. \qquad (6.117)$$

Using (6.116) and (6.117) we can obtain the grid-point values of $\partial\psi/\partial\lambda$ and $(1-\mu^2)\partial\psi/\partial\mu$. Similarly, the grid-point values of $(1-\mu^2)\partial\zeta/\partial\mu$ and $\partial\zeta/\partial\lambda$ are calculated from the spectral coefficients of vorticity ζ_n^m. Having obtained $(1-\mu^2)\partial\psi/\partial\mu$, $\partial\psi/\partial\lambda$, $(1-\mu^2)\partial\zeta/\partial\mu$, and $\partial\zeta/\partial\lambda$ in grid space, we obtain $F(\lambda,\mu)$ from (6.113) in grid space. The Fourier-Legendre transform is then applied to $F(\lambda,\mu)$ to obtain F_n^m.

Knowing the spectral coefficients ψ_n^m and F_n^m, one can calculate the term on the right-hand side of the spectral form of vorticity equation (6.112). This gives us $d\psi_n^m/dt$, the time tendency of ψ_n^m, which, along with the value of ψ_n^m at the previous time step, can be used to obtain ψ_n^m at a future time step. In practice, the following steps are needed to integrate the barotropic vorticity equation:

Step 1: From the coefficients ψ_n^m and ζ_n^m, obtain the grid-point values of $(1-\mu^2)\partial\psi/\partial\mu$ and $(1-\mu^2)\partial\zeta/\partial\mu$ along a latitude circle using (6.116).

Step 2: Similarly, using (6.117), obtain grid-point values of $\partial\psi/\partial\lambda$ and $\partial\zeta/\partial\lambda$ along the latitude circle.

Step 3: Multiply $(1-\mu^2)\partial\psi/\partial\mu$, $\partial\zeta/\partial\lambda$, $\partial\psi/\partial\lambda$ and $(1-\mu^2)\partial\zeta/\partial\mu$ to calculate

$$\begin{aligned} F(\lambda,\mu) &= \frac{1}{a^2}\left(\frac{\partial\psi}{\partial\mu}\frac{\partial\zeta}{\partial\lambda} - \frac{\partial\psi}{\partial\lambda}\frac{\partial\zeta}{\partial\mu}\right) \\ &= \frac{1}{a^2(1-\mu^2)}\left((1-\mu^2)\frac{\partial\psi}{\partial\mu}\frac{\partial\zeta}{\partial\lambda} - \frac{\partial\psi}{\partial\lambda}(1-\mu^2)\frac{\partial\zeta}{\partial\mu}\right) \end{aligned}$$

on the grid points (λ,μ).

Step 4: Perform the Fourier transform of $F(\lambda,\mu)$ along the latitude circle to obtain the Fourier components $F^m(\mu)$. This is done by using the FFT.

Step 5: Perform the Legendre transform of $F^m(\mu)$ at the various latitude to obtain F_n^m, the spherical harmonic amplitudes of $F(\lambda,\mu)$. This is done by using Gaussian quadrature.

Step 6: From ψ_n^m and F_n^m, calculate the right-hand side of the spectral vorticity equation and obtain $d\psi_n^m/dt$.

Step 7: From $\psi_n^m(t-\Delta t)$ and $d\psi_n^m/dt$, obtain $\psi_n^m(t+\Delta t)$ at the next time step as

$$\psi_n^m(t+\Delta t) = \psi_n^m(t-\Delta t) + 2\Delta t\frac{d\psi_n^m}{dt}.$$

6.9 Shallow-Water Spectral Model

Earlier we discussed the integration of the shallow-water equations using finite differences in both time and space. We now structure the shallow-water system to a similar time-differencing scheme, but the space differentials will be calculated spectrally. In this system, we use the momentum and the mass continuity equations. In a shallow-water model, we consider a homogeneous incompressible fluid with a rigid lower boundary and an upper free surface. The horizontal velocity is taken to be invariant with height.

The continuity equation for an incompressible fluid is

$$\frac{\partial u}{\partial x}+\frac{\partial v}{\partial y}+\frac{\partial w}{\partial z}=0. \tag{6.118}$$

Integrating (6.118) vertically with a lower boundary condition w_0 at $z_0=0$, we obtain

$$\int\frac{\partial w}{\partial z}dz=-\int\left(\frac{\partial u}{\partial x}+\frac{\partial v}{\partial y}\right)dz,$$

$$w_z-w_0=-z\left(\frac{\partial u}{\partial x}+\frac{\partial v}{\partial y}\right),$$

$$w_z=-z\left(\frac{\partial u}{\partial x}+\frac{\partial v}{\partial y}\right), \tag{6.119}$$

where z is the height of the free surface and w_z is the vertical velocity of the free surface. Also,

$$w_z=\frac{dz}{dt}=\frac{\partial z}{\partial t}+\vec{V}_H\cdot\nabla z. \tag{6.120}$$

From (6.119) and (6.120) we obtain

$$\frac{\partial z}{\partial t}+\vec{V}_H\cdot\nabla z=-z\left(\frac{\partial u}{\partial x}+\frac{\partial v}{\partial y}\right),$$

$$\frac{\partial z}{\partial t}=-\vec{V}_H\cdot\nabla z-z\nabla\cdot\vec{V}_H,$$

which on multiplication by g gives

$$\frac{\partial \phi}{\partial t}=-\vec{V}_H\cdot\nabla\phi-\phi\nabla\cdot\vec{V}_H, \tag{6.121}$$

where ϕ is the geopotential at the free surface.

Writing $\phi=\overline{\phi}+\phi'$ as the sum of the mean and its deviation from the mean, where $\overline{\phi}$ is the time-variant area mean, we obtain

$$\frac{\partial \phi'}{\partial t}=-\vec{V}_H\cdot\nabla(\overline{\phi}+\phi')-(\overline{\phi}+\phi')\nabla\cdot\vec{V}_H=-\vec{V}_H\cdot\nabla\phi'-\phi'\nabla\cdot\vec{V}_H-\overline{\phi}\nabla\cdot\vec{V}_H,$$

or

$$\frac{\partial \phi'}{\partial t} = -\nabla \cdot \left(\phi' \vec{V}_H \right) - \bar{\phi} \cdot \nabla \vec{V}_H. \tag{6.122}$$

Equation (6.122) now represents the continuity equation.

The horizontal momentum equation for the shallow-water model is

$$\frac{\partial \vec{V}_H}{\partial t} = -\left(\vec{V}_H \cdot \nabla \right) \vec{V}_H - f \vec{k} \times \vec{V}_H - \nabla \phi. \tag{6.123}$$

Using the vector identity

$$\left(\vec{V}_H \cdot \nabla \right) \vec{V}_H = \nabla \left(\frac{\vec{V}_H \cdot \vec{V}_H}{2} \right) + \zeta \vec{k} \times \vec{V}_H,$$

the horizontal momentum equation may be written as

$$\frac{\partial \vec{V}_H}{\partial t} = -(\zeta + f) \vec{k} \times \vec{V}_H - \nabla \left(\phi + \frac{\vec{V}_H \cdot \vec{V}_H}{2} \right),$$

or

$$\frac{\partial \vec{V}_H}{\partial t} = -(\zeta + f) \vec{k} \times \vec{V}_H - \nabla \left(\phi' + \frac{\vec{V}_H \cdot \vec{V}_H}{2} \right), \tag{6.124}$$

because $\nabla \bar{\phi} = 0$. We express the momentum equation (6.124) in terms of the vorticity and divergence equations, which are more suitable for a spectral model.

The vorticity equation is obtained by operating on (6.124) with the vector operator $\vec{k} \cdot \nabla \times$. Thus

$$\frac{\partial \left(\vec{k} \cdot \nabla \times \vec{V}_H \right)}{\partial t} = -\vec{k} \cdot \nabla \times \left[(\zeta + f) \vec{k} \times \vec{V}_H \right] - \vec{k} \cdot \nabla \times \nabla \left(\phi' + \frac{\vec{V}_H \cdot \vec{V}_H}{2} \right).$$

After simplifying, we obtain

$$\frac{\partial \zeta}{\partial t} = -\nabla \cdot (\zeta + f) \vec{V}_H \tag{6.125}$$

as the desired vorticity equation. Similarly, the divergence equation is obtained by applying the vector operator $\nabla \cdot$ to (6.124). Thus

$$\frac{\partial \left(\nabla \cdot \vec{V}_H \right)}{\partial t} = -\nabla \cdot \left[(\zeta + f) \vec{k} \times \vec{V}_H \right] - \nabla \cdot \nabla \left(\phi' + \frac{\vec{V}_H \cdot \vec{V}_H}{2} \right).$$

After further simplification we obtain the divergence equation as

$$\frac{\partial D}{\partial t} = \frac{\partial v(\zeta + f)}{a \cos\theta \, \partial \lambda} - \frac{\partial u \cos\theta (\zeta + f)}{a \cos\theta \, \partial \theta} - \nabla^2 \left(\phi' + \frac{\vec{V}_H \cdot \vec{V}_H}{2} \right). \tag{6.126}$$

The vorticity, divergence, and continuity equations given by (6.125), (6.126), and (6.122) form the shallow-water model on a sphere. To remove the discontinuity in the wind field at the poles, we replace u and v by the Robert functions defined by (6.73). Using these functions, the vorticity, divergence, and continuity equations take the following form in spherical coordinates:

$$\frac{\partial \zeta}{\partial t} = -\frac{1}{\cos^2\theta}\left(\frac{\partial U(\zeta+f)}{\partial \lambda} + \cos\theta \frac{\partial V(\zeta+f)}{\partial \theta}\right), \tag{6.127}$$

$$\frac{\partial D}{\partial t} = \frac{1}{\cos^2\theta}\left(\frac{\partial V(\zeta+f)}{\partial \lambda} - \cos\theta \frac{\partial U(\zeta+f)}{\partial \theta}\right)$$

$$-\nabla^2\left(\frac{U^2+V^2}{2\cos^2\theta} + \phi'\right), \tag{6.128}$$

$$\frac{\partial \phi'}{\partial t} = -\frac{1}{\cos^2\theta}\left(\frac{\partial (U\phi')}{\partial \lambda} + \cos\theta \frac{\partial (V\phi')}{\partial \theta}\right) - \bar{\phi}D. \tag{6.129}$$

These can be written in terms of ψ, χ, and ϕ' as

$$\frac{\partial \nabla^2\psi}{\partial t} = -\frac{1}{\cos^2\theta}\left(\frac{\partial U(\nabla^2\psi+f)}{\partial \lambda} + \cos\theta \frac{\partial V(\nabla^2\psi+f)}{\partial \theta}\right), \tag{6.130}$$

$$\frac{\partial \nabla^2\chi}{\partial t} = \frac{1}{\cos^2\theta}\left(\frac{\partial V(\nabla^2\psi+f)}{\partial \lambda} - \cos\theta \frac{\partial U(\nabla^2\psi+f)}{\partial \theta}\right)$$

$$-\nabla^2\left(\frac{U^2+V^2}{2\cos^2\theta} + \phi'\right), \tag{6.131}$$

$$\frac{\partial \phi'}{\partial t} = -\frac{1}{\cos^2\theta}\left(\frac{\partial (U\phi')}{\partial \lambda} + \cos\theta \frac{\partial (V\phi')}{\partial \theta}\right) - \bar{\phi}D. \tag{6.132}$$

It can be shown that the above set of vorticity, divergence, and continuity equations ensures the conservation of vorticity, kinetic energy, and potential energy during the course of computations. Because the first term on the right-hand side in these equations is of the same form, the equations can be calculated by an identical computation algorithm.

Our aim is to solve (6.130), (6.131), and (6.132) spectrally using a fully explicit method. This is achieved by expanding ψ, χ, ϕ', U, and V spectrally, that is,

$$\psi = a^2 \sum_m \sum_n \psi_n^m Y_n^m, \tag{6.133}$$

and similarly for χ, ϕ', U, and V. In (6.130) to (6.132), ψ, χ, and ϕ' are the prognostic variables. In addition to these, we need U and V to calculate the right-hand side of these equations. To obtain U_n^m and V_n^m, the spectral components of U and V, from the spectral components of ψ and χ, we make use of (6.88) and (6.89).

Substituting the spectral expansions of ψ, χ, ϕ', U, and V into (6.130) to (6.132), the spectral form of the shallow-water equations takes the following form:

$$-n(n+1)\frac{\partial \psi_n^m}{\partial t} = -\alpha_n^m(A,B), \tag{6.134}$$

$$-n(n+1)\frac{\partial \chi_n^m}{\partial t} = \alpha_n^m(B,-A) - \nabla^2\left(\frac{U^2+V^2}{2\cos^2\theta}+\phi'\right)_n^m, \tag{6.135}$$

$$\frac{\partial \phi'^m_n}{\partial t} = -\alpha_n^m(U\phi, V\phi) - \bar{\phi}D_n^m, \tag{6.136}$$

where $A = U(\nabla^2\psi + f)$, $B = V(\nabla^2\psi + f)$, $D = \nabla\cdot\vec{V}$, $\vec{V} = U\vec{i} + V\vec{j}$, and the operator α is defined as

$$\alpha(A,B) = \frac{1}{\cos^2\theta}\left(\frac{\partial A}{\partial \lambda} + \cos\theta\frac{\partial B}{\partial \theta}\right).$$

The spectral amplitudes of the nonlinear terms on the right-hand side of (6.134) to (6.136) are calculated by first projecting the values of ψ, χ, ϕ', U, and V from the spectral domain onto the space domain on a Gaussian grid, and then multiplying them in the space domain to obtain the grid-point values of the nonlinear terms. The Fourier-Legendre transform is then used to convert these grid-point values to the spectral amplitudes of the nonlinear terms. The sum of the nonlinear terms and linear terms forms the tendencies of the various spectral amplitudes in the shallow-water model. These tendencies, along with the values of ψ_n^m, χ_n^m, and ϕ'^m_n at the previous time step, are used to obtain the future values of these functions using the following finite-difference analog:

$$\psi_n^m(t+\Delta t) = \psi_n^m(t-\Delta t) + \left(\frac{\partial \psi_n^m}{\partial t}\right)2\Delta t, \tag{6.137}$$

$$\chi_n^m(t+\Delta t) = \chi_n^m(t-\Delta t) + \left(\frac{\partial \chi_n^m}{\partial t}\right)2\Delta t, \tag{6.138}$$

$$\phi'^m_n(t+\Delta t) = \phi'^m_n(t-\Delta t) + \left(\frac{\partial \phi'^m_n}{\partial t}\right)2\Delta t. \tag{6.139}$$

To march forward in time, we make use of centered time differencing except for the first time step, where forward time differencing is applied. Marching forward in time over the forecast period, we obtain the final forecast values of ψ_n^m, χ_n^m, and ϕ'^m_n. These amplitudes are then projected onto the space domain to obtain the forecast fields of stream function (or vorticity), velocity potential (or divergence), and the geopotential (or height) at the free surface.

This shallow water spectral model has gravity waves and Rossby waves. Gravity waves require small computational time steps while Rossby waves can take larger computational time steps. If we are using centered time differencing we need to use small time steps for this model which is computationally expensive. For this reason, the fully explicit method is rarely used and the semi-implicit shallow water model is employed instead.

6.10 Semi-implicit Shallow-Water Spectral Model

We next address the formulation of a shallow-water model based on the semi-implicit time-integration scheme. In this formulation, the linear terms are integrated implicitly, while the nonlinear terms are integrated explicitly. As the vorticity equation has only nonlinear terms on its right-hand side which do not excite gravity waves, it is integrated explicitly. The divergence equation and the continuity equation have both linear and nonlinear terms on their right-hand sides and therefore are integrated implicitly.

Separating the linear and nonlinear terms, the divergence and continuity equations can be written as

$$\frac{\partial}{\partial t}D = F_1(\psi) - \nabla^2\phi' \tag{6.140}$$

and

$$\frac{\partial}{\partial t}\phi' = F_2(\phi') - \bar{\phi}D, \tag{6.141}$$

where

$$F_1(\psi) = \frac{1}{\cos^2\theta}\left(\frac{\partial}{\partial\lambda}V(\nabla^2\psi + f) - \cos\theta\frac{\partial}{\partial\theta}U(\nabla^2\psi + f)\right) - \nabla^2\left(\frac{U^2+V^2}{2\cos^2\theta}\right)$$

and

$$F_2(\phi') = -\frac{1}{\cos^2\theta}\left(\frac{\partial}{\partial\lambda}(U\phi') + \cos\theta\frac{\partial}{\partial\theta}(V\phi')\right)$$

are the nonlinear terms.

The reason for doing such a partitioning is that we want to treat the gravitational modes differently from the Rossby waves. We integrate the linear terms responsible for the fast-moving gravitational modes implicitly and the nonlinear terms explicitly.

Let us define a time-average operator $\overline{\overline{()}}^t$ as

$$\overline{\overline{D}}^t = \frac{D^{t+\Delta t} + D^{t-\Delta t}}{2}, \tag{6.142}$$

noting that from the centered time differencing scheme we have

$$\frac{\partial D}{\partial t} = \frac{D^{t+\Delta t} - D^{t-\Delta t}}{2\Delta t}.$$

Solving equation (6.142) for $D^{t+\Delta t}$ and substituting it into the above equation we can rewrite (6.140) using the above averaged value as

$$\frac{D^{t+\Delta t} - D^{t-\Delta t}}{2\Delta t} = \frac{\overline{\overline{D}}^t - D^{t-\Delta t}}{\Delta t} = F_1^t(\psi) - \nabla^2\overline{\overline{\phi'}}^t. \tag{6.143}$$

Similarly, equation (6.141) can be rewritten as

$$\frac{\phi'^{t+\Delta t}-\phi'^{t-\Delta t}}{2\Delta t}=\frac{\overline{\overline{\phi'^{t}}}-\phi'^{t-\Delta t}}{\Delta t}=F_2^t(\phi')-\overline{\phi}\,\overline{\overline{D^t}}\,. \tag{6.144}$$

It should be noted that the values of ϕ' and D in the linear terms on the right-hand side of (6.143) and (6.144) are averages of their values at time levels $t-\Delta t$ and $t+\Delta t$, while the nonlinear terms are calculated at time level t.

Equations (6.143) and (6.144) may be written as

$$\overline{\overline{D^t}}=D^{t-\Delta t}+F_1^t(\psi)\Delta t-\nabla^2\overline{\overline{\phi'^t}}\Delta t\,, \tag{6.145}$$

$$\overline{\overline{\phi'^t}}=\phi'^{t-\Delta t}+F_2^t(\phi')\Delta t-\overline{\phi}\,\overline{\overline{D^t}}\Delta t\,. \tag{6.146}$$

In order to solve this system of equations, we eliminate one of the unknown variables ($\overline{\overline{D^t}}$ or $\overline{\overline{\phi'^t}}$) from these equations in order to obtain an equation in a single unknown variable. We eliminate $\overline{\overline{\phi'^t}}$ to get an equation in the single variable $\overline{\overline{D^t}}$. This is achieved by applying the ∇^2 operator to (6.146) and multiplying the resultant equation by Δt

$$\nabla^2\overline{\overline{\phi'^t}}\Delta t=\Delta t\nabla^2\phi'^{t-\Delta t}+(\Delta t)^2\,\nabla^2F_2^t(\phi')-(\Delta t)^2\,\overline{\phi}\nabla^2\overline{\overline{D^t}}\,,$$

then substituting for $\nabla^2\overline{\overline{\phi'^t}}\Delta t$ into (6.145) to obtain

$$\overline{\overline{D^t}}=D^{t-\Delta t}+F_1^t(\psi)\Delta t-\Delta t\nabla^2\phi'^{t-\Delta t}-(\Delta t)^2\,\nabla^2F_2^t(\phi')+(\Delta t)^2\,\overline{\phi}\nabla^2\overline{\overline{D^t}}\,,$$

or, on rearranging, we obtain a Helmholtz equation, that is,

$$(\Delta t)^2\,\overline{\phi}\nabla^2\overline{\overline{D^t}}-\overline{\overline{D^t}}=F_3\left(D,\phi',\psi\right)^{t,t-\Delta t}, \tag{6.147}$$

where

$$F_3\left(D,\phi',\psi\right)^{t,t-\Delta t}=(\Delta t)^2\,\nabla^2F_2^t(\phi')-\Delta tF_1^t(\psi)+(\Delta t)\nabla^2\phi'^{t-\Delta t}-D^{t-\Delta t}\,.$$

We can write (6.147) in spectral form as

$$-(\Delta t)^2\,\overline{\phi}\frac{n(n+1)}{a^2}\overline{\overline{D_n^{m^t}}}-\overline{\overline{D_n^{m^t}}}=F_{3n}^{\;m}\,,$$

or

$$\overline{\overline{D_n^{m^t}}}=-\frac{F_{3n}^{\;m}}{1+\left(\dfrac{n(n+1)}{a^2}(\Delta t)^2\overline{\phi}\right)}\,. \tag{6.148}$$

Hence we can obtain the unknown variable $\overline{\overline{D_n^{m^t}}}$ using the above equation. The value of D_n^m at time level $t+\Delta t$ is then obtained from

$$D_n^{m^{t+\Delta t}}=2\overline{\overline{D_n^{m^t}}}-D_n^{m^{t-\Delta t}}\,. \tag{6.149}$$

By substituting $\overline{\overline{D_n^{m^t}}}$ into (6.146), one also obtains the spectrum of $\overline{\overline{\phi^{tt}}}$ and hence $\phi^{t+\Delta t}$. The spectrum of $\zeta^{t+\Delta t}$ is obtained explicitly from the spectral form of the vorticity equation (6.134). This completes a one-time-step forecast of the shallow-water model using a semi-implicit time-integration scheme.

6.11 Inclusion of Bottom Topography

It is relatively easy to cast a spectral shallow-water model that includes the effects of mountains. The only changes that are involved in the basic equations are in the pressure gradient force, where

$$\frac{\partial \phi}{\partial \lambda} \quad \text{is replaced by} \quad \frac{\partial}{\partial \lambda}(\phi + gh)$$

and

$$\frac{\partial \phi}{\partial \theta} \quad \text{is replaced by} \quad \frac{\partial}{\partial \theta}(\phi + gh).$$

The mass continuity equation remains unaltered. Here h is the height of the mountain (i.e., the bottom topography of the shallow-water model) and $\phi + gh$ defines the geopotential of the free surface. The derivation of the Helmholtz equations for the shallow-water equations with bottom topography is left as an exercise for the student.

6.12 Exercises

6.1. In Section 6.2, Rodrigues' formula,

$$P_n(\mu) = \frac{1}{2^n n!} \frac{d^n}{d\mu^n}(\mu^2 - 1)^n, \qquad n = 0,\ 1,\ 2,\ 3,\ \dots, \qquad |\mu| \le 1,$$

was described. Using (6.13), try to prove the above formula. *Hint*: Apply a binomial expansion of $(1-\mu^2)^n$.

6.2. We define the generating function of the ordinary Legendre polynomial as follows:

$$\frac{1}{(1 - 2\mu x + x^2)^{1/2}} = \sum_{h=0}^{\infty} P_n(\mu) x^n, \qquad |\mu| < 1, \qquad |x| < 1.$$

Prove the above equality. *Hint*: Use the binomial expansion of $1/(1-y)^{1/2}$, set $y = 2\mu x - x^2$, and multiply the powers of $2\mu x - x^2$ out and collect terms involving x^n.

6.3. Prove the recurrence relation

$$(n+1) P_{n+1}(\mu) = (2n+1)\mu P_n(\mu) - n P_{n-1}(\mu), \qquad n = 1,\ 2,\ 3,\ \dots$$

Hint: Differentiate the generating function with respect to x. The above equation is called *Bonnet's recursion*.

6.4. Represent the following polynomials in terms of Legendre polynomials: 1, μ, μ^2, and μ^3; i.e., find the matrix A such that

$$A\begin{bmatrix} P_0(\mu) \\ P_1(\mu) \\ P_2(\mu) \\ P_3(\mu) \end{bmatrix} = \begin{bmatrix} 1 \\ \mu \\ \mu^2 \\ \mu^3 \end{bmatrix}.$$

Describe the most salient feature of matrix A. What does this feature imply?

6.5. Show that $|P_n(\mu)| \le 1$ for $-1 \le \mu \le 1$.

Chapter 7

Multilevel Global Spectral Model

7.1 Introduction

Since the 1970s, the spectral method has become an increasingly popular technique for global numerical weather prediction. Global numerical models formulated using the spectral technique are used worldwide for both research and operational purposes. The success of the spectral technique can be attributed to the spectral transform technique developed independently by Eliasen et al. (1970) and Orszag (1970), and later refined by Bourke (1972).

Prior to the introduction of the transform technique, the nonlinear terms were computed using a very tedious process called the *interaction coefficients method.* This method required large amounts of computer resources as well as enormous bookkeeping. The transform technique facilitates the computation of the nonlinear terms, as discussed later in Section 7.4. Furthermore, the Galerkin method discussed in Chapter 4 is widely used in most spectral models and provides us with alias-free computation of the nonlinear terms.

The transform technique enables the current spectral models to be competitive in terms of computational overhead with respect to their grid-point counterparts. The transform technique is also well suited for incorporating the terms dealing with physics in the prediction scheme. There are a number of advantages to using the spectral technique over the conventional grid-point method. However, we will not get into this discussion here. It should be noted that the model truncation limit specifies the scale of the shortest wavelength that can be resolved by the model. In the following section, we discuss the two most widely used truncations in a spectral model.

7.2 Truncation in a Spectral Model

As we saw earlier, any variable field (say ζ, ψ, etc.) can be represented by a series of spherical harmonics as

$$\zeta(\lambda,\mu)=\sum_{m}\sum_{n}\zeta_n^m Y_n^m(\lambda,\mu), \tag{7.1}$$

$$\psi(\lambda,\mu)=\sum_{m}\sum_{n}\psi_n^m Y_n^m(\lambda,\mu), \tag{7.2}$$

where the double summation is infinite in general, in other words, $-\infty \leq m \leq \infty$ and $|m| \leq n \leq \infty$. However, in practice, representation in terms of an infinite series is not possible and one needs to truncate the series at some point. It can be shown that once we decide on the maximum number of spectral components to be present in the double summation, then we can achieve a consistent energy, momentum, and vorticity conservation for the spectrally truncated equations. In doing so, we neglect waves outside the spectrum (i.e., beyond the designated highest wavenumber). In practice, one truncates smaller-scale waves in the spectral truncation.

In a global spectral model, two types of truncation schemes are generally used. They are called the *triangular truncation* and the *rhomboidal truncation.* In the case of triangular truncation, the highest degree of the various spherical harmonics Y_n^m (m denotes the order of a spherical harmonic and n the degree) is fixed and is set equal to the highest order of these waves. Thus, this truncation is represented mathematically by

$$\psi(\lambda,\mu) = \sum_{m=-N}^{N} \sum_{n=|m|}^{N} \psi_n^m Y_n^m(\lambda,\mu), \tag{7.3}$$

and is schematically represented in Fig. 7.1.

Under this truncation, the number of zeroes $n-m$ for different wavenumbers m vary, and thus different wavenumbers m have different degrees of freedom along a latitude. The two-dimensional wave index has the same maximum value for all waves, and all waves are truncated at the same two-dimensional scale.

The rhomboidal truncation has a fixed value of the maximum number of zeroes for every wavenumber m, thus giving equal degrees of freedom along a latitude for all waves. This truncation is represented as

$$\psi(\lambda,\mu) = \sum_{m=-N}^{N} \sum_{n=|m|}^{|m|+J} \psi_n^m Y_n^m. \tag{7.4}$$

If $J = N$, the truncation is called rhomboidal. When $J \neq N$, it may be called *parallelogramic truncation.* Schematically, this can be represented as in Fig. 7.2. The triangular truncation at wavenumber N is often abbreviated as $T-N$; likewise, rhomboidal truncation at wavenumber N is abbreviated as $R-N$.

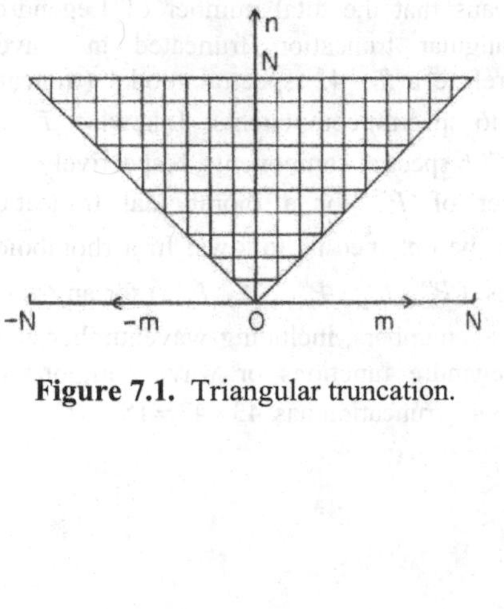

Figure 7.1. Triangular truncation.

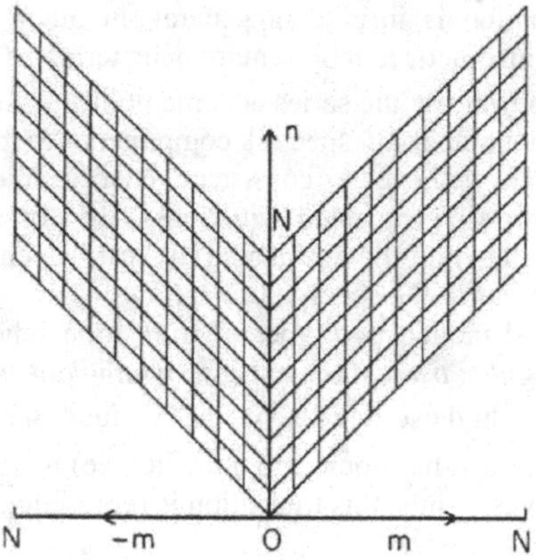

Figure 7.2. Rhomboidal truncation.

There has been considerable discussion as to which of the two truncations is better suited for different problems. Both of these representations have their advantages and disadvantages. The triangular truncation assumes that the variance contained in waves shorter than a particular two-dimensional wave is zero. However, the representation in the meridional direction does not have the same number of degrees of freedom for all waves. In general, it is seen that for the same number of wave components, triangular truncation contains more variance than the rhomboidal one.

The total number of P_n^m for a triangular truncation that is truncated at wavenumber N is equal to $(N+1)(N+2)/2$ (see Fig. 7.1). This can be obtained as follows: In a triangular truncation with maximum wavenumber N, there are $N+1$ Legendre functions with P_0^0, P_1^0, P_2^0, ..., P_N^0 zonal wavenumbers. Likewise, the number of Legendre functions containing wavenumbers 1, 2, 3, ..., N is N, $N-1$, $N-2$, ..., 1, respectively. Thus on average there are $(N+2)/2$ Legendre functions per zonal wavenumber. The total number of zonal wavenumbers, including wavenumber zero, equals $N+1$. This means that the total number of Legendre functions, or spectral components, in a triangular truncation truncated at wavenumber N is set at $(N+1)(N+2)/2$. Therefore a $T-42$ spectral model (truncated at wavenumber 42) contains $(43\times44)/2=946$ spectral components. Likewise, $T-106$ and $T-170$ models would have 5778 and 14706 spectral components, respectively.

The total number of P_n^m for a rhomboidal truncation $R-N$ is equal to $(N+1)(N+1)$. This can be obtained as follows: In a rhomboidal truncation, there are $N+1$ Legendre functions (P_n^m, P_{n+1}^m, P_{n+2}^m, ..., P_{n+J}^m) for any zonal wavenumber m. The total number of zonal wavenumbers, including wavenumber zero, is $N+1$. Therefore, the total number of Legendre functions or wave components in this truncation is $(N+1)(N+1)$. Thus $R-42$ truncation has $43\times43=1849 P_n^m$.

7.3 Aliasing

Given some data at $2N+1$ points along a line, we can analyze them into a maximum of N-Fourier (N-sine and N-cosine) components. If one tries to calculate more than N components, *a false representation, or aliasing*, of lower wavenumbers into higher wavenumbers occurs. To show this, let

$$y(x)=\sum_{m=0}^{M} a_m e^{imx}, \tag{7.5}$$

where y has values $y_1, y_2, \ldots, y_{2N+1}$ at $2N+1$ equally spaced points $x_1, x_2, \ldots, x_{2N+1}$, with $x_1=0$ and $x_{2N+1}=2\pi$. Multiplying both sides of (7.5) by e^{-imx} and summing over all possible grid points j, we obtain

$$a_m=\frac{1}{2N+1}\sum_{j=1}^{2N+1} y_j e^{-imx_j}=\frac{1}{2N+1}\sum_{j=1}^{2N+1} y_j e^{-im(\pi/N)(j-1)}. \tag{7.6}$$

If $m>N$, then we can write $m=2N-m'$, where $m'<N$,

$$\begin{aligned} a_m &= \sum_{j=1}^{2N+1} y_j e^{-i(2N-m')(\pi/N)(j-1)} \\ &= \sum_{j=1}^{2N+1} y_j e^{im'(\pi/N)(j-1)}, \qquad \text{because } e^{-i2\pi(j-1)}=1 \\ &= a_{m'}, \qquad \text{or} \qquad a_m=a_{m'}. \end{aligned} \tag{7.7}$$

Thus we see that the amplitude a_m is actually reflected as $a_{m'}$, which is the amplitude of the lower wavenumber m', where $m'<N$. Thus if one tries to analyze a data set at $2N+1$ equally spaced points into wavenumbers greater than N, one runs into aliasing problems. The minimum number of points needed to analyze a field into N-Fourier components is $2N+1$. When we analyze the product terms, we need still more points for alias-free calculating, as will be seen in the next section.

7.4 Transform Method

The nonlinear terms in a spectral model truncated at the model's spectral resolution can be calculated by one of the following methods: (a) the interaction coefficients method or (b) the transform method. The *interaction coefficients method* consists of obtaining the spectral amplitudes of the nonlinear products directly from the spectral amplitudes of the individual members of the product terms using relevant trigonometric or Legendre function formulas. This procedure is exact and free from any computational errors. If the number of spectral components in the model is small, say four to six, as is the case in many theoretical studies, this method is very efficient and elegant. However, when dealing with a large number of spectral components, the number of interaction coefficients becomes too large and their storage , bookkeeping, and off and on retrieval becomes very cumbersome. For a large number of wave components, the transform method is simpler and more efficient.

In the *transform method,* the nonlinear terms are calculated in the space domain and then transformed to the spectral domain. The nonlinear terms in an atmospheric model are generally quadratic products of the form $u\partial u/\partial\lambda$, $v\partial u/\partial\mu$, etc., where u and v are the spectrally truncated variables. We are required to spectrally analyze these products truncated at the model's resolution.

To spectrally analyze the product of variables $A(\lambda,\mu)$ and $B(\lambda,\mu)$, the transform method consists of the following three steps, which are repeated at every time step during model integration:

1. Perform a spectral to grid-point transform of the model variables.
   (a) This involves the calculation of the Fourier amplitudes from the spherical harmonic amplitudes at each latitude using the relations

$$A^m(\mu) = \sum_{n=|m|}^{N} a_n^m P_n^m(\mu),$$

$$B^m(\mu) = \sum_{n=|m|}^{N} b_n^m P_n^m(\mu),$$

   where a_n^m and b_n^m are the spherical harmonic amplitudes of the two variables involved in the nonlinear term, while $A^m(\mu)$ and $B^m(\mu)$ are their Fourier amplitudes at a particular latitude μ.

   (b) Calculate the grid-point values from the Fourier components as

$$A(\lambda,\mu) = \sum_{m=-M}^{M} A^m(\mu)e^{im\lambda},$$

$$B(\lambda,\mu) = \sum_{m=-M}^{M} B^m(\mu)e^{im\lambda}.$$

   Processes 1(a) and 1(b) are the inverse Legendre and inverse Fourier transforms, respectively.

2. Perform a calculation of the nonlinear products on grid points. The components $A(\lambda,\mu)$ and $B(\lambda,\mu)$ at each grid point are multiplied to get the nonlinear product $C(\lambda,\mu)$, that is,

$$C(\lambda,\mu) = A(\lambda,\mu)B(\lambda,\mu).$$

   Here, A and B may be the grid point values of variables like u, v, $\partial u/\partial\lambda$, $\partial v/\partial\mu$, and so on.

3. Perform a spectral transform of the nonlinear products at grid points. This process is again achieved in two transform steps:

(a) A Fourier transform of grid-point products on latitude circles.

(b) A Legendre transform of the Fourier components obtained in step 3(a).

Now we consider the requirements for alias-free calculations. For this, let $A(\lambda,\mu)$ and $B(\lambda,\mu)$ be the space values of the two variables spectrally truncated at order (zonal wavenumber) M and degree N of the spherical harmonics series. Then the product $C(\lambda,\mu)$ may be written as

$$C(\lambda,\mu)=\sum_{m_1=-M}^{M}\sum_{n_1=|m_1|}^{N} a_{n_1}^{m_1}Y_{n_1}^{m_1}(\lambda,\mu)\sum_{m_2=-M}^{M}\sum_{n_2=|m_2|}^{N} b_{n_2}^{m_2}Y_{n_2}^{m_2}(\lambda,\mu)$$
$$=\sum_{m_1=-M}^{M}\sum_{m_2=-M}^{M}\sum_{n_1=|m_1|}^{N}\sum_{n_2=|m_2|}^{N} a_{n_1}^{m_1}b_{n_2}^{m_2}P_{n_1}^{m_1}P_{n_2}^{m_2}e^{i(m_1+m_2)\lambda}. \tag{7.8}$$

The two-dimensional spectral amplitude C_n^m of a spherical harmonic wave is given by

$$C_n^m=\frac{1}{4\pi}\int_{-1}^{1}\int_{0}^{2\pi}C(\lambda,\mu)Y_n^{m*}(\lambda,\mu)\,d\lambda\,d\mu$$
$$=\frac{1}{4\pi}\int_{-1}^{1}\left(\int_{0}^{2\pi}C(\lambda,\mu)e^{-im\lambda}d\lambda\right)P_n^m(\mu)\,d\mu.$$

Substituting for $C(\lambda,\mu)$ from (7.8), we get

$$C_n^m=\frac{1}{2}\int_{-1}^{1}\left(\frac{1}{2\pi}\int_{0}^{2\pi}\sum_{m_1=-M}^{M}\sum_{m_2=-M}^{M}\sum_{n_1=|m_1|}^{N}\sum_{n_2=|m_2|}^{N}\right.$$
$$\left.a_{n_1}^{m_1}b_{n_2}^{m_2}P_{n_1}^{m_1}P_{n_2}^{m_2}e^{i(m_1+m_2)\lambda}e^{-im\lambda}d\lambda\right)P_n^m(\mu)\,d\mu. \tag{7.9}$$

The integral with respect to λ on the right-hand side of (7.9) is nonzero only for cases when $m=m_1+m_2$. Thus the zonal wavenumber m of the nonlinear product results from the interaction of those wavenumbers m_1 and m_2 of the individual components of the product for which $m=m_1+m_2$. The integration with respect to λ results in the Fourier analysis of $C(\lambda,\mu)$ along latitude circles. Also, if C^m is the Fourier amplitude of wavenumber m, then

$$C_n^m=\frac{1}{2}\int_{-1}^{1}C^m(\mu)P_n^m(\mu)d\mu. \tag{7.10}$$

Comparing (7.9) and (7.10), we have

$$C_n^m=\frac{1}{2\pi}\int_{0}^{2\pi}\sum_{m_1=-M}^{M}\sum_{m_2=-M}^{M}\sum_{n_1=|m_1|}^{N}\sum_{n_2=|m_2|}^{N} a_{n_1}^{m_1}b_{n_2}^{m_2}P_{n_1}^{m_1}(\mu)P_{n_2}^{m_2}(\mu)$$
$$\times e^{i(m_1+m_2-m)\lambda}d\lambda$$

$$= \sum_{m_1=-M}^{M} \sum_{m_2=-M}^{M} \sum_{n_1=|m_1|}^{N} \sum_{n_2=|m_2|}^{N} a_{n_1}^{m_1} b_{n_2}^{m_2} P_{n_1}^{m_1} P_{n_2}^{m_2} , \quad (7.11)$$

where $m_1 + m_2 = m$. Thus

$$C^m = \sum_{m_1=-M}^{M} \sum_{m_2=-M}^{M} \sum_{n_1=|m_1|}^{N} \sum_{n_2=|m_2|}^{N} \frac{1}{2} a_{n_1}^{m_1} b_{n_2}^{m_2} \int_{-1}^{1} P_{n_1}^{m_1} P_{n_2}^{m_2} P_n^m d\mu . \quad (7.12)$$

This calculation of spectral components C_n^m of the product $C = AB$ involves the evaluation of an integral of the form $\int_{-1}^{1} P_{n_1}^{m_1} P_{n_2}^{m_2} P_n^m d\mu$, where $m_1 + m_2 = m$. During the Legendre transform, it is necessary that this integral be evaluated exactly. We examine the requirements of numerical quadrature to achieve such exact evaluation of the integral.

We may write a normalized Legendre function,

$$P_n^m(\mu) = \frac{1}{2^n n!} \left(\frac{(n-m)!}{(n+m)!} \right)^{1/2} \left(\frac{2n+1}{2} \right)^{1/2} (1-\mu^2)^{m/2} \frac{d^{n+m}(\mu^2-1)^n}{d\mu^{n+m}}$$
$$= \left(1-\mu^2\right)^{m/2} p_{n-m} , \quad (7.13)$$

where p_{n-m} is a polynomial of degree $n-m$. With this, we can write

$$\begin{aligned} P_{n_1}^{m_1}(\mu) P_{n_2}^{m_2}(\mu) P_n^m(\mu) &= (1-\mu^2)^{m_1/2} p_{n_1-m_1} \\ &\times (1-\mu^2)^{m_2/2} p_{n_2-m_2} (1-\mu^2)^{m/2} p_{n-m} \\ &= (1-\mu^2)^{(m_1+m_2+m)/2} p_{n_1-m_1} p_{n_2-m_2} p_{n-m} \\ &= p_{2m} p_{n_1-m_1} p_{n_2-m_2} p_{n-m} . \end{aligned} \quad (7.14)$$

Because $m = m_1 + m_2$,

$$(1-\mu^2)^{(m_1+m_2+m)/2} - (1-\mu^2)^m = p_{2m}$$

is a polynomial of degree $2m$. Furthermore, $p_{n_1-m_1}$, $p_{n_2-m_2}$, and p_{n-m}, are polynomials in μ of degree $n_1 - m_1$, $n_2 - m_2$, and $n - m$, respectively. Consider first the rhomboidal or the parallelogramic truncation of the form

$$\sum_{m=-N}^{N} \sum_{n=|m|}^{|m|+J} .$$

The highest degree of p_{2m} is $2N$ and that of $p_{n_1-m_1}$, $p_{n_2-m_2}$, and p_{n-m} is J.

Therefore, $P_{n_1}^{m_1}(\mu) P_{n_2}^{m_2}(\mu) P_n^m(\mu)$ has the highest degree $2N + 3J$. By using a Gaussian quadrature, we can integrate a polynomial of highest degree $\leq 2k-1$ from its values available at k Gaussian quadrature points. In this case, the highest degree of the triple-product $P_{n_1}^{m_1} P_{n_2}^{m_2} P_n^m$ is $2N + 3J$. Therefore

$$\int_{-1}^{1} P_{n_1}^{m_1}(\mu) P_{n_2}^{m_2}(\mu) P_n^m(\mu) d\mu \quad (7.15)$$

can be evaluated exactly if $2k-1=2N+3J$ or if $k=(2N+3J+1)/2$. In a rhomboidal truncation $J=N$, therefore $k=(5N+1)/2$ is the minimum number of Gaussian latitudes for the Legendre transform.

Consider next the triangular truncation, which has the form

$$\sum_{m=-N}^{N}\sum_{n=|m|}^{N}.$$

For triangular truncation, the product $P_{n_1}^{m_1}(\mu)P_{n_2}^{m_2}(\mu)P_n^m(\mu)$ can be represented by a polynomial of the form (7.14). However, in this case the highest degree of $P_{n_1}^{m_1}(\mu)$, $P_{n_2}^{m_2}(\mu)$, and $P_n^m(\mu)$ is fixed at *N*. The highest degree of the product $P_{n_1}^{m_1}(\mu)P_{n_2}^{m_2}(\mu)P_n^m(\mu)$ in this case may therefore be represented by the polynomial

$$\begin{aligned}(1-\mu^2)^{\frac{m_1+m_2+m}{2}} p_{N-m_1}p_{N-m_2}p_{N-m} &= p_{2m}p_{N-m_1}p_{N-m_2}p_{n-m} \\ &= p_{3N}, \end{aligned} \tag{7.16}$$

since $m=m_1+m_2$. The highest degree of the triple product $P_{n_1}^{m_1}(\mu)P_{n_2}^{m_2}(\mu)P_n^m(\mu)$ in this case is 3*N*, and it can be integrated exactly using Gaussian quadrature with a minimum of *k* Gaussian latitudes, where $2k-1=3N$ or $k=(3N+1)/2$.

For the Fourier transform in an $R-42$ resolution model, the minimum number of grid points in the zonal direction is $3\times 42+1=127$. Because the FFT needs these points to be multiples of 2, 3, or 5, the minimum number of points is taken as 128. The same number of points in the zonal direction are required for $T-42$ truncation. However, the minimum number of Gaussian latitudes for an $R-42$ resolution is 106 [$\geq(5\times 42+1)/2$], and for $T-42$ it is 64 [$\geq(3\times 42+1)/2$].

7.5 The *x*-*y*-σ Coordinate System

In most applications of the hydrodynamic equations, the horizontal coordinate system is either the Cartesian coordinate system or the spherical coordinate system. However, the vertical coordinate system has more possible variations. The most common vertical coordinates are height (*z*) and pressure (*p*). One of the disadvantages of using *z* or *p* as a vertical coordinate is that the lower coordinate surfaces are frequently below the ground in regions of high elevation (Andes, Himalayas, Rockies), which may cause difficulties when incorporating the effect of surface topography into the model equations.

To avoid this, use is made of the sigma (σ) coordinate (Phillips 1957), in which the lowest σ surface follows the earth's surface. The σ vertical coordinate is defined as

$$\sigma=\frac{p}{p_s}, \tag{7.17}$$

where p_s is the surface pressure. Then

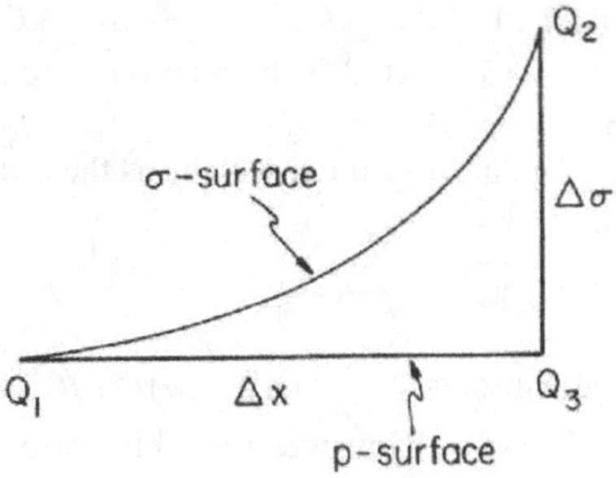

Figure 7.3. Location of points on pressure and sigma surfaces.

$$\sigma = \begin{cases} 1 & \text{at the earth's surface} \\ 0 & \text{at the top of the atmosphere.} \end{cases}$$

In a quasi-static system, σ varies monotonically with height and has the desired properties of a vertical coordinate.

The transform from an x-y-p coordinate system to an x-y-σ coordinate will be presented here. As shown in Fig. 7.3 let Q_1 and Q_2 be the values of variable Q on a σ surface at two points a zonal distance Δx apart from each other. Let Q_3 be the value at the projection of Q_2 on the p surface so that the zonal distance between Q_1 and Q_3 is also Δx. Let the vertical distance between Q_2 and Q_3 in the σ coordinate be $\Delta\sigma$. With this, we may write

$$\frac{Q_3 - Q_1}{\Delta x} = \frac{Q_2 - Q_1}{\Delta x} + \frac{Q_3 - Q_2}{\Delta\sigma}\frac{\Delta\sigma}{\Delta x}. \tag{7.18}$$

Taking the limits $\Delta x \to 0$ and $\Delta\sigma \to 0$, we obtain

$$\left.\frac{\partial Q}{\partial x}\right|_p = \left.\frac{\partial Q}{\partial x}\right|_\sigma + \left.\frac{\partial Q}{\partial \sigma}\frac{\partial \sigma}{\partial x}\right|_p. \tag{7.19}$$

Similarly, one can write an expression for the derivatives with respects to y and t.

However, along vertical we have

$$\frac{\partial Q}{\partial p} = \frac{\partial Q}{\partial \sigma}\frac{\partial \sigma}{\partial p}. \tag{7.20}$$

If we assume that σ is a single-valued function of pressure, then the above expression is justified. In a non-hydrostatic, vigorous, small-scale weather system such as a tornado, pressure may not decrease monotonically with height. Therefore (7.20) will not be applicable in such cases.

The following transformation laws describe the conversion from the p to the σ coordinate system:

$$\left.\frac{\partial Q}{\partial x}\right|_p = \left.\frac{\partial Q}{\partial x}\right|_\sigma + \left.\frac{\partial Q}{\partial \sigma}\frac{\partial \sigma}{\partial x}\right|_p, \tag{7.21}$$

$$\left.\frac{\partial Q}{\partial y}\right|_p = \left.\frac{\partial Q}{\partial y}\right|_\sigma + \left.\frac{\partial Q}{\partial \sigma}\frac{\partial \sigma}{\partial y}\right|_p, \tag{7.22}$$

$$\left.\frac{\partial Q}{\partial t}\right|_p = \left.\frac{\partial Q}{\partial t}\right|_\sigma + \left.\frac{\partial Q}{\partial \sigma}\frac{\partial \sigma}{\partial t}\right|_p, \tag{7.23}$$

$$\frac{\partial Q}{\partial p} = \frac{\partial Q}{\partial \sigma}\frac{\partial \sigma}{\partial p}. \tag{7.24}$$

Now that we have the template for conversion from the p to the σ coordinate system, in order to form a closed system of equations for the full spectral model, we first need to transform the substantial derivative, the pressure gradient term, the hydrostatic equation, and the mass continuity equation from pressure coordinates to sigma coordinates. This is done in the following sections.

7.5.1 *Substantial Time Derivative* (d/dt) *in* σ *Coordinates*

The substantial derivative (i.e., dQ/dt), in the (x, y, p) system is given by

$$\frac{dQ}{dt} = \left.\frac{\partial Q}{\partial t}\right|_p + u\left.\frac{\partial Q}{\partial x}\right|_p + v\left.\frac{\partial Q}{\partial y}\right|_p + \omega\frac{\partial Q}{\partial p}. \tag{7.25}$$

On making use of (7.21) to (7.24), this can be written as

$$\frac{dQ}{dt} = \left.\frac{\partial Q}{\partial t}\right|_\sigma + u\left.\frac{\partial Q}{\partial x}\right|_\sigma + v\left.\frac{\partial Q}{\partial y}\right|_\sigma + \left(\left.\frac{\partial \sigma}{\partial t}\right|_p + u\left.\frac{\partial \sigma}{\partial x}\right|_p + v\left.\frac{\partial \sigma}{\partial y}\right|_p + \omega\frac{\partial \sigma}{\partial p}\right)\frac{\partial Q}{\partial \sigma}.$$

Thus we obtain

$$\frac{dQ}{dt} = \left.\frac{\partial Q}{\partial t}\right|_\sigma + u\left.\frac{\partial Q}{\partial x}\right|_\sigma + v\left.\frac{\partial Q}{\partial y}\right|_\sigma + \frac{\partial Q}{\partial \sigma}\frac{d\sigma}{dt},$$

or

$$\frac{dQ}{dt} = \left.\frac{\partial Q}{\partial t}\right|_\sigma + u\left.\frac{\partial Q}{\partial x}\right|_\sigma + v\left.\frac{\partial Q}{\partial y}\right|_\sigma + \dot{\sigma}\frac{\partial Q}{\partial \sigma}, \tag{7.26}$$

where $\dot{\sigma} = d\sigma/dt$, the vertical velocity in σ coordinates.

7.5.2. *Pressure Gradient Term in* σ *Coordinates*

We start from the terms $-g\partial z/\partial x|_p$ and $-g\partial z/\partial y|_p$ in the pressure coordinate system. Using transformation rule (7.21), we can write

$$-g\frac{\partial z}{\partial x}\bigg|_p = -g\frac{\partial z}{\partial x}\bigg|_\sigma - g\frac{\partial \sigma}{\partial x}\bigg|_p \frac{\partial z}{\partial \sigma}.$$

Using the hydrostatic equation, $\partial p/\partial z = -g\rho = -g(p/RT)$, we can rearrange $g\,\partial z/\partial\sigma$ from the above equation to obtain

$$g\frac{\partial z}{\partial \sigma} = g\frac{\partial z}{\partial p}\frac{\partial p}{\partial \sigma} = -\frac{RT}{p}\frac{\partial p}{\partial \sigma}.$$

Hence

$$-g\frac{\partial z}{\partial x}\bigg|_p = -g\frac{\partial z}{\partial x}\bigg|_\sigma + \frac{RT}{p}\frac{\partial \sigma}{\partial x}\bigg|_p \frac{\partial p}{\partial \sigma}. \tag{7.27}$$

If we let $Q = p$ in (7.21) and noting that $\partial p/\partial x|_p = 0$, we obtain

$$0 = \frac{\partial p}{\partial x}\bigg|_\sigma + \frac{\partial p}{\partial \sigma}\frac{\partial \sigma}{\partial x}\bigg|_p. \tag{7.28}$$

From (7.27) and (7.28) we obtain

$$-g\frac{\partial z}{\partial x}\bigg|_p = -g\frac{\partial z}{\partial x}\bigg|_\sigma - \frac{RT}{p}\frac{\partial p}{\partial x}\bigg|_\sigma. \tag{7.29}$$

Similarly,

$$-g\frac{\partial z}{\partial y}\bigg|_p = -g\frac{\partial z}{\partial y}\bigg|_\sigma - \frac{RT}{p}\frac{\partial p}{\partial y}\bigg|_\sigma. \tag{7.30}$$

Equations (7.29) and (7.30) give the conversion of the horizontal pressure gradient terms from p to σ coordinates.

It is often convenient to introduce the *Exner function*

$$\pi = \left(\frac{p}{p_0}\right)^{R/c_p}, \tag{7.31}$$

where $p_0 = 1000$ mb. From the above definition, by taking the natural logarithm of both sides of (7.31) and differentiating with respect to x it follows that

$$\ln\pi = \ln\left(\frac{p}{p_0}\right)^{R/c_p},$$

$$\ln\pi = \frac{R}{c_p}\ln p - \ln p_0,$$

$$\frac{\partial}{\partial x}\ln\pi = \frac{\partial}{\partial x}\left(\frac{R}{c_p}\ln p - \ln p_0\right),$$

$$\frac{1}{\pi}\frac{\partial \pi}{\partial x}\bigg|_\sigma = \frac{R}{c_p}\frac{\partial}{\partial x}(\ln p - \ln p_0)\bigg|_\sigma,$$

$$\left.\frac{1}{\pi}\frac{\partial \pi}{\partial x}\right|_\sigma = -\frac{R}{pc_p}\left.\frac{\partial p}{\partial x}\right|_\sigma . \tag{7.32}$$

Now (7.29) may be written as

$$-g\left.\frac{\partial z}{\partial x}\right|_p = -g\left.\frac{\partial z}{\partial x}\right|_\sigma - c_p\left.\frac{T}{\pi}\frac{\partial \pi}{\partial x}\right|_\sigma .$$

Furthermore, using $T = \theta\pi$, we obtain

$$-g\left.\frac{\partial z}{\partial x}\right|_p = -g\left.\frac{\partial z}{\partial x}\right|_\sigma - c_p\theta\left.\frac{\partial \pi}{\partial x}\right|_\sigma . \tag{7.33}$$

Similarly,

$$-g\left.\frac{\partial z}{\partial y}\right|_p = -g\left.\frac{\partial z}{\partial y}\right|_\sigma - c_p\theta\left.\frac{\partial \pi}{\partial y}\right|_\sigma . \tag{7.34}$$

It should be noted that the individual terms on the right-hand side of (7.29), (7.30), (7.33), and (7.34), like $-g\partial z/\partial x|_\sigma$ and $-(RT/p)\partial p/\partial x|_\sigma$, are very large in magnitude but have opposite signs. Small errors in the calculation of these terms can lead to larger errors in the estimation of the pressure gradient force, which is a small sum of these two large terms with opposite signs. This problem becomes particularly serious over regions covered by steep mountains.

7.5.3 *Hydrostatic Equation in σ Coordinates*

The hydrostatic equation may be expressed as

$$\frac{\partial \phi}{\partial p} = -\frac{1}{\rho} = -\frac{RT}{p},$$

or

$$\frac{\partial \phi}{\partial \sigma}\frac{\partial \sigma}{\partial p} = -\frac{RT}{p}.$$

Using the relation $\sigma = p/p_s$ and noting that

$$\frac{\partial \sigma}{\partial p} = \frac{\partial}{\partial p}\left(\frac{p}{p_s}\right) = \frac{1}{p_s}\frac{\partial p}{\partial p} = \frac{1}{p_s},$$

we get

$$\frac{p}{p_s}\frac{\partial \phi}{\partial \sigma} = -RT,$$

or

$$\sigma\frac{\partial \phi}{\partial \sigma} = -RT. \tag{7.35}$$

This is the hydrostatic equation in σ coordinates. An alternative form of the hydrostatic equation can be obtained from the Exner function (7.31). We take the natural logarithm of both sides of (7.31) and differentiate with respect to σ to obtain

$$\frac{1}{\pi}\frac{\partial \pi}{\partial \sigma}=\frac{R}{c_p p}\frac{\partial p}{\partial \sigma}. \tag{7.36}$$

Using the hydrostatic equation, $\partial p/\partial z=-g\rho=-g(p/RT)$, we obtain

$$\frac{1}{\pi}\frac{\partial \pi}{\partial \sigma}=-\frac{1}{c_p T}g\frac{\partial z}{\partial \sigma},$$

or

$$g\frac{\partial z}{\partial \sigma}=-c_p\theta\frac{\partial \pi}{\partial \sigma}, \qquad \text{as} \qquad T=\theta\pi. \tag{7.37}$$

This is also an hydrostatic equation in (x, y, σ) coordinates.

7.5.4 *Mass Continuity Equation in σ Coordinates*

The mass continuity equation in the (x, y, p) system is

$$\left.\frac{\partial u}{\partial x}\right|_p+\left.\frac{\partial v}{\partial y}\right|_p+\frac{\partial \omega}{\partial p}=0, \tag{7.38}$$

where $\omega=dp/dt$ is the vertical velocity in the pressure coordinate system. Making use of (7.21) and (7.22), we may write

$$\left.\frac{\partial u}{\partial x}\right|_p=\left.\frac{\partial u}{\partial x}\right|_\sigma+\left.\frac{\partial u}{\partial \sigma}\frac{\partial \sigma}{\partial x}\right|_p, \tag{7.39}$$

$$\left.\frac{\partial v}{\partial y}\right|_p=\left.\frac{\partial v}{\partial y}\right|_\sigma+\left.\frac{\partial v}{\partial \sigma}\frac{\partial \sigma}{\partial y}\right|_p. \tag{7.40}$$

The term $\partial\omega/\partial p$ may also be written as

$$\begin{aligned}\frac{\partial \omega}{\partial p}&=\frac{\partial}{\partial p}\left(\frac{dp}{dt}\right)=\frac{\partial}{\partial \sigma}\left(\frac{dp}{dt}\right)\frac{\partial \sigma}{\partial p}\\ &=\frac{\partial}{\partial \sigma}\left(\frac{\partial p}{\partial t}+u\frac{\partial p}{\partial x}+v\frac{\partial p}{\partial y}+\dot{\sigma}\frac{\partial p}{\partial \sigma}\right)\frac{\partial \sigma}{\partial p}\\ &=\left[\left(\frac{\partial}{\partial t}+u\frac{\partial}{\partial x}+v\frac{\partial}{\partial y}+\dot{\sigma}\frac{\partial}{\partial \sigma}\right)\frac{\partial p}{\partial \sigma}\right.\\ &\qquad\left.+\left.\frac{\partial u}{\partial \sigma}\frac{\partial p}{\partial x}\right|_\sigma+\left.\frac{\partial v}{\partial \sigma}\frac{\partial p}{\partial y}\right|_\sigma+\frac{\partial \dot{\sigma}}{\partial \sigma}\frac{\partial p}{\partial \sigma}\right]\frac{\partial \sigma}{\partial p},\end{aligned} \tag{7.41}$$

or

$$\frac{\partial \omega}{\partial p} = \frac{\partial \sigma}{\partial p}\frac{d}{dt}\left(\frac{\partial p}{\partial \sigma}\right) + \left(\frac{\partial u}{\partial \sigma}\frac{\partial p}{\partial x}\bigg|_{\sigma} + \frac{\partial v}{\partial \sigma}\frac{\partial p}{\partial y}\bigg|_{\sigma} + \frac{\partial \dot{\sigma}}{\partial \sigma}\frac{\partial p}{\partial \sigma}\right)\frac{\partial \sigma}{\partial p}. \tag{7.42}$$

Letting $Q = p$ in (7.21) and (7.22), we obtain

$$\frac{\partial p}{\partial x}\bigg|_{p} = \frac{\partial p}{\partial x}\bigg|_{\sigma} + \frac{\partial p}{\partial \sigma}\frac{\partial \sigma}{\partial x}\bigg|_{p}$$

and

$$\frac{\partial p}{\partial y}\bigg|_{p} = \frac{\partial p}{\partial y}\bigg|_{\sigma} + \frac{\partial p}{\partial \sigma}\frac{\partial \sigma}{\partial y}\bigg|_{p}.$$

Since $\partial p / \partial x\big|_p = \partial p / \partial y\big|_p = 0$, we obtain

$$\frac{\partial p}{\partial x}\bigg|_{\sigma} = -\frac{\partial p}{\partial \sigma}\frac{\partial \sigma}{\partial x}\bigg|_{p}$$

and

$$\frac{\partial p}{\partial y}\bigg|_{\sigma} = -\frac{\partial p}{\partial \sigma}\frac{\partial \sigma}{\partial y}\bigg|_{p}. \tag{7.43}$$

Substituting for $\partial p / \partial x\big|_\sigma$ and $\partial p / \partial y\big|_\sigma$ into (7.42), we obtain

$$\frac{\partial \omega}{\partial p} = \frac{\partial \sigma}{\partial p}\frac{d}{dt}\left(\frac{\partial p}{\partial \sigma}\right) - \frac{\partial p}{\partial \sigma}\left(\frac{\partial \sigma}{\partial x}\bigg|_{p}\frac{\partial u}{\partial \sigma} + \frac{\partial \sigma}{\partial y}\bigg|_{p}\frac{\partial v}{\partial \sigma} - \frac{\partial \dot{\sigma}}{\partial \sigma}\right)\frac{\partial \sigma}{\partial p}. \tag{7.44}$$

From (7.39), (7.40) and (7.44), the continuity equation (7.38) may be written in the (x, y, σ) coordinate system as

$$\frac{\partial u}{\partial x}\bigg|_{\sigma} + \frac{\partial u}{\partial \sigma}\frac{\partial \sigma}{\partial x}\bigg|_{p} + \frac{\partial v}{\partial y}\bigg|_{\sigma} + \frac{\partial v}{\partial \sigma}\frac{\partial \sigma}{\partial y}\bigg|_{p} + \frac{\partial \sigma}{\partial p}\frac{d}{dt}\left(\frac{\partial p}{\partial \sigma}\right)$$

$$-\frac{\partial p}{\partial \sigma}\left(\frac{\partial \sigma}{\partial x}\bigg|_{p}\frac{\partial u}{\partial \sigma} + \frac{\partial \sigma}{\partial y}\bigg|_{p}\frac{\partial v}{\partial \sigma} - \frac{\partial \dot{\sigma}}{\partial \sigma}\right)\frac{\partial \sigma}{\partial p} = 0,$$

or

$$\frac{\partial u}{\partial x}\bigg|_{\sigma} + \frac{\partial v}{\partial y}\bigg|_{\sigma} + \frac{\partial \dot{\sigma}}{\partial \sigma} + \frac{\partial \sigma}{\partial p}\frac{d}{dt}\left(\frac{\partial p}{\partial \sigma}\right) = 0. \tag{7.45}$$

As $\sigma = p / p_s$ and $\partial p / \partial \sigma = p_s$, (7.45) can be written as

$$\frac{1}{p_s}\frac{dp_s}{dt} + \frac{\partial u}{\partial x}\bigg|_{\sigma} + \frac{\partial v}{\partial y}\bigg|_{\sigma} + \frac{\partial \dot{\sigma}}{\partial \sigma} = 0, \tag{7.46}$$

or

$$\frac{d \ln p_s}{dt} + \left.\frac{\partial u}{\partial x}\right|_\sigma + \left.\frac{\partial v}{\partial y}\right|_\sigma + \frac{\partial \dot{\sigma}}{\partial \sigma} = 0. \tag{7.47}$$

This is the mass continuity equation in the (x, y, σ) coordinate system.

Often it is useful to write the continuity equation in flux form. This can be done by multiplying (7.46) by p_s to obtain

$$\frac{dp_s}{dt} + p_s\left(\left.\frac{\partial u}{\partial x}\right|_\sigma + \left.\frac{\partial v}{\partial y}\right|_\sigma + \frac{\partial \dot{\sigma}}{\partial \sigma}\right) = 0,$$

or

$$\frac{\partial p_s}{\partial t} + u\left.\frac{\partial p_s}{\partial x}\right|_\sigma + v\left.\frac{\partial p_s}{\partial y}\right|_\sigma + \dot{\sigma}\frac{\partial p_s}{\partial \sigma} + p_s\left.\frac{\partial u}{\partial x}\right|_\sigma + p_s\left.\frac{\partial v}{\partial y}\right|_\sigma + p_s\frac{\partial \dot{\sigma}}{\partial \sigma} = 0,$$

or

$$\frac{\partial}{\partial t}p_s + \frac{\partial}{\partial x}(up_s) + \frac{\partial}{\partial y}(vp_s) + \frac{\partial}{\partial \sigma}(\dot{\sigma}p_s) = 0. \tag{7.48}$$

This is the flux form of the continuity equation in the (x, y, σ) system.

7.6 A Closed System of Equations in σ Coordinates on a Sphere

In Section 7.5 we transformed the substantial time derivative, the pressure gradient term, and the mass continuity equation from the (x, y, p) to the (x, y, σ) coordinate system. We make use of these to write a closed system of equations for a global model with σ as the vertical coordinate. The appropriate coordinate system for a global domain is the spherical curvilinear coordinates. The position of any point in this coordinate system is given by (λ, θ, r), where λ is the longitude, θ is the latitude, and r is the distance from the center of the earth.

It is convenient to regard the earth as a sphere with radius a, so that $r = a + z$, where z is the vertical height of the point from the earth's surface. With this, we obtain

$$\delta x = r\cos\theta\delta\lambda, \qquad \delta y = r\delta\theta, \qquad \text{and} \quad \delta z = \delta r.$$

Because $z \ll a$, $r \approx a$, so that we may write

$$\delta x = a\cos\theta\delta\lambda, \qquad \delta y = a\delta\theta, \qquad \text{and} \quad \delta z = \delta r. \tag{7.49}$$

If instead of height we take pressure as the vertical coordinate, then the position of any point is given by (λ, θ, p) and the distance increments are

$$\delta x = a\cos\theta\delta\lambda, \qquad \delta y = a\delta\theta, \qquad \text{and} \quad \delta z = -\frac{\delta p}{g\rho}. \tag{7.50}$$

The basic equations governing the atmosphere are the three momentum equations, the mass continuity equation, and the thermodynamic equation. The large-scale atmospheric flow is quasi-horizontal with very small vertical acceleration. For such an atmosphere, the vertical momentum equation is approximated by the hydrostatic assumption:

$$\frac{\partial p}{\partial z} = -g\rho ,$$

where p is the pressure, z is the height, g is the acceleration due to gravity, and ρ is the density of air. With this, the equations for a global model in (λ, θ, p) coordinates are as follows:

- The horizontal momentum equation

$$\frac{\partial \vec{V}}{\partial t} = -\left(\vec{V}\cdot\nabla\right)\vec{V} - \omega\frac{\partial \vec{V}}{\partial p} - f\vec{k}\times\vec{V} - \nabla\phi + \vec{F} . \tag{7.51}$$

- The hydrostatic equation

$$\frac{\partial p}{\partial \phi} = -\rho . \tag{7.52}$$

- The thermodynamic equation

$$\frac{\partial T}{\partial t} = -\vec{V}\cdot\nabla T - \omega\left(\frac{\partial T}{\partial p} - \frac{RT}{c_p p}\right) + \frac{\dot{Q}}{c_p} . \tag{7.53}$$

- The mass continuity equation

$$\nabla\cdot\vec{V} + \frac{\partial \omega}{\partial p} = 0 . \tag{7.54}$$

- The moisture equation

$$\frac{\partial r}{\partial t} = -u\frac{\partial r}{\partial x} - v\frac{\partial r}{\partial y} - \omega\frac{\partial r}{\partial p} + E - P . \tag{7.55}$$

Here $\vec{V} = u\vec{i} + v\vec{j}$ is the horizontal wind vector, t is time, $\phi = gz$ is the geopotential, f is the Coriolis parameter, T is the temperature, ω is the vertical velocity dp/dt, $\vec{F}$ is the frictional force vector, $\dot{Q}$ represents all the diabatic sources and sinks of heat, r is the specific humidity, E is the evaporation, and P is the precipitation.

If we take the vector equation (7.51) as two scalar equations, (7.51) to (7.55) are the six equations in the six dependent variables u, v, ω, ϕ, T, and r (x, λ, θ, and p are independent variables) that form a closed system.

We now write the above set of equations with σ as the vertical coordinate. With σ as the vertical coordinate, we have $\sigma = 0$ at the top $(p = 0)$ and $\sigma = 1$ at the bottom $(p = p_s)$ of the atmosphere. Therefore, $\dot{\sigma} = d\sigma/dt = 0$ at $\sigma = 0$ and $\sigma = 1$. These are the top and bottom boundary conditions for $\dot{\sigma}$. From (7.29) and (7.30), we have

$$-\nabla_p\phi = -\nabla_\sigma\phi - \frac{RT}{p}\nabla_\sigma p = -\nabla_\sigma\phi - RT\nabla_\sigma \ln p , \tag{7.56}$$

where ∇ is a two-dimensional operator. With the fact that $p=\sigma p_s$ and that σ is constant on σ surfaces, (7.56) may be written as

$$-\nabla_p \phi = -\nabla_\sigma \phi - RT\nabla_\sigma \ln p_s . \tag{7.57}$$

Thus the momentum equation in the σ coordinate system becomes

$$\frac{\partial \vec{V}}{\partial t} = -\left(\vec{V}\cdot\nabla\right)\vec{V} - \dot{\sigma}\frac{\partial \vec{V}}{\partial \sigma} - f\vec{k}\times\vec{V} - (\nabla\phi + RT\nabla \ln p_s) + \vec{F} .$$

Furthermore, making use of the identity

$$\left(\vec{V}\cdot\nabla\right)\vec{V} = \nabla\left(\frac{\vec{V}\cdot\vec{V}}{2}\right) + (\nabla\times\vec{V})\times\vec{V} = \nabla\left(\frac{\vec{V}\cdot\vec{V}}{2}\right) + \zeta\vec{k}\times\vec{V} ,$$

the momentum equation takes the form

$$\frac{\partial \vec{V}}{\partial t} = -(\zeta + f)\vec{k}\times\vec{V} - \dot{\sigma}\frac{\partial \vec{V}}{\partial \sigma} - \nabla\left(\phi + \frac{\vec{V}\cdot\vec{V}}{2}\right) + RT\nabla \ln p_s + \vec{F} . \tag{7.58}$$

The hydrostatic equation takes the form

$$\sigma\frac{\partial \phi}{\partial \sigma} = -RT . \tag{7.59}$$

The continuity equation (7.47) is then

$$\frac{\partial \ln p_s}{\partial t} = -\vec{V}\cdot\nabla \ln p_s - \nabla\cdot\vec{V} - \frac{\partial \dot{\sigma}}{\partial \sigma} . \tag{7.60}$$

It should be noted that since $\ln p_s$ is a function of λ and θ only, then $\partial \ln p_s / \partial\sigma = 0$. Thus the vertical advection term $\dot{\sigma}\partial \ln p_s / \partial\sigma$ does not appear in (7.47). The thermodynamic equation, or the first law of thermodynamics, is

$$c_p\frac{dT}{dt} - \frac{1}{\rho}\frac{dp}{dt} = \dot{Q} . \tag{7.61}$$

From the definition of σ, we obtain $dp/dt = \dot{\sigma}p_s + \sigma\dot{p}_s$. Then (7.61) takes the form

$$\begin{aligned}
\frac{\partial T}{\partial t} &= -\vec{V}\cdot\nabla T - \dot{\sigma}\frac{\partial T}{\partial \sigma} + \frac{RT}{c_p p}(\dot{\sigma}p_s + \dot{p}_s\sigma) + H_T \\
&= -\vec{V}\cdot\nabla T - \dot{\sigma}\left(\frac{\partial T}{\partial \sigma} - \frac{RT}{c_p\sigma}\right) + \frac{RT}{c_p p_s}\frac{dp_s}{dt} + H_T \\
&= -\vec{V}\cdot\nabla T - \dot{\sigma}\left(\frac{\partial T}{\partial \sigma} - \frac{RT}{c_p\sigma}\right) + \frac{RT}{c_p}\frac{d\ln p_s}{dt} + H_T .
\end{aligned} \tag{7.62}$$

From (7.62) and (7.60) we get

$$\frac{\partial T}{\partial t} = -\vec{V}\cdot\nabla T + \dot{\sigma}\gamma - \frac{RT}{c_p}\left(\nabla\cdot\vec{V} + \frac{\partial\dot{\sigma}}{\partial\sigma}\right) + H_T, \tag{7.63}$$

where $\gamma = RT/c_p\sigma - \partial T/\partial\sigma$ is the static stability and $H_T = \dot{Q}/c_p$ represents the diabatic heat sources and sinks. The moisture equation is

$$\frac{\partial r}{\partial t} = -\vec{V}\cdot\nabla r - \dot{\sigma}\frac{\partial r}{\partial\sigma} + E - P. \tag{7.64}$$

Thus (7.58), (7.59), (7.60), (7.63), and (7.64) represent a closed system of equations in (λ, θ, σ) coordinates.

For a global spectral model, it is convenient to write the horizontal momentum equation (7.58) as vorticity and divergence equations. If we operate on (7.58) by $\vec{k}\cdot\nabla\times$, we obtain the vorticity equation. Operating on the same equation by $\nabla\cdot$ results in the divergence equation. Furthermore, it is advantageous to write these equations in flux form. With this, the desired closed system of equations takes the following form as given by Daley et al. (1976):

$$\frac{\partial\zeta}{\partial t} = -\nabla\cdot(\zeta+f)\vec{V} - \vec{k}\cdot\nabla\times\left(RT\nabla q + \dot{\sigma}\frac{\partial\vec{V}}{\partial\sigma} - \vec{F}\right), \tag{7.65}$$

$$\frac{\partial D}{\partial t} = \vec{k}\cdot\nabla\times(\zeta+f)\vec{V}$$

$$-\nabla\cdot\left(RT\nabla q + \dot{\sigma}\frac{\partial\vec{V}}{\partial\sigma} - \vec{F}\right) - \nabla^2\left(\phi + \frac{\vec{V}\cdot\vec{V}}{2}\right), \tag{7.66}$$

$$\frac{\partial T}{\partial t} = -\nabla\cdot\left(T\vec{V}\right) + TD + \dot{\sigma}\gamma - \frac{RT}{c_p}\left(D + \frac{\partial\dot{\sigma}}{\partial\sigma}\right) + H_T, \tag{7.67}$$

$$\frac{\partial q}{\partial t} = -\vec{V}\cdot\nabla_q - D - \frac{\partial\dot{\sigma}}{\partial\sigma}, \tag{7.68}$$

$$\sigma\frac{\partial\phi}{\partial\sigma} = -RT, \tag{7.69}$$

$$\frac{\partial r}{\partial t} = -\nabla\cdot\left(r\vec{V}\right) + rD - \dot{\sigma}\frac{\partial r}{\partial\sigma} + E - P. \tag{7.70}$$

A summary of the basic equations can be found in Appendix B.

In the above equations, ζ is the vorticity, D is the divergence, f is the Coriolis parameter, $\vec{V}$ is the horizontal wind vector $u\vec{i} + v\vec{j}$, T is the temperature, $q = \ln p_s$, r is the specific humidity, $\gamma = RT/c_p\sigma - \partial T/\partial\sigma$ is the static stability parameter, $\vec{F}$ is the frictional force, H_T is the diabatic heat sources and sinks, E is the evaporation, and P is the precipitation.

There are different ways to treat the moisture variables in a model. Most of the models treat the moisture in terms of specific humidity, as in (7.70). In the Florida State University Global Spectral Model, however, it is the dew-point depression $T - T_d$ that is used as the moisture variable. We present below a prediction equation for the dew-point depression also.

The dew-point temperature T_d is a function of pressure p and specific humidity r, so that

$$\frac{dT_d}{dt} = \left(\frac{\partial T_d}{\partial p}\right)_r \frac{dp}{dt} + \left(\frac{\partial T_d}{\partial r}\right)_p \frac{dr}{dt}. \tag{7.71}$$

The first term on the right-hand side gives the change in T_d due to changes in pressure, keeping specific humidity constant. This change occurs due to atmospheric dynamics. The second term represents the change in T_d due to changes in specific humidity at constant pressure. This results from external moisture sources or sinks (evaporation and precipitation, etc.) and may be denoted by H_M.

To determine a suitable expression for $\partial T_d / \partial p|_r$, consider the Clausius-Clapeyron equation,

$$\frac{1}{e_s}\frac{\partial e_s}{\partial p} = \frac{\in L(T_d)}{R}\frac{1}{T_d^2}\frac{\partial T_d}{\partial p}, \tag{7.72}$$

where e_s is the saturation vapor pressure, $\in = 0.622$ is the ratio of the molecular weight of water vapor to the molecular weight of dry air, and R is the gas constant for dry air, and L is the latent heat of phase change and is a function of T_d. The specific humidity is defined as

$$r = \frac{\in e_s}{p}. \tag{7.73}$$

Taking the natural logarithm of (7.73) and differentiating the resulting equation with respect to p, we obtain

$$\ln r = \ln\left(\frac{\in e_s}{p}\right) = \left(\ln \in e_s - \ln p\right) = \left(\ln \in + \ln e_s - \ln p\right),$$

$$\frac{\partial}{\partial p}\ln r = \left(\frac{\partial}{\partial p}\ln \in + \frac{\partial}{\partial p}\ln e_s - \frac{\partial}{\partial p}\ln p\right),$$

$$\frac{\partial \ln r}{\partial p} = \frac{1}{e_s}\frac{\partial e_s}{\partial p} - \frac{1}{p}. \tag{7.74}$$

From (7.72) and (7.74) we get

$$\frac{\partial \ln r}{\partial p} = \frac{\in L(T_d)}{R}\frac{\partial T_d}{T_d^2 \partial p} - \frac{1}{p}. \tag{7.75}$$

For constant r,

$$\left.\frac{\partial \ln r}{\partial p}\right|_{r=\text{constant}} = 0.$$

Therefore

$$\frac{\in L(T_d)}{R}\frac{\partial T_d}{\partial p}\frac{1}{T_d^2}-\frac{1}{p}=0,$$

or

$$\left.\frac{\partial T_d}{\partial p}\right|_{r=\text{constant}}=\frac{RT_d^2}{\in pL(T_d)}. \tag{7.76}$$

Thus from (7.71)

$$\frac{dT_d}{dt}=\frac{RT_d^2}{\in pL(T_d)}\omega+H_M. \tag{7.77}$$

Vertical velocity ω is given by

$$\omega=\dot{\sigma}p_s+\sigma\dot{p}_s=\sigma p_s\left(\frac{\dot{\sigma}}{\sigma}+\frac{\dot{p}_s}{p_s}\right)$$

$$=p\left(\frac{\dot{\sigma}}{\sigma}+\frac{d\ln p_s}{dt}\right)=p\left(\frac{\dot{\sigma}}{\sigma}-D-\frac{\partial\dot{\sigma}}{\partial\sigma}\right). \tag{7.78}$$

With this, (7.77) takes the form

$$\frac{\partial T_d}{\partial t}=-\nabla\cdot\left(T_d\vec{V}\right)+T_dD-\dot{\sigma}\frac{\partial T_d}{\partial\sigma}-\frac{RT_d^2}{\in L(T_d)}\left(D+\frac{\partial\dot{\sigma}}{\partial\sigma}-\frac{\dot{\sigma}}{\sigma}\right)+H_M. \tag{7.79}$$

This is a prediction equation for dew-point temperature, T_d. Subtracting (7.79) from (7.67), we obtain a prediction equation for dew-point depression as

$$\frac{\partial S}{\partial t}=-\nabla\cdot\left(\vec{V}S\right)+SD-\dot{\sigma}\frac{\partial S}{\partial\sigma}$$

$$-\left(\frac{RT}{c_p}-\frac{RT_d^2}{\in L(T_d)}\right)\left(D+\frac{\partial\dot{\sigma}}{\partial\sigma}-\frac{\dot{\sigma}}{\sigma}\right)+H_T-H_M, \tag{7.80}$$

where $S=T-T_d$ is the dew-point depression.

We have used $\dot{\sigma}$, the vertical velocity in σ coordinates, in the model equations, but have not discussed the procedure to calculate it. We now derive a diagnostic equation to obtain $\dot{\sigma}$.

We define the vertical integral operators (^) and ($^{\wedge\sigma}$) as

$$\hat{F}=\int_0^1 Fd\sigma \qquad \text{and} \qquad \hat{F}^{\sigma}=\int_\sigma^1 Fd\sigma. \tag{7.81}$$

Integrating the continuity equation (7.68) from $\sigma=0$ to $\sigma=1$, we obtain

$$\int_0^1\frac{\partial q}{\partial t}d\sigma=-\int_0^1\vec{V}\cdot\nabla q\,d\sigma-\int_0^1\nabla\cdot\vec{V}\,d\sigma-\int_0^1\frac{\partial\dot{\sigma}}{\partial\sigma}d\sigma.$$

As $q=\ln p_s$ is not a function σ, and $\dot{\sigma}$ vanishes at $\sigma=0$ and $\sigma=1$, we obtain

$$\frac{\partial q}{\partial t}=-\hat{\vec{V}}\cdot\nabla q-\nabla\cdot\hat{\vec{V}}, \tag{7.82}$$

where $\hat{\vec{V}} = \int_0^1 \vec{V} d\sigma$. If we integrate the continuity equation from $\sigma = \sigma$ to $\sigma = 1$, we obtain

$$\int_\sigma^1 \frac{\partial q}{\partial t} d\sigma = -\int_\sigma^1 \vec{V} \cdot \nabla q \, d\sigma - \int_\sigma^1 \nabla \cdot \vec{V} \, d\sigma - \int_\sigma^1 \frac{\partial \dot{\sigma}}{\partial \sigma} d\sigma ,$$

or

$$(1-\sigma)\frac{\partial q}{\partial t} = -\hat{\vec{V}}^{\sigma} \cdot \nabla q - \nabla \cdot \hat{\vec{V}}^{\sigma} + \dot{\sigma} . \tag{7.83}$$

Eliminating $\partial q / \partial t$ from (7.82) and (7.83), we obtain

$$\dot{\sigma} = (\sigma - 1)\left(\nabla \cdot \hat{\vec{V}} + \hat{\vec{V}} \cdot \nabla q\right) + \left(\nabla \cdot \hat{\vec{V}}^{\sigma} + \hat{\vec{V}}^{\sigma} \cdot \nabla q\right),$$

or

$$\dot{\sigma} = (\sigma - 1)\left(\hat{D} + \hat{\vec{V}} \cdot \nabla q\right) + \left(\hat{D}^{\sigma} + \hat{\vec{V}}^{\sigma} \cdot \nabla q\right). \tag{7.84}$$

This is the diagnostic equation for $\dot{\sigma}$, where $D = \nabla \cdot \vec{V}$.

Before we transform the global model equations into their spectral form, we separate their linear and nonlinear parts. This is necessary in order to integrate the model using a semi-implicit time-integration scheme.

The pressure gradient term is divided into linear and nonlinear parts assuming $T = T^* + T'$, where T^* is the horizontal mean at a level, which is a function of σ only, and T' is the deviation from T^*. Likewise, $\gamma = \gamma^* + \gamma'$. To remove the wind singularity at the poles, we define the Robert functions as

$$U = \frac{u \cos\theta}{a} \quad \text{and} \quad V = \frac{v \cos\theta}{a}. \tag{7.85}$$

We also define an operator α as

$$\alpha(A,B) = \frac{1}{\cos^2\theta}\left(\frac{\partial A}{\partial \lambda} + \cos\theta \frac{\partial B}{\partial \theta}\right). \tag{7.86}$$

We now make use of this operator α, the Robert functions, and the definition of T^* to recast the model equations.

If we now consider the terms on the right-hand side of the vorticity equation (7.65), we have

$$\begin{aligned} -\nabla \cdot (\zeta + f)\vec{V} = -\frac{1}{\cos^2\theta}\Bigg(&\frac{\partial(\zeta + f)U}{\partial \lambda} \\ &+ \cos\theta \frac{\partial(\zeta + f)V}{\partial \theta}\Bigg), \end{aligned} \tag{7.87}$$

$$\begin{aligned} -\hat{k} \cdot \nabla \times (RT\nabla q) &= -\hat{k} \cdot \nabla \times (RT'\nabla q) - \hat{k} \cdot \nabla \times (RT^*\nabla q) \\ &= -\hat{k} \cdot \nabla \times (RT'\nabla q) \end{aligned}$$

$$= -\frac{1}{\cos^2\theta}\left[\frac{\partial}{\partial\lambda}\left(\frac{RT'}{a^2}\cos\theta\frac{\partial q}{\partial\theta}\right)\right.$$

$$\left.-\cos\theta\frac{\partial}{\partial\theta}\left(\frac{RT'}{a^2}\frac{\partial q}{\partial\lambda}\right)\right], \tag{7.88}$$

$$-\hat{\vec{k}}\cdot\nabla\times\left(\dot{\sigma}\frac{\partial\vec{V}}{\partial\sigma}\right) = -\frac{1}{\cos^2\theta}\left[\frac{\partial}{\partial\lambda}\left(\dot{\sigma}\frac{\partial V}{\partial\sigma}\right)\right.$$

$$\left.-\cos\theta\frac{\partial}{\partial\theta}\left(\dot{\sigma}\frac{\partial U}{\partial\sigma}\right)\right], \tag{7.89}$$

$$\hat{\vec{k}}\cdot\nabla\times\vec{F} = \frac{1}{\cos\theta}\frac{\partial}{\partial\lambda}\frac{F_\theta}{a} + \frac{\partial}{\partial\theta}\frac{F_\lambda}{a}. \tag{7.90}$$

Thus using (7.87), (7.88), (7.89) and (7.90), the vorticity equation takes the form

$$\frac{\partial\zeta}{\partial t} = -\frac{1}{\cos^2\theta}\left[\frac{\partial}{\partial\lambda}\left((\zeta+f)U + \dot{\sigma}\frac{\partial V}{\partial\sigma}\right.\right.$$

$$\left. + \frac{RT'}{a^2}\cos\theta\frac{\partial q}{\partial\theta} - \cos\theta\frac{F_\theta}{a}\right) + \cos\theta\frac{\partial}{\partial\theta}\left((\zeta+f)V\right.$$

$$\left.\left. -\dot{\sigma}\frac{\partial U}{\partial\sigma} - \frac{RT'}{a^2}\frac{\partial q}{\partial\lambda} + \cos\theta\frac{F_\lambda}{a}\right)\right]$$

$$= -\frac{1}{\cos^2\theta}\left(\frac{\partial A}{\partial\lambda} + \cos\theta\frac{\partial B}{\partial\theta}\right). \tag{7.91}$$

In a more compact form, we may write the vorticity equation as

$$\frac{\partial\zeta}{\partial t} = -\alpha(A,B). \tag{7.92}$$

Similarly, the divergence equation may be written as

$$\frac{\partial D}{\partial t} + \nabla^2\left(\phi + RT^* q\right) = \alpha(B,-A) - a^2\nabla^2 E. \tag{7.93}$$

The thermodynamic equation may be expressed as

$$\frac{\partial T}{\partial t} - \gamma^*\dot{\sigma} - \frac{RT^*}{c_p}\frac{\partial q}{\partial t} = -\alpha(UT', VT') + B_T. \tag{7.94}$$

The continuity equation can be written as

$$\frac{\partial q}{\partial t} + \hat{G} + \hat{D} = 0. \tag{7.95}$$

The hydrostatic equation is given as

$$\sigma\frac{\partial\phi}{\partial\sigma} = -RT. \tag{7.96}$$

Equations (7.92) to (7.96) along with the equation of state form a closed set of equations for the dry atmosphere. Finally, the moisture (dew-point depression) equation takes the form

$$\frac{\partial S}{\partial t} = -\alpha\left(US, VS\right) + B_S \, . \tag{7.97}$$

In the above set of equations,

$$A = (\zeta + f)U + \dot{\sigma}\frac{\partial V}{\partial \sigma} + \frac{RT'}{a^2}\cos\theta\frac{\partial q}{\partial \theta} - \cos\theta\frac{F_\theta}{a},$$

$$B = (\zeta + f)V - \dot{\sigma}\frac{\partial U}{\partial \sigma} - \frac{RT'}{a^2}\frac{\partial q}{\partial \lambda} + \cos\theta\frac{F_\lambda}{a},$$

$$G = \frac{1}{\cos^2\theta}\left(U\frac{\partial q}{\partial \lambda} + V\cos\theta\frac{\partial q}{\partial \theta}\right),$$

$$E = \frac{U^2 + V^2}{2\cos^2\theta},$$

$$\dot{\sigma} = (\sigma - 1)\left(\hat{G} + \hat{D}\right) + \hat{G}^\sigma + \hat{D}^\sigma,$$

$$B_T = T'D + \gamma'\dot{\sigma} - \frac{RT'}{c_p}\left(\hat{G} + \hat{D}\right) + \frac{RT}{c_p}G + H_T + \gamma^*\left(\hat{G}^\sigma - \hat{G}\right),$$

$$B_S = SD - \dot{\sigma}\frac{\partial S}{\partial \sigma} + \left(\frac{RT}{c_p} - \frac{RT_d^2}{\in L(T_d)}\right)\left(\frac{\dot{\sigma}}{\sigma} + G - \hat{G} - \hat{D}\right) + H_T - H_M \, .$$

We next define a pseudopressure function P as

$$P = \phi + RT^* q \, . \tag{7.98}$$

This function enables the pressure gradient term to be split into linear and nonlinear parts. Differentiating (7.98) by σ, we get

$$\sigma\frac{\partial P}{\partial \sigma} = \sigma\frac{\partial \phi}{\partial \sigma} + \sigma R\frac{\partial T^*}{\partial \sigma}q \, .$$

Making use of the hydrostatic equation $(\sigma\partial\phi/\partial\sigma = -RT)$, we can write

$$\sigma\frac{\partial P}{\partial \sigma} = -RT + \sigma R\frac{\partial T^*}{\partial \sigma}q \, . \tag{7.99}$$

If we take $\partial/\partial t$ of the above equation and substitute for $\partial T/\partial t$ from (7.94), we obtain

$$\frac{\partial}{\partial t}\left(\sigma\frac{\partial P}{\partial \sigma}\right) = -R\left(\gamma^*\dot{\sigma} + \frac{RT^*}{c_p}\frac{\partial q}{\partial t} - \alpha(UT', VT') + B_T\right)$$
$$+ R\sigma\frac{\partial T^*}{\partial \sigma}\frac{\partial q}{\partial t} + R\frac{\partial \sigma}{\partial t}\frac{\partial T^*}{\partial \sigma}q + R\sigma q\frac{\partial}{\partial t}\frac{\partial T^*}{\partial \sigma} \, . \tag{7.100}$$

Since T^* is independent of time, the term $\partial/\partial t(\partial T^*/\partial\sigma)$ is zero. Furthermore, the local time rate of change of σ on a σ surface is zero. Thus

$$\frac{\partial}{\partial t}\left(\sigma\frac{\partial P}{\partial\sigma}\right)=-R\left(\gamma^*\dot{\sigma}+\frac{RT^*}{c_p}\frac{\partial q}{\partial t}-\alpha(UT',VT')+B_T\right)$$

$$+R\sigma\frac{\partial T^*}{\partial\sigma}\frac{\partial q}{\partial t}. \qquad (7.101)$$

From (7.95) and (7.101), we get

$$\frac{\partial}{\partial t}\left(\sigma\frac{\partial P}{\partial\sigma}\right)=-R\left[\gamma^*\dot{\sigma}-\alpha(UT',VT')+B_T\right]$$

$$-R\left(\frac{\partial T^*}{\partial\sigma}-\frac{RT^*}{c_p\sigma}\right)\sigma\left(\hat{G}+\hat{D}\right). \qquad (7.102)$$

We next introduce a *pseudo-vertical velocity function* W by the expression $W=\dot{\sigma}-\sigma(\hat{G}+\hat{D})$, where $\hat{G}$ is the vertically integrated advection of surface pressure and $\hat{D}$ is the vertically integrated divergence. W may also be expressed as

$$W=\dot{\sigma}+\sigma\frac{\partial q}{\partial t}. \qquad (7.103)$$

The surface value of W is given by $W_s=W$ (at $\sigma=1$) $=-(\hat{G}+\hat{D})$. With this and noting that $\gamma^*=(RT^*)/c_p\sigma-\partial T^*/\partial\sigma$, (7.102) becomes

$$\frac{\partial}{\partial t}\left(\sigma\frac{\partial P}{\partial\sigma}\right)+R\gamma^*W=R\alpha(UT',VT')-RB_T. \qquad (7.104)$$

The divergence equation takes the form

$$\frac{\partial D}{\partial t}+\nabla_\sigma{}^2P=\alpha(B,-A)-a^2\nabla_\sigma{}^2E. \qquad (7.105)$$

Furthermore, the vertically integrated continuity equation (7.95) can be expressed as

$$\frac{\partial q}{\partial t}=-\left(\hat{G}+\hat{D}\right)=W_s. \qquad (7.106)$$

Relating the four variables *P, W, D*, and *q*, we have three equations. Thus, we need an additional equation to complete this system. For this purpose, we write the continuity equation in the form

$$\frac{\partial q}{\partial t}=-(D+G)-\frac{\partial\dot{\sigma}}{\partial\sigma} \quad \text{at any level.} \qquad (7.107)$$

Since $W=\dot{\sigma}-\sigma\left(\hat{G}+\hat{D}\right)$, we obtain

$$\frac{\partial W}{\partial\sigma}=\frac{\partial\dot{\sigma}}{\partial\sigma}-(\hat{G}+\hat{D})=\frac{\partial\dot{\sigma}}{\partial\sigma}+\frac{\partial q}{\partial t}. \qquad (7.108)$$

Substituting for $\partial q/\partial t$, we obtain

$$\frac{\partial W}{\partial \sigma} + D = B_W, \tag{7.109}$$

where $B_W = -G$.

We have four equations for the four unknowns *P*, *W*, *D*, and *q*. A summary of the multi-level spectral model equations can be found in Appendix B. We will be using a semi-implicit time-differencing scheme for integrating (7.104) to (7.106) and (7.109). It is customary to treat the vorticity and moisture equations explicitly. Along with the other model equations, we transform the above four equations to spectral form, and then we combine them to form a single spectral equation for semi-implicit integration.

7.7 Spectral Form of the Primitive Equations

The model equations are transformed into their spectral form similar to Daley et al. (1976) by (a) writing each of the model variables as a truncated series of spherical harmonics and (b) multiplying both sides of the equations by the complex conjugate of the spherical harmonics and integrating, making use of the orthogonality relationship

$$\int_{-1}^{1}\int_{0}^{2\pi} Y_{l_1}^{m_1} Y_{l_2}^{m_2 *} d\lambda du = \begin{cases} 1 & \text{if } l_1 = l_2 \text{ and } m_1 = m_2 \\ 0 & \text{if } l_1 \neq l_2 \text{ and/or } m_1 \neq m_2 \end{cases}.$$

This enables us to obtain the spectral form of the equations for each of the spherical harmonic amplitudes. The spectral amplitudes of the nonlinear parts of the equations are, however, obtained by the transform method.

The spectral amplitude of the horizontal component of wind in terms of Robert functions is obtained from the relations

$$U_l^m = \frac{1}{a^2}\left[im\chi_l^m + (l-1)\in_l^m \psi_{l-1}^m - (l+2)\in_{l+1}^m \psi_{l+1}^m\right] \tag{7.110}$$

and

$$V_l^m = \frac{1}{a^2}\left[im\psi_l^m - (l-1)\in_l^m \chi_{l-1}^m + (l+2)\in_{l+1}^m \chi_{l+1}^m\right], \tag{7.111}$$

where

$$\in_l^m = \left(\frac{l^2 - m^2}{4l^2 - 1}\right)^{1/2}.$$

There are also relationships between U_l^m, V_l^m, ζ_l^m, and D_l^m, where ζ_l^m and D_l^m are coefficients of ζ and *D*, respectively. Namely,

$$l(l+1)U_l^m = -imD_l^m - (l+1)\in_l^m \zeta_{l-1}^m + l\in_{l+1}^m \zeta_{l+1}^m, \tag{7.112}$$

$$l(l+1)V_l^m = (l+1)\in_l^m D_{l-1}^m - l\in_{l+1}^m D_{l+1}^m - im\zeta_l^m. \tag{7.113}$$

Following the above procedure, we get the spectral form of the equations:

Vorticity equation:

$$\frac{\partial \zeta_l^m}{\partial t} = -\left[\alpha(A,B)\right]_l^m. \tag{7.114}$$

Divergence equation:

$$\frac{\partial D_l^m}{\partial t} - a^{-2} l(l+1) P_l^m = \left[\alpha(B,-A)\right]_l^m + l(l+1) E_l^m . \tag{7.115}$$

Thermodynamic equation:

$$\sigma \frac{\partial^2 P_l^m}{\partial t \partial \sigma} + R\gamma^* W_l^m = R[\alpha(UT',VT') - B_T]_l^m . \tag{7.116}$$

Hydrostatic equation:

$$\sigma \frac{\partial \phi_l^m}{\partial \sigma} = -RT_l^m . \tag{7.117}$$

Pseudo-vertical velocity equation:

$$\frac{\partial W_l^m}{\partial \sigma} + D_l^m = \left(B_W\right)_l^m . \tag{7.118}$$

Mass continuity equation:

$$\frac{\partial q_l^m}{\partial t} = \left(W_s\right)_l^m . \tag{7.119}$$

Moisture (dew-point depression) equation:

$$\frac{\partial S_l^m}{\partial t} = \left[-\alpha\left(US,VS\right) + B_S\right]_l^m . \tag{7.120}$$

The left-hand side of the above equations contain the linear parts, while the nonlinear parts are put on the right-hand side. This is to facilitate their integration using the semi-implicit time-integration scheme. Note that in (7.116), P_l^m is a spectral component of $P = \phi + RT^* q$. It should not be confused with the Legendre polynomial $P_n^m(\mu)$.

Semi-implicit Time-Differencing Scheme. The primitive equation model has both slow-moving atmospheric waves and fast-moving gravity waves as its solution. Gravity waves have phase speeds of the order of 300 ms^{-1}, which is more than an order of magnitude higher than the phase speeds of atmospheric waves. A centered time-differencing scheme used to accommodate the gravity waves will need very small time steps to avoid violating the CFL criterion. This makes the integration of a primitive equation model computationally very expensive, particularly for a high-resolution model.

To overcome this problem, the gravity wave part of the primitive equation solution is integrated using an implicit time-integration scheme, which is an unconditionally stable scheme at time steps much larger than those permitted by the CFL criterion. The atmospheric wave solution is integrated using an explicit time-differencing scheme. Implicit integration is used only for the linear parts of the equations. The nonlinear parts are integrated explicitly. This time-differencing scheme, where part of the equations are integrated explicitly and part implicitly, is called a semi-implicit time-differencing scheme. The time step needed for this scheme will be that needed by the fastest-moving atmospheric waves to satisfy the CFL criterion.

The gravity waves are excited in the divergent part of flow by the force of gravity. The divergence equation, the thermodynamic equation, and the mass continuity equation contain the gravity wave solution as well as the atmospheric wave solution. They will be integrated using the semi-implicit time-integration scheme. The vorticity equation and the moisture equation are integrated explicitly.

To develop semi-implicit time differencing, we define

$$\overline{F}^t = \frac{F(t+\Delta t)+F(t-\Delta t)}{2}, \tag{7.121}$$

so that

$$\frac{\partial F}{\partial t} = \frac{F(t+\Delta t)-F(t-\Delta t)}{2\Delta t} = \frac{\overline{F}^t - F(t-\Delta t)}{\Delta t}. \tag{7.122}$$

The terms in the model equations that are treated explicitly are calculated at time t, while those that are treated implicitly are calculated as means at time $t+\Delta t$ and $t-\Delta t$.

With this, the spectral forms of the divergence equation may be written as

$$\frac{\overline{D_l^m}^t - D_l^m(t-\Delta t)}{\Delta t} - a^{-2}l(l+1)\overline{P_l^m}^t = [\alpha(B,-A)]_l^m + l(l+1)E_l^m,$$

or

$$\overline{D_l^m}^t - a^{-2}\Delta t l(l+1)\overline{P_l^m}^t \tag{7.123}$$
$$= \Delta t\{[\alpha(B,-A)]_l^m + l(l+1)E_l^m\} + D_l^m(t-\Delta t).$$

The thermodynamic equation is written as

$$\frac{\sigma\frac{\partial \overline{P_l^m}^t}{\partial\sigma} - \sigma\frac{\partial P_l^m(t-\Delta t)}{\partial\sigma}}{\Delta t} + R\gamma^*\overline{W_l^m}^t = R[\alpha(UT', VT') - B_T]_l^m,$$

or

$$\sigma\frac{\partial \overline{P_l^m}^t}{\partial\sigma} + \Delta t R\gamma^*\overline{W_l^m}^t = \Delta t R[\alpha(UT', VT') - B_T]_l^m,$$
$$+\sigma\frac{\partial P_l^m(t-\Delta t)}{\partial\sigma}. \tag{7.124}$$

The equation for pseudo-vertical velocity is written as

$$\frac{\partial \overline{W_l^m}^t}{\partial\sigma} + \overline{D_l^m}^t = \left(B_W\right)_l^m, \tag{7.125}$$

and the mass continuity equation becomes

$$\frac{\overline{q_l^m}^t - q_l^m(t-\Delta t)}{\Delta t} = \left(W_s\right)_l^m,$$

or

$$\overline{q_l^m}^t = \Delta t\left(W_s\right)_l^m + q_l^m(t-\Delta t). \tag{7.126}$$

After eliminating $\overline{D_l^m}^t$ between the divergence equation (7.123) and the pseudo-vertical velocity equation (7.125), we get

$$-\frac{\partial \overline{W_l^m}^t}{\partial \sigma} - a^2 \Delta t l(l+1)\overline{P_l^m}^t \tag{7.127}$$
$$= \Delta t\left\{\left[\alpha\left(B,-A\right)\right]_l^m + l(l+1)E_l^m\right\} - \left(B_W\right)_l^m + D_l^m\left(t-\Delta t\right).$$

Now eliminating $\overline{W_l^m}^t$ from the above equation and the thermodynamic equation (7.124), we get

$$\frac{\partial}{\partial \sigma}\left(\frac{\sigma}{\gamma^*}\frac{\partial}{\partial \sigma}\overline{P_l^m}^t\right) - a^{-2}R\Delta t^2 l(l+1)\overline{P_l^m}^t$$
$$= \frac{\partial}{\partial \sigma}\left(\frac{\Delta t R}{\gamma^*}[\alpha(UT',VT') - B_T]_l^m + \frac{\sigma}{\gamma^*}\frac{\partial P_l^m(t-\Delta t)}{\partial \sigma}\right)$$
$$+R\Delta t\left\{\Delta t\left[\alpha\left(B,-A\right) + l\left(l+1\right)E\right]_l^m\right.$$
$$\left.-\left(B_W\right)_l^m + D_l^m\left(t-\Delta t\right)\right\},$$

which may be written as

$$\frac{\partial}{\partial \sigma}\left(\frac{\sigma}{\gamma^*}\frac{\partial}{\partial \sigma}\overline{P_l^m}^t\right) - a^{-2}R\Delta t^2 l(l+1)\overline{P_l^m}^t = \frac{\partial}{\partial \sigma}\left(\frac{C_T}{\gamma^*}\right)_l^m + (C_D)_l^m, \tag{7.128}$$

where

$$\left(C_T\right)_l^m = R\Delta t\left[\alpha\left(UT',VT'\right) - B_T\right]_l^m + \sigma\frac{\partial P_l^m(t-\Delta t)}{\partial \sigma}$$

and

$$\left(C_D\right)_l^m = R\Delta t\left\{\Delta t\left[\alpha\left(B,-A\right) + l\left(l+1\right)E\right]_l^m - \left(B_W\right)_l^m + D_l^m\left(t-\Delta t\right)\right\}.$$

Equation (7.128) is a Helmholtz-type second-order differential equation in $\overline{P_l^m}^t$. The term on the right-hand side of this equation is a function of time t or $t-\Delta t$ and can be calculated via the transform method.

One can solve this equation using a finite-difference analog. For this, consider the nth level of an N-level model (i.e., $1 \le n \le N$). Let σ_n be the sigma value at the layer interface and let $\tilde{\sigma}_n$ be the sigma value in the center of the layer, where $\tilde{\sigma}_n = (\sigma_n \sigma_{n+1})^{1/2}$. Also, the logarithmic spacing between two adjacent levels is defined as

$$d_n = \ln \sigma_{n+1} - \ln \sigma_n = \ln\left(\frac{\sigma_{n+1}}{\sigma_n}\right), \tag{7.129}$$

and the vertical spacing between adjacent $\tilde{\sigma}$ levels is $\tilde{\delta}_n = \tilde{\sigma}_{n+1} - \tilde{\sigma}_n$ for $1 \le n \le N-2$. Considering the top of the model as the top of the atmosphere (where $\sigma = 0$) and the

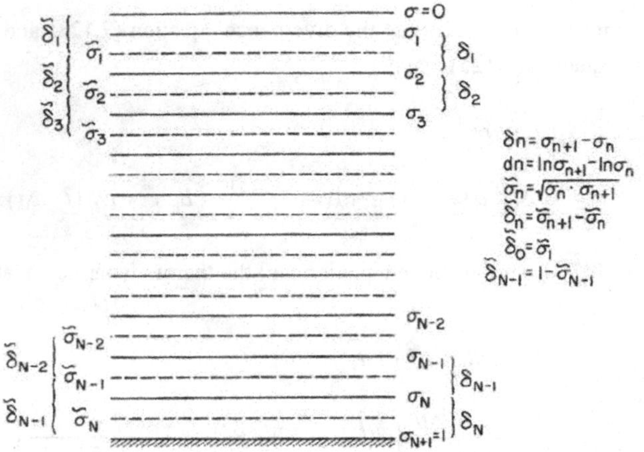

Figure 7.4. Vertical structure of the model.

bottom of the model as the earth's surface (where $\sigma = 1$), we get $\tilde{\delta}_0 = \tilde{\sigma}_1$ and $\tilde{\delta}_{N-1} = 1 - \tilde{\sigma}_{N-1}$. Figure 7.4 shows the vertical structure of the model.

One can then write the finite-difference analog of (7.128) as

$$\frac{1}{\tilde{\delta}_{n-1}}\left(\frac{\overline{P}_{n+1}^t - \overline{P}_n^t}{\tilde{\gamma}_n^* d_n} - \frac{\overline{P}_n^t - \overline{P}_{n-1}^t}{\tilde{\gamma}_{n-1}^* d_{n-1}}\right) - \frac{R\Delta t^2}{a^2} l(l+1)\overline{P}_n^t \tag{7.130}$$

$$= \frac{1}{\tilde{\delta}_{n-1}}\left(\frac{1}{\tilde{\gamma}_n^*}(\tilde{C}_T)_n - \frac{1}{\tilde{\gamma}_{n-1}^*}(\tilde{C}_T)_{n-1}\right) + (C_D)_n,$$

where the tilde (˜) indicates the value in the middle of the layer. For simplicity, we have dropped the wavenumber index *m*, *l* in the above equation.

Equation (7.130) leads to $N-2$ algebraic equations for the *N* unknowns $\overline{P}_n^t$. Two equations can be obtained by using the boundary conditions; thus this leads to a closed set of *N* algebraic equations for the *N* unknowns. At the top boundary ($\sigma = 0$), $\dot{\sigma} = 0$ and thus $W = 0$. Then from (7.125) we get

$$\frac{\overline{W}_1^t - 0}{\tilde{\delta}_0} + \overline{D}^t = (B_W)_1,$$

or

$$\frac{\overline{W}_1^t}{\tilde{\delta}_0} + \overline{D}^t = (B_W)_1 \qquad \text{at} \qquad \sigma = \sigma_1.$$

Substituting the above value of $\overline{W}_1^t$ into the finite-difference analog of (7.124), we get

$$\frac{1}{\tilde{\delta}_0}\left(\frac{\overline{P}_2^t-\overline{P}_1^t}{\tilde{\gamma}_1^* d_1}\right)+R\Delta t\left((B_W)_1-\overline{D}_1^t\right)=\frac{1}{\tilde{\delta}_0}\frac{(C_T)_1}{\tilde{\gamma}_1^*}. \tag{7.131}$$

Eliminating $\overline{D}_1^t$ between (7.131) and (7.123) at level one gives

$$\frac{1}{\tilde{\delta}_0}\left(\frac{\overline{P}_2^t-\overline{P}_1^t}{\tilde{\gamma}_1^* d_1}\right)-\frac{R\Delta t^2}{a^2}l(l+1)\overline{P}_1^t=\frac{(\tilde{C}_T)_1}{\tilde{\delta}_0\tilde{\gamma}_1^*}+(C_D)_1. \tag{7.132}$$

At the lower boundary, the finite-difference analog of (7.125) becomes

$$\frac{\overline{P}_{N+1}^t-\overline{P}_N^t}{d_N}+R\Delta t\,\tilde{\gamma}_N^*\overline{W}_N^t=\left(\tilde{C}_T\right)_N.$$

As $\overline{P}_{N+1}^t=\overline{P}_s^t=\phi_s+RT_s^*\overline{q}^t$, the above equation may be written as

$$-\frac{\overline{P}_N^t}{d_N}+\frac{\phi_s}{d_N}+\frac{RT_s^*}{d_N}\overline{q}^t+R\Delta t\,\tilde{\gamma}_N^*\overline{W}_N^t=\left(\tilde{C}_T\right)_N. \tag{7.133}$$

Assuming that

$$\tilde{\gamma}_{N-1}^*=\tilde{\gamma}_N^*=\frac{1}{\sigma_N}\left(\frac{R\tilde{T}_N^*}{c_p}-\frac{T_{N+1}^*-T_N^*}{d_N}\right),$$

we get

$$T_s^*=T_{N+1}^*=d_N\left(\frac{R}{c_p}\tilde{T}_N^*-\sigma_N\tilde{\gamma}_{N-1}^*\right). \tag{7.134}$$

Also, from the definition of pseudo-vertical velocity in layers $N-1$ and N, we get

$$\overline{\tilde{W}}_{N-1}^t=\tilde{\dot{\sigma}}_{N-1}+\tilde{\sigma}_{N-1}\overline{W}_s^t \tag{7.135}$$

and

$$\overline{\tilde{W}}_N^t=\tilde{\dot{\sigma}}_N+\tilde{\sigma}_N\overline{W}_s^t. \tag{7.136}$$

Combining (7.135) and (7.136) gives

$$\overline{\tilde{W}}_N^t=h\overline{\tilde{W}}_{N-1}^t-(h\tilde{\sigma}_{N-1}-\tilde{\sigma}_N)\overline{W}_s^t, \tag{7.137}$$

where $h=\tilde{\dot{\sigma}}_N/\tilde{\dot{\sigma}}_{N-1}$.

Substituting for T_s^* from (7.133) and $\overline{\tilde{W}}_N^t$ from (7.136) into (7.132) and with some further simplification, we get the finite-difference analog of the thermodynamic equation at level $\tilde{\sigma}_N$ as

$$\frac{-\overline{P}_N^t}{\tilde{\lambda}_N^*}d_N+\frac{R\Delta T}{\tilde{\lambda}_N^*}h\tilde{\gamma}_{N-1}^*\overline{\tilde{W}}_{N-1}^t+R\Delta T\overline{W}_s^t=\frac{\left(\tilde{C}_T\right)_N}{\tilde{\lambda}_N^*}, \tag{7.138}$$

where

$$\tilde{\lambda}_N^* = \frac{R\tilde{T}_N^*}{c_p} + \frac{T_N^*}{d_N} - h\tilde{\gamma}_{N-1}^*\tilde{\sigma}_{N-1}.$$

Here $0 \le h \le 1$ and $\tilde{\tilde{\sigma}}_N = h\tilde{\tilde{\sigma}}_{N-1}$.

A relation can be obtained between $\overline{W}_s^t$, $\overline{\tilde{W}}_{N-1}^t$, and $\overline{P}_N^t$ using finite-difference analogs of the divergence equation and the pseudo-vertical-velocity equation applied to the σ_N level. Furthermore, a relation between $\overline{P}_{N-1}^t$, $\overline{P}_N^t$, and $\overline{\tilde{W}}_{N-1}^t$ can be obtained by the finite-difference analog of the thermodynamic equation at the σ_{N-1} level.

Thus we have four unknowns, namely $\overline{P}_{N-1}^t$, $\overline{P}_N^t$, $\overline{\tilde{W}}_{N-1}^t$, and $\overline{W}_s^t$, and three equations. These three equations can be simplified to eliminate $\overline{\tilde{W}}_{N-1}^t$ and $\overline{W}_s^t$ to obtain a single equation involving $\overline{P}_{N-1}^t$ and $\overline{P}_N^t$, that is,

$$\frac{1}{\tilde{\delta}_{N-1}}\left(\frac{-\overline{P}_N^t}{\tilde{\lambda}_N^* d_N} - \frac{\overline{P}_N^t - \overline{P}_{N-1}^t}{\tilde{\lambda}_{N-1}^* d_{N-1}}\right) - \frac{R\Delta t^2 l(l+1)\overline{P}_N^t}{a^2} \tag{7.139}$$

$$= \frac{1}{\tilde{\delta}_{N-1}}\left(\frac{1}{\tilde{\lambda}_N^*}\left(\tilde{C}_T\right)_N - \frac{1}{\tilde{\lambda}_{N-1}^*}\left(\tilde{C}_T\right)_{N-1}\right) + \left(\tilde{C}_D\right)_N,$$

where

$$\tilde{\lambda}_{N-1}^* = \frac{\tilde{\gamma}_{N-1}^*\tilde{\lambda}_N^*}{\tilde{\lambda}_N^* + h\tilde{\gamma}_{N-1}^*}.$$

Finally, (7.130), (7.132), and (7.139) define a system of N equations in N unknowns. The coefficient matrix is an $N \times N$ tridiagonal matrix and the unknowns are $\overline{P}_n^t$ for $1 \le n \le N$. Once $\overline{P}_n^t$ is obtained, one can find $\overline{D}_n^t$, $\overline{\tilde{W}}_n^t$, $\overline{\tilde{W}}_{N-1}^t$, and $\overline{W}_s^t$. Furthermore, q can be obtained from $\overline{W}_s^t$ using the mass continuity equation. A summary of the spectral equations can be found in Appendix B.

7.8 Examples

We show here some examples of the quality of forecasts produced by the multilevel spectral model. Overall performance of the model is determined by anomaly correlation. Also we demonstrate the prediction of hurricanes by such models.

Anomaly Correlation. We start with an anomaly which is defined as $\nabla Q(t) = Q(t) - Q_n$, where $Q(t)$ is a variable and Q_n is its climatological mean value. The anomaly correlation, as shown in Figure 7.5, is defined by

$$A = \frac{1}{N}\sum_{n=1}^{N}\left(\frac{\left[\Delta Q_p(t) - \overline{\Delta Q_p(t)}\right]\left[\Delta Q_o(t) - \overline{\Delta Q_o(t)}\right]}{\left\{\left[\Delta Q_p(t) - \overline{\Delta Q_p(t)}\right]^2\right\}^{1/2}\left\{\left[\Delta Q_o(t) - \overline{\Delta Q_o(t)}\right]^2\right\}^{1/2}}\right).$$

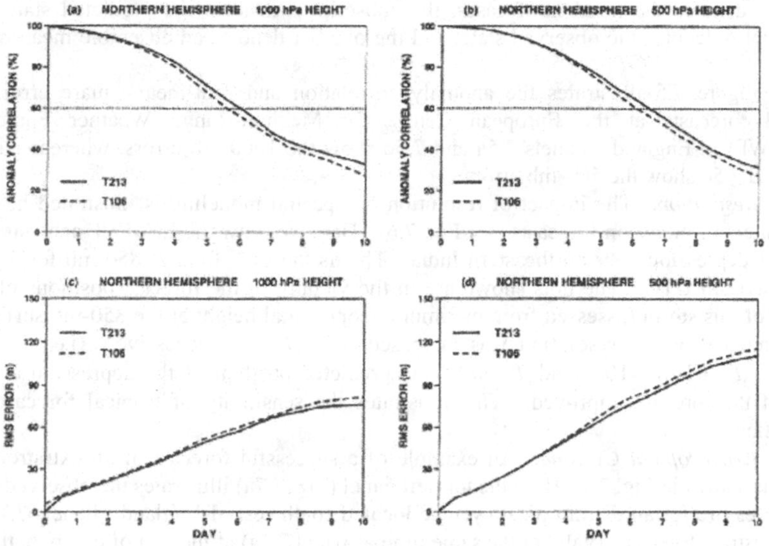

Figure 7.5. (a) Geopotential height anomaly correlation at (a) 1000 mb and (b) 500 mb and root-mean-square error at (c) 1000 mb and (d) 500 mb for the Northern Hemisphere at model resolutions *T*-213 and *T*-106. Source: ECMWF.

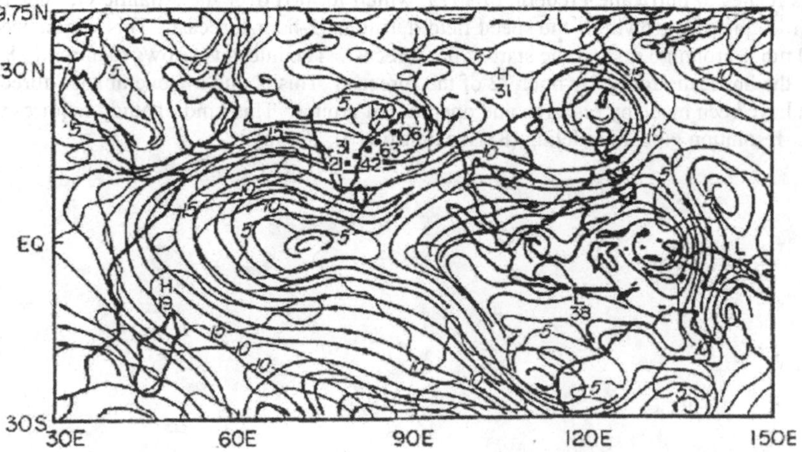

Figure 7.6. Predicted positions of a monsoon depression on day five of the forecast are shown by dots. Each dot is the forecast at a particular horizontal resolution ($T-21$, $T-31$, $T-42$, $T-63$, $T-106$, and $T-170$) of the global spectral model. The flow field shows the verification chart on day five, based on ECMWF analysis.

Here N denotes the number of cases, the subscript p denotes the predicted state, the subscript o denotes the observed state, and the overbar denotes an ensemble mean of all cases.

Figure 7.5 illustrates the anomaly correlation and root-mean-square errors of typical forecasts at the European Centre for Medium-Range Weather Forecasts (ECMWF) in England. Panels 7.5a and 7.5c show the 1000-mb errors, whereas panels 7.5b and 7.5d show the 500-mb errors.

Resolution. The impact of resolution on spectral modeling is illustrated here in the context of monsoon forecasts (see Fig. 7.6). Here we show the ECMWF analysis of a tropical depression over northeastern India. This is the flow field at 850 mb for 7 July 1979 at 1200 UTC. The dots shown are in the vicinity of the forecast positions of the center of this storm (assessed from minimum geopotential height at the 850-mb surface). We find that as the resolution was increased from $T-21$ successively (i.e., $T-31$, $T-42$, $T-63$, $T-106$, and $T-170$), the predicted position of the depression at day five of the forecast improved. This illustrates the sensitivity of tropical forecasts to resolution.

Extratropical Cyclone. An example of a successful forecast of an extratropical storm is shown in Fig. 7.7. Here the top-left panel (Fig. 7.7a) illustrates the observed sea-level pressure for an extratropical cyclone located southwest of England. Panels 7.7b to 7.7f illustrate forecasts (valid at the same time as panel 7.7a) at the end of day two, three, four, five, and six. These forecasts were experimental and based on a new advancement in the analysis procedure called *four-dimensional assimilation* (Rabier et al. 1993). It is interesting to note that all of these forecasts were successful in predicting the intense cyclone over southwestern Europe through day six of the forecasts.

Hurricane Forecast. An example of hurricane formation is illustrated in Fig. 7.8. This relates to hurricane Frederic of 1979, which formed over the Atlantic Ocean. Here we illustrate the growth of the speed field during a 72-hour forecast. This wind speed at 850 mb is around 6 ms^{-1} at the start of the forecast. The intensity grows to around 38 ms^{-1} as the hurricane forms by hour 72 of the forecast. This is one of several such forecasts that have been made by research and operational groups. This kind of study requires very high-resolution global modeling (Krishnamurti et al. 1994a).

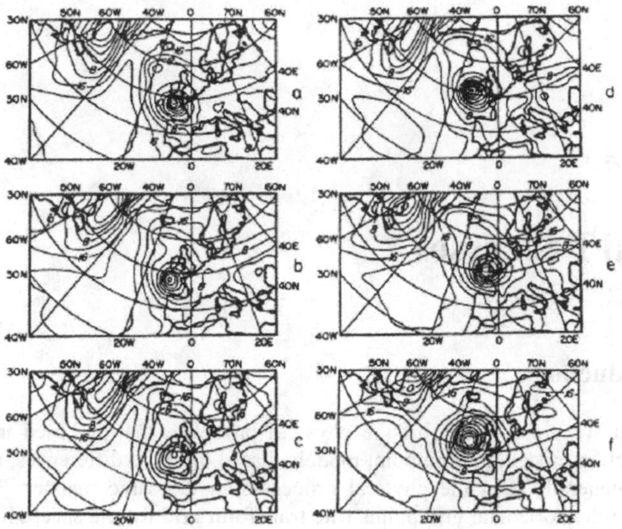

Figure 7.7. An example of the prediction of an extratropical cyclone. (a) Observed field of surface pressure (mb). Days two (b), three (c), four (d), five (e), and six (f) of an ECMWF model prediction are valid at the same time as (a).

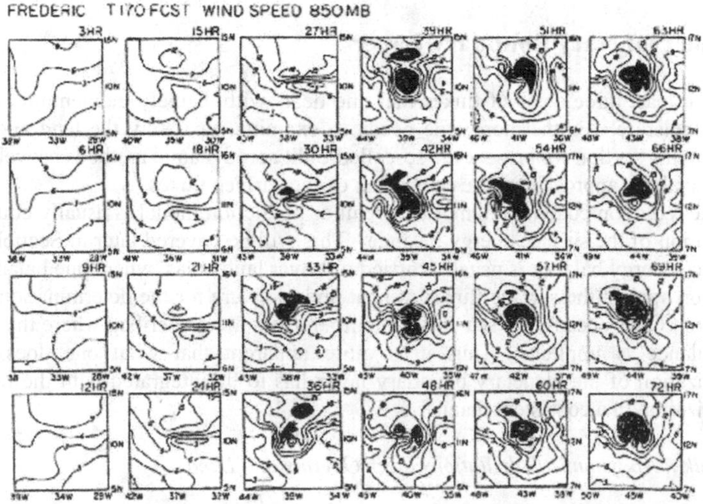

Figure 7.8. Forecasts of the wind speed (ms^{-1}) at 850 mb made with a global spectral model for hurricane Frederic of 1979 over the Atlantic Ocean. The three-hourly fields cover the period between hours 3 and 72 of a high-resolution ($T-170$) forecast.

Chapter 8

Physical Processes

8.1 Introduction

In this chapter we present some of the physical processes that are used in numerical weather prediction modeling. Grid-point models, based on finite differences, and spectral models both generally treat the physical processes in the same manner. The vertical columns above the horizontal grid points (the transform grid for the spectral models) are the ones along which estimates of the effects of the physical processes are made. In this chapter we present a treatment of the planetary boundary layer, including a discussion on the surface similarity theory. Also covered is the cumulus parameterization problem in terms of the Kuo scheme and the Arakawa Schubert scheme. Large-scale condensation and radiative transfer in clear and cloudy skies are the final topics reviewed.

8.2 The Planetary Boundary Layer

There are at least three types of fluxes that one deals with, namely momentum, sensible heat, and moisture. Furthermore, one needs to examine separately the land and ocean regions. In this section we present the so-called bulk aerodynamic methods as well as the similarity analysis approach for the estimation of the surface fluxes.

The radiation code in a numerical weather prediction model is usually coupled to the calculation of the surface energy balance. This will be covered later in Section 8.5.6. This surface energy balance is usually carried out over land areas, where one balances the net radiation against the surface fluxes of heat and moisture for the determination of soil temperature. Over oceans, the sea-surface temperatures are prescribed where the surface energy balance is implicit. Thus it is quite apparent that what one does in the parameterization of the planetary boundary layer has to be integrated with the radiative parameterization in a consistent manner.

8.2.1 *Bulk Aerodynamic Calculations Over Oceans and Land*

Here the surface fluxes of sensible heat, water vapor, and momentum are expressed by relations of the type

$$F_H = \rho C_p C_H |V_a| (T_w - T_a) \qquad \text{sensible heat,}$$

$$F_q = \rho L_q C_q \left|V_a\right|(q_w - q_a) \qquad \text{latent heat,} \qquad (8.1)$$
$$F_M = \rho C_D \left|V_a\right|\left|V_a\right| \qquad \text{momentum.}$$

V_a, T_a, and q_a are the wind speed, the air temperature, and the specific humidity, respectively, at the anemometer level. T_w and q_w denote the surface temperature and the saturation specific humidity at the surface. The currently accepted values of the dimensionless exchange coefficients are

$$C_H = 1.4\times10^{-3}, \quad C_q = 1.6\times10^{-3}, \quad \text{and} \quad C_D = 1.1\times10^{-3}.$$

These were determined experimentally during the Global Atmospheric Research Program (GARP) Atlantic Tropical Experiment (GATE).

The choice of units is quite important for the calculation of these fluxes. For most meteorological purposes it is desirable to express F_H and F_q in units of W m^{-2}, while F_M is usually expressed in the familiar units of dynes cm^{-2}. For these units, one can simplify the bulk formulas (using $\rho = 1.23\times10^{-3}$ g cm^{-3}) to read

$$\begin{aligned} F_H &= 1.72\left|V_a\right|(T_w - T_a), \\ F_q &= 4.9\times10^3\left|V_a\right|(q_w - q_a), \\ F_M &= 1.35\times10^{-2} V_a^2. \end{aligned} \qquad (8.2)$$

Here the wind speed is measured in m s^{-1}, temperature in K or °C and specific humidity in g kg^{-1}.

In situations of strong wind speed, the ocean waves exert a large drag on the air, and one finds it desirable to allow for a variation of drag coefficient as a function of wind speed. The variation of drag as a function of wind speed has been expressed by

$$\begin{aligned} C_D = C_{DO} &= 1.1\times10^{-3} && \text{for} && V < 5.8 \text{ m s}^{-1}, \\ &= C_{DO}(0.74 + 0.046V) && \text{for} && 5.8 \le V \le 16.8 \text{ m s}^{-1}, \\ &= C_{DO}(0.94 + 0.034V) && \text{for} && V > 16.8 \text{ m s}^{-1}. \end{aligned}$$

In dealing with tropical storms and hurricanes, a variation of drag as a function of wind speed is generally invoked. The variation of the surface drag coefficient as a function of wind speed is also formally described in the section on the roughness parameter, z_0, described below.

8.2.2. *The Roughness Parameter* z_0

Although z_0 varies in space, it should be regarded as a function of time as well. Over oceans under conditions of strong winds, the wave drag can be quite large. Given a domain of calculation $0 \le x \le L$ and $0 \le y \le M$, one needs a tabulation of a land-ocean matrix *IL* and a tabulation of orographic height *h*. Let us suppose $IL = 0$ over oceans and $IL = 1$ over land. The tabulation of *h* can be obtained from data centers that provide such information.

Over Oceans. The well-known *Charnock formula* (Charnock 1955) is an accepted method for defining z_0 over oceans, that is,

$$z_0 = M \frac{u^{*2}}{g}, \tag{8.3}$$

where M is a constant and has a value around 0.04. One cannot calculate z_0 from this formula until u^*, the friction velocity, is known. That, as we will show, appears in the similarity analysis and requires a knowledge of z_0. Thus an iteration procedure is required to solve for z_0 and u^* successively. Note that

$$u^{*2} = -u'w' = \frac{\tau_0}{\rho_0}, \tag{8.4}$$

where τ_0 is the surface stress and ρ_0 is the density of air. A first-guess field of u^* can be obtained from the bulk aerodynamic representation, noting that $u'w'$ represents the vertical eddy flux of momentum. Given a first-guess field of z_0, one next goes to the similarity approach (described below) to determine the surface fluxes which define u^*, and in turn define the final value of z_0, the roughness parameter.

Over Land. Over land, an empirical method described by Manobianco (1989) allows for the variation of the roughness parameter as a function of the elevation based on the mesoscale variance of the mountain heights. In its simplest form, it is expressed by the relation

$$z_0 = 15 + (473.6 + 0.0368h) \times 10^{-6}, \tag{8.5}$$

where units of the grid-scale mountain height h and of the roughness parameter z_0 are cm. For numerical weather prediction, it is desirable to restrict the upper limit of z_0 to 4000 cm.

8.2.3 *Surface Similarity Theory*

The basis of the similarity analysis presented here follows planetary boundary-layer observations, e.g., Businger et al. (1971). According to these observations, nondimensionalized vertical gradients of large-scale quantities such as wind, potential temperature, and specific humidity can be expressed as universal functions of a nondimensional height z/L, where z is the height above the earth's surface and L is the *Monin-Obukhov length,* which is defined by

$$L = -\frac{u^{*2}}{\kappa\beta\theta^*}. \tag{8.6}$$

Here $\beta = g/\theta_0$, where θ_0 is a reference temperature, u^* is the friction velocity, θ^* is a characteristic temperature, and κ is the von Karman constant. These are usually expressed by the following relations:
Nondimensional shear:

$$\frac{\kappa z}{u^*}\frac{\partial \overline{u}}{\partial z} = \phi_m(z/L). \tag{8.7}$$

Nondimensional vertical gradient of potential temperature:

$$\frac{\kappa z}{\theta^*}\frac{\partial \overline{\theta}}{\partial z} = \phi_h(z/L). \tag{8.8}$$

Nondimensional vertical gradient of specific humidity:

$$\frac{\kappa z}{q^*}\frac{\partial \overline{q}}{\partial z} = \phi_q(z/L), \tag{8.9}$$

where q^* is a characteristic specific humidity. Here u^*, θ^*, and q^* are related to the surface fluxes by the expressions:

$$u^{*2} = -\overline{u'w'}|_0, \tag{8.10}$$

$$u^*\theta^* = \overline{\theta'w'}|_0, \tag{8.11}$$

$$u^*q^* = \overline{q'w'}|_0. \tag{8.12}$$

The right-hand sides of the above equations are eddy correlations estimated over a period of time. This defines the variables u^*, θ^*, and q^*.

The empirical fits of the boundary-layer observations are usually separated in terms of stability. Stability is usually expressed as a function of the sign of the Monin-Obukhov length *L*, or of the bulk Richardson number Ri_B. The definition of stability is based on surface heat flux. Unstable (heat flux up), stable (heat flux down), or neutral (heat flux = 0) correspond to $L<0$, $L>0$, and $L=0$, respectively. Note that u^* is a velocity and is always positive definite. Hence, this implies that for an unstable case ($L<0$) θ^* is always positive and vice versa for the stable case.

We next express this in terms of the *bulk Richardson number*, which is the ratio of stability to the square of wind shear. Here

$$\mathrm{Ri}_B = \beta \frac{\dfrac{d\overline{\theta}}{dz}}{\left(\dfrac{du}{dz}\right)^2}. \tag{8.13}$$

For a stable surface layer, the heat flux is downward, in other words, $L>0$ or $Ri_B>0$. For an unstable surface layer, the heat flux is upward, that is, $L<0$ or $Ri_B<0$. For a neutral surface layer, there is no vertical heat flux, that is $L=0$ or $Ri_B=0$. Since *a priori* the Monin-Obukhov length is an unknown quantity, stability is assessed from the sign of the bulk Richardson number Ri_B. Empirical fits of the boundary-layer observations (Businger et al. 1971) for the unstable and stable cases may be expressed following Chang (1978) by nondimensional relations such as

$$\frac{\kappa z}{u^*}\frac{\partial \overline{u}}{\partial z}=\left(1-15\frac{z}{L}\right)^{-1/4}, \tag{8.14}$$

$$\frac{\kappa z}{\theta^*}\frac{\partial \overline{\theta}}{\partial z}=0.74\left(1-9\frac{z}{L}\right)^{-1/2}, \tag{8.15}$$

$$\frac{\kappa z}{q^*}\frac{\partial \overline{q}}{\partial z}=0.74\left(1-9\frac{z}{L}\right)^{-1/2} \tag{8.16}$$

for the unstable case and

$$\frac{\kappa z}{u^*}\frac{\partial \overline{u}}{\partial z}=1.0+4.7\frac{z}{L}, \tag{8.17}$$

$$\frac{\kappa z}{\theta^*}\frac{\partial \overline{\theta}}{\partial z}=0.74+4.7\frac{z}{L}, \tag{8.18}$$

$$\frac{\kappa z}{q^*}\frac{\partial \overline{q}}{\partial z}=0.74+4.7\frac{z}{L} \tag{8.19}$$

for the stable case. The above six relations were obtained from a least-squares fit of observations.

For both the stable and unstable cases, the definition of the Monin-Obukhov length yields a fourth equation,

$$\frac{z}{L}=-\frac{z\kappa\beta\theta^*}{u^{*2}}. \tag{8.20}$$

The four equations (for the stable or the unstable case) need to be solved for the four variables u^*, θ^*, q^*, and the Monin–Obukhov length L. The surface fluxes of momentum, heat, and moisture are defined by

$$\begin{aligned} F_M &= u^{*2} \equiv -\overline{u'w'}|_0, \\ F_H &= u^*\theta^* \equiv \overline{\theta'w'}|_0, \\ F_q &= u^*q^* \equiv \overline{q'w'}|_0. \end{aligned} \tag{8.21}$$

The solution procedures followed by different investigators for the aforementioned equations vary somewhat. We illustrate here a method developed in our studies with a global spectral model (Krishnamurti et al. 1983, 1984). A two-level surface-layer representation is convenient for the evaluation of surface fluxes.

Let z_1 and z_2 denote two levels, where $z_1=z_0$ (the roughness length), z_2 is the top of the surface layer, and $\Delta z = z_2 - z_1$ is roughly 10 to 50 meters, which is considered the depth of the constant flux layer. At these two levels, the respective winds, potential temperature, and specific humidity may be denote by $\left(\overline{u_1},\overline{v_1}\right)$, $\left(\overline{u_2},\overline{v_2}\right)$, $\left(\overline{\theta_1},\overline{\theta_2}\right)$, and $\left(\overline{q_1},\overline{q_2}\right)$. These are known quantities.

We next write the bulk Richardson number in finite-difference form, that is,

$$\mathrm{Ri}_B = \beta \frac{\left(\overline{\theta}_2 - \overline{\theta}_1\right)\Delta z}{u_2^{\,2} + v_2^{\,2}}. \tag{8.22}$$

Note that $u_1 = v_1 = 0$. The solution for the unstable case is more complicated because of the fractional negative powers in (8.14) to (8.16). Following Chang (1978), we can express the two-level representation of these flux relations for the unstable case by

$$\frac{z_2}{L} = \mathrm{Ri}_B \frac{z_2}{\Delta z}\left(\frac{\ln(z_2/z_1) - \psi_1}{0.74\left[\ln(z_2/z_1) - \psi_2\right]}\right), \tag{8.23}$$

where ψ_1 and ψ_2 are functions of L and are defined as

$$\psi_1(z_2/L, z_1/L) = \ln\left[\left(\frac{1+\epsilon_2}{1+\epsilon_1}\right)^2 \frac{1+\epsilon_2^{\,2}}{1+\epsilon_1^{\,2}}\right] - 2\tan^{-1}\epsilon_2 + 2\tan^{-1}\epsilon_1, \tag{8.24}$$

$$\psi_2(z_2/L, z_1/L) = 2\ln\frac{1+\gamma_2}{1+\gamma_1}. \tag{8.25}$$

Here

$$\epsilon_1 = \left(1 - 15\frac{z_2/L}{z_2/z_1}\right)^{1/4},$$

$$\epsilon_2 = \left(1 - 15\frac{z_2}{L}\right)^{1/4},$$

$$\gamma_1 = \left(1 - 9\frac{z_2/L}{z_2/z_1}\right)^{1/2},$$

and

$$\gamma_2 = \left(1 - 9\frac{z_2}{L}\right)^{1/2}.$$

Integrating (8.14) from z_1 to z_2, we obtain

$$\int_{\overline{u}_1}^{\overline{u}_2} \frac{\kappa}{u^*} du = \int_{z_1}^{z_2} \frac{1}{z}\left(1 - \frac{15z}{L}\right)^{-1/4} dz,$$

or

$$\frac{\kappa}{u^*}(u_2 - u_1) = \int_{z_1}^{z_2}\left[\frac{1}{z} - \frac{1}{z} + \frac{1}{z}\left(1 - \frac{15z}{L}\right)^{-1/4}\right] dz$$

$$= \int_{z_1}^{z_2} \frac{1}{z} dz - \int_{z_1}^{z_2} \frac{1}{z}\left[1 - \left(1 - \frac{15z}{L}\right)^{-1/4}\right] dz$$

$$= \ln\left(\frac{z_2}{z_1}\right) - \psi_1\left(\frac{z_2}{L}, \frac{z_2}{z_1}\right). \tag{8.26}$$

Similar vertical integration of (8.15) and (8.16) results in

$$\frac{\kappa\left(\overline{\theta_2} - \overline{\theta_1}\right)}{0.74\theta^*} = \ln\left(\frac{z_2}{z_1}\right) - \psi_2\left(\frac{z_2}{L}, \frac{z_2}{z_1}\right), \tag{8.27}$$

$$\frac{\kappa\left(\overline{q_2} - \overline{q_1}\right)}{0.74q^*} = \ln\left(\frac{z_2}{z_1}\right) - \psi_2\left(\frac{z_2}{L}, \frac{z_2}{z_1}\right). \tag{8.28}$$

Here

$$\psi_2 = \int_{z_1}^{z_2} \frac{1}{z}\left[1 - \left(1 - \frac{9z}{L}\right)^{-1/2}\right] dz,$$

$$\psi_1 = \int_{z_1}^{z_2} \frac{1}{z}\left[1 - \left(1 - \frac{15z}{L}\right)^{-1/4}\right] dz.$$

Equations (8.14), (8.15), (8.16), and (8.20) are solved for the variables L, u^*, θ^*, and q^* rather simply. The variation of the Monin-Obukhov length is monotonic with respect to u^* and θ^* (Businger et al. 1971). A simple linear incremental search of L in (8.20) provides a rapid solution to the desired degree of accuracy. After substitution for L on the right-hand side of (8.14), (8.15), and (8.16), one obtains the corresponding solution for u^*, θ^*, and q^*.

The solution for the stable case is relatively straightforward and requires a sequential solution of four linear algebraic equations. One of these is

$$\frac{z_2 - z_1}{L} = \ln\left(\frac{z_2}{z_1}\right)\frac{9.4\mathrm{Ri}_B - 0.74 + (4.89\mathrm{Ri}_B + 0.55)^{1/2}}{9.4 - 44.18\mathrm{Ri}_B}. \tag{8.29}$$

Equation (8.29) is obtained from eliminating u^* and θ^* from vertically integrated forms of (8.17), (8.18), and (8.20) within the constant flux layer. The finite-difference forms of these three vertically integrated equations are

$$\frac{\kappa\left(\overline{u_2} - \overline{u_1}\right)}{u^*} = \ln\left(\frac{z_2}{z_1}\right) + 4.7\frac{z_2 - z_1}{L}, \tag{8.30}$$

$$\frac{\kappa\left(\overline{\theta_2} - \overline{\theta_1}\right)}{\theta^*} = 0.47\ \ln\left(\frac{z_2}{z_1}\right) + 4.7\frac{z_2 - z_1}{L}, \tag{8.31}$$

$$\frac{z_2 - z_1}{L} = \frac{(z_2 - z_1)\kappa\beta\theta^*}{u^{*2}}. \tag{8.32}$$

The lower level z_1 is identified with the roughness length z_0. At these two levels, one needs to define wind, potential temperature, and specific humidity in order to carry out the desired computations. At the lower level z_1 we set all wind (u_1, v_1) to zero. The temperature T_1 at the lower level is set to the sea-surface temperature. Over land surfaces, T_1 is determined from surface energy balance, which is described below. Furthermore one sets $\theta_1 = T_1$, since the surface pressure is close to 1000 mb. The specific humidity q_1 at the lower level over the ocean is set to the saturation value at temperature T_1. The upper-level height z_2 is set to a value of 10 meters above the surface. The wind components u_2 and v_2 are usually interpolated using a log-linear profile, while the temperature T_2 and the specific humidity q_2 are linearly interpolated between 1000 and 850 mb from the analyzed data.

The numerical model algorithm requires information at the top of the constant flux layer, which is usually obtained by interpolation of large-scale atmospheric information between 1000 and 850 mb. The von Karman constant κ has a value of 0.35. β stands for g/θ_0 where θ_0 is a constant reference potential temperature. Since Ri_B is know from large-scale data sets, this sequence of calculations yields L, u^*, θ^*, and q^*.

A rigorous comparison of the aforementioned method was made with respect to direct observations of fluxes obtained from the so-called eddy correlation method. This was done using GATE observations, and results were in close agreement. The virtue of this method over the bulk method lies in the dependence of the implicit bulk coefficients on stability and shear. These are expressed by the following equivalent bulk coefficients:

$$C_D = \frac{\kappa^2}{\left[\ln(z_2/z_1) + (4.7\Delta z)/L\right]^2},$$

$$C_\theta = C_q = \frac{-\kappa^2}{\left[\ln(z_2/z_1) + (4.7\Delta z)/L\right]} \frac{1}{\left[0.74\ln(z_2/z_1) + (4.7\Delta z)/L\right]}$$

for the stable case and

$$C_D = \frac{\kappa^2}{\left[\ln(z_2/z_1) - \psi_1\right]^2},$$

$$C_\theta = C_q = \frac{-\kappa^2}{0.74\left[\ln(z_2/z_1) - \psi_2\right]\left[\ln(z_2/z_1) - \psi_1\right]}$$

for the unstable case. The neutral cases are usually calculated in the same manner as the stable situations.

8.2.4 *Vertical Disposition of Surface Fluxes*

The vertical disposition of eddy fluxes of momentum, heat, and moisture above the surface layer is based on *K-theory*. This theory uses an eddy diffusion coefficient K that depends on the mixing length l, the vertical wind shear, and the stability of the atmosphere as determined by the bulk Richardson number Ri_B.

The eddy diffusion coefficient for heat and moisture is

$$K_H = K_Q = l^2 \frac{\partial |V|}{\partial z} F_h(\mathrm{Ri})_B, \tag{8.33}$$

and for momentum it is given by

$$K_M = l^2 \frac{\partial |V|}{\partial z} F_m(\mathrm{Ri})_B. \tag{8.34}$$

Here, l is the mixing length based on a formula given by Blackadar (1962), i.e.,

$$l = \frac{\kappa z}{1 + \kappa z / \lambda},$$

where κ is the von Karman constant, taken as 0.35, and λ is the asymptotic mixing length, which is set to 450 m for heat and moisture exchange and 150 m for momentum exchange.

Following Louis (1979), F_h and F_m are given as

$$F_h = F_m = \frac{1}{(1 + 5\mathrm{Ri}_B)^2}, \qquad \mathrm{Ri}_B \geq 0 \tag{8.35}$$

for the stable case and

$$F_h = \frac{1 + 1.286 |\mathrm{Ri}_B|^{1/2} - 8\mathrm{Ri}_B}{1 + 1.286 |\mathrm{Ri}_B|^{1/2}} \qquad \mathrm{Ri}_B < 0, \tag{8.36}$$

$$F_m = \frac{1 + 1.746 |\mathrm{Ri}_B|^{1/2} - 8\mathrm{Ri}_B}{1 + 1.746 |\mathrm{Ri}_B|^{1/2}} \qquad \mathrm{Ri}_B < 0 \tag{8.37}$$

for the unstable case. Here, the bulk Richardson number over an atmospheric layer is given by

$$\mathrm{Ri}_B = \frac{g}{\theta} \frac{\frac{\partial \bar{\theta}}{\partial z}}{\left| \frac{\partial |V|}{\partial z} \right|^2}. \tag{8.38}$$

It can be shown that $\mathrm{Ri}_B \leq 0.212$ is a physically valid value.

The time tendency due to diffusive fluxes at any level is given by

$$\frac{\partial \tau}{\partial t} = \frac{1}{\rho} \frac{\partial}{\partial z} \left(\rho K \frac{\partial \tau}{\partial z} \right), \tag{8.39}$$

where $\tau = u,\ v,\ \theta,$ or q, and ρ is density of air. In σ coordinates, (8.39) takes the form

$$\frac{\partial \tau}{\partial t} = \frac{g^2}{p_s^2} \frac{\partial}{\partial \sigma} \left(\rho^2 K \frac{\partial \tau}{\partial \sigma} \right). \tag{8.40}$$

The upper and lower boundary conditions for (8.40) are $\partial \tau / \partial \sigma = 0$ at the top and $\partial \tau / \partial \sigma =$ the boundary-layer fluxes at the top of the constant flux layer at the bottom.

Using these boundary conditions, (8.40) is solved implicitly, where it takes the finite-difference form

$$\frac{\tau(t+\Delta t)-\tau^*(t+\Delta t)}{2\Delta t}=\frac{g^2}{p_s^2}\frac{\partial}{\partial\sigma}\left(\rho K\frac{\partial\tau(t+\Delta t)}{\partial\sigma}\right), \tag{8.41}$$

where $\tau^*(t+\Delta t)$ is the value of u, v, θ, or q before vertical disposition of the fluxes at time $(t+\Delta t)$, and $\tau(t+\Delta t)$ is the value after vertical disposition of the fluxes. The solution of (8.41) is obtained by writing the equation in the form $|A|\tau(t+\Delta t)=\tau^*(t+\Delta t)$, where $|A|$ is a tridiagonal coefficient matrix and $\tau(t+\Delta t)$ and $\tau^*(t+\Delta t)$ are vectors defining the values of $\tau(t+\Delta t)$ and $\tau^*(t+\Delta t)$ at different levels.

8.3 Cumulus Parameterization

The scale of cumulus clouds is much smaller than the scale of the grid squares (or the smallest resolvable scale) of a numerical weather prediction model. The individual cumulus cloud has a scale of a few kilometers, whereas the model grid square is more like a few hundred kilometers. Within larger synoptic-scale disturbances often reside organized mesoscale convective systems with scales of the order of a few hundred kilometers. These are the mesoscales over which the clouds show organization. Thus the interaction of cumulus clouds with the broad synoptic scale appears to be an important multiscale problem. Cumulus parameterization addresses the effects of the cumulus scale on the resolvable scale, given the information on the latter scales.

Minimally, one needs to derive three parameters from a cumulus parameterization scheme. These are the vertical distribution of both heating and moistening, and the rainfall rates. We first outline a version of Kuo's scheme which has been modified in recent years by Krishnamurti et al. (1983) to provide successful forecasts of the life cycle of tropical cyclones. We follow that with a scheme based on the work of Arakawa and Schubert (1974), which has also been used successfully by several research scientists and numerical weather prediction centers.

8.3.1 *Kuo's Scheme*

According to Kuo, organized cumulus convection requires the presence of conditional instability and a net supply of moisture. In the earlier formulation of Kuo (1965), in those regions where the two conditions were met the following form of the moisture equation was used, i.e.,

$$\frac{\partial q}{\partial t}=a\frac{q_s-q}{\Delta\tau}, \tag{8.42}$$

where $\Delta\tau$ is a cloud time-scale parameter and a denotes the fraction area of the grid box that would be covered by the newly formed convective clouds. The parameter a is defined by the relation

$$a = \frac{-\frac{1}{g}\int_{p_T}^{p_B}\left(\nabla\cdot q\vec{V} + \frac{\partial}{\partial p}q\omega\right)dp}{\frac{1}{g}\int_{p_T}^{p_B}\left(\frac{c_p(T_s - T)}{L\Delta\tau} + \frac{q_s - q}{\Delta\tau}\right)dp}. \tag{8.43}$$

In the following we use the symbol Q for the denominator of (8.43). In this relation the denominator may be interpreted as the amount of moisture supply needed to cover the entire grid-square area by a model cloud (a local moist adiabat T_s, q_s). The numerator, on the other hand, is a measure of the available moisture supply. The total convective rainfall rate is expressed by

$$P_T = \frac{1}{g}\int_{p_T}^{p_B}\frac{ac_p(T_s - T)}{L\Delta\tau}dp. \tag{8.44}$$

It should be noted that the definition of a as given by (8.43) and the parameterization of the moisture in (8.42) are consistent with the principle of conservation of moisture. This can be shown by considering the moisture conservation law in the form

$$\frac{\partial q}{\partial t} = -\nabla\cdot q\vec{V} - \frac{\partial}{\partial p}qw + E - P. \tag{8.45}$$

After integration of (8.45) from p_T to p_B with the assumption that evaporation E occurs only at the air-sea interface (i.e., below p_B), we obtain

$$\frac{1}{g}\int_{p_T}^{p_B}\frac{\partial q}{\partial t}dp = I - P_T, \tag{8.46}$$

where P_T is the total precipitation rate for the vertical column and I is the moisture convergence rate at the grid point. Noting that

$$I = aQ = a\left(\frac{1}{g}\int c_p\frac{T_s - T}{L\Delta\tau}dp + \frac{1}{g}\int\frac{q_s - q}{\Delta\tau}dp\right) \tag{8.47}$$

and substituting for P_T from (8.44) into (8.46), we obtain

$$\frac{1}{g}\int_{p_T}^{p_B}\frac{\partial q}{\partial t}dp = \frac{1}{g}a\int_{p_T}^{p_B}\frac{q_s - q}{\Delta\tau}dp. \tag{8.48}$$

Hence the use of (8.43) is consistent with the moisture conservation law (8.45).

We furthermore note that part of the moisture convergence I is being used for raising the level of moisture, and part of it goes into condensational heating. This partitioning may be expressed by the relation

$$I = I_q + I_\theta. \tag{8.49}$$

We note that in general I_q is larger than I_θ initially. However, as q tends to q_s, Q becomes small; as a consequence, a becomes very large. From then on, I_θ starts to become very large and I_q tends to zero. This leads to a slight computational instability at a few grid points due to the very large heat release and very strong resulting vertical motions. This defect is primarily related to a disproportionate partitioning of I into I_q and I_θ.

Kuo applied this early formulation of the cumulus parameterization scheme to hurricanes for which it worked well. However, it was found that it did not succeed for modeling other large-scale disturbances. This scheme produces too much moistening, too little rain, and too little heating. In a high-resolution model, such as used to model a hurricane, there is a great deal of moisture convergence and hence more rain and heating. However, in a coarser resolution model, there is less moisture convergence and therefore less heating and rain.

Now we consider the parameterization of deep, moist convection with a modified Kuo scheme. The large-scale convergence of the flux of moisture is expressed by the relation

$$C_M = -\nabla \cdot \overline{\vec{V}}_q - \frac{\partial \overline{\omega q}}{\partial p},$$

or in advective form,

$$C_M = -\overline{\vec{V}} \cdot \nabla \overline{q} - \overline{\omega} \frac{\partial \overline{q}}{\partial p}. \tag{8.50}$$

In the Kuo type of cumulus parameterization schemes, the supply of moisture is usually defined from a vertical integral of the above expression. The supply is then used to define the moist adiabatic cloud elements in various versions of Kuo's scheme.

Krishnamurti et al. (1983) and Anthes (1977) have noted that the supply of moisture for the definition of clouds may be expressed simply by the second term in (8.50). The first term, namely the horizontal advection, is used for a direct moistening of the air as a large-scale advective term. Thus we define the *supply of moisture* by

$$I_L = \frac{1}{g} \int_{p_T}^{p_B} \omega \frac{\partial q}{\partial p} dp. \tag{8.51}$$

Here p_T and p_B denote the cloud top and the cloud base, respectively. They are defined in terms of the vanishing buoyancy level and the lifting condensation level, respectively.

From our experience (Krishnamurti et al. 1980, 1983), we have noted that the above definition for the large-scale supply is a close measure of the rainfall rate, and thus sufficient supply is not available to account for the observed moistening of the vertical columns. These statements are based on semiprognostic studies made with GATE observations. This research leads us to propose a *mesoscale convergence parameter* η and a *moistening parameter* b

$$I = I_L (1 + \eta), \tag{8.52}$$

where $I_L \eta$ denotes the net mesoscale moisture supply.

The total moisture supply is partitioned into a precipitation part and a moistening part via the following relations:

$$\begin{aligned} R &= I(1-b) = I_L(1+\eta)(1-b), \\ M &= Ib \quad = I_L(1+\eta)b. \end{aligned} \tag{8.53}$$

Following Krishnamurti et al. (1983), we define the total supply of moisture required to produce a grid-scale moist-adiabatic sounding by

$$Q = \frac{1}{g}\int_{p_T}^{p_B} \frac{q_s - q}{\Delta\tau}\,dp + \frac{1}{g}\int_{p_T}^{p_B}\left(\frac{c_p T(\theta_s - \theta)}{L\theta\Delta\tau} + \omega\frac{c_p T}{L\theta}\frac{\partial\theta}{\partial p}\right)dp. \tag{8.54}$$

Here $\Delta\tau$ denotes a cloud time-scale approximately equal to 30 minutes. The two respective terms are denoted by Q_q and Q_θ, so that

$$Q = Q_q + Q_\theta. \tag{8.55}$$

Note that equation (8.54) differs from the definition of Q in the classic Kuo scheme. The last term on the right-hand side has been added and it is the amount of energy needed to overcome adiabatic cooling. In the classic Kuo scheme when saturation was reached $Q \to 0$ so that $a = I/Q$ would blow up. The new term was added so that if saturation is reached Q would tend toward the value of this term.

The total supply I may likewise be split into moistening and heating parts by

$$I_q = Ib = I_L b(1+\eta) \tag{8.56}$$

and

$$I_\theta = I(1-b) = I_L(1-b)(1+\eta). \tag{8.57}$$

The temperature and moisture equations are expressed by

$$\frac{\partial\theta}{\partial t} + \vec{V}\cdot\nabla\theta + \omega\frac{\partial\theta}{\partial p} = a_\theta\left(\frac{\theta_s - \theta}{\Delta\tau} + \omega\frac{\partial\theta}{\partial p}\right), \tag{8.58}$$

$$\frac{\partial q}{\partial t} + \vec{V}\cdot\nabla q = a_q\frac{q_s - q}{\Delta\tau}, \tag{8.59}$$

where a_θ, the fractional area of a cloud that releases latent heat, and a_q, the fractional area of a cloud that moistens, are defined by

$$a_\theta = \frac{I_\theta}{Q_\theta} = \frac{I(1-b)}{Q_\theta} = \frac{I_L(1+\eta)(1-b)}{Q_\theta}, \tag{8.60}$$

$$a_q = \frac{I_q}{Q_q} = \frac{Ib}{Q_q} = \frac{I_L b(1+\eta)}{Q_q}. \tag{8.61}$$

The parameterization is closed if b and η are somehow determined. Then a_θ and a_q may be evaluated from (8.60) and (8.61). Note that Q_θ and Q_q are known quantities.

Krishnamurti et al. (1983) proposed a closure for b and η based on a screening multiregression analysis of GATE observations. Here they regressed normalized heating R/I_L and moistening M/I_L against a number of large-scale variables. From GATE observations they noted significant correlations for the heating and moistening from the following relations.

$$\frac{M}{I_L} = a_1\zeta + b_1\bar{\omega} + c_1, \tag{8.62}$$

$$\frac{R}{I_L} = a_2\zeta + b_2\bar{\omega} + c_2. \tag{8.63}$$

Here ζ is the relative vorticity of the lower troposphere, $\bar{\omega}$ is a vertically averaged vertical velocity, and a_1, b_1, c_1, a_2, b_2, and c_2 are regression constants whose magnitudes may be found in Krishnamurti et al. (1983).

Thus in numerical weather prediction, ζ and ω determine M/I_L and R/I_L since

$$\frac{M}{I_L} = b(1+\eta) \tag{8.64}$$

and

$$\frac{R}{I_L} = (1+\eta)(1-b). \tag{8.65}$$

These two relations determine b and η. We can find a_θ and a_q from

$$a_\theta = \frac{I_L(1-b)(1+\eta)}{Q_\theta} = \frac{R}{Q_\theta} \tag{8.66}$$

and

$$a_q = \frac{I_L b(1+\eta)}{Q_q} = \frac{M}{Q_q}. \tag{8.67}$$

It is also of interest to note that the *apparent heat source* Q_1 and the *apparent moisture sink* Q_2 for this formulation are expressed by

$$Q_1 = a_\theta\left(c_p\frac{T}{\theta}\frac{\theta_s-\theta}{\Delta\tau} + \omega c_p\frac{T}{\theta}\frac{\partial\theta}{\partial p}\right) + c_p\frac{T}{\theta}(H_R + H_s), \tag{8.68}$$

$$Q_2 = -La_q\left(\frac{q_s-q}{\Delta\tau} + \omega\frac{\partial q}{\partial p}\right), \tag{8.69}$$

where H_R is the total radiative potential temperature rate of change and H_s is the vertical sensible heat flux by subgrid-scale motions. The total rainfall is given by

$$P = \frac{1}{g}\int_{p_T}^{p_B} a_\theta\frac{c_p}{L}\frac{T}{\theta}\left(\frac{\theta_s-\theta}{\Delta\tau} + \omega\frac{\partial\theta}{\partial p}\right)dp. \tag{8.70}$$

8.3.2 *The Arakawa–Schubert Cumulus Parameterization*

This scheme is considered to be the best scheme in terms of relating the physics and dynamics of the cloud system to the large-scale environment. The Arakawa-Schubert (hereafter AS) cumulus parameterization scheme (Arakawa and Schubert 1974) attempts to quantify the effect of cumulus convection on the large-scale environment. From a modeling perspective, this parameterization scheme breaks down into static control, dynamic control, and the feedback. However, this scheme is somewhat complicated and requires a large amount of computational overhead. In this section we present a brief outline of the basic concepts and formulation of the AS scheme.

In the AS scheme, we have a cloud ensemble consisting of various subensembles. Each cloud type is characterized by a parameter λ. Furthermore, it is assumed that λ can take values between zero and $\lambda_{\max}$. By doing so, the whole cloud ensemble is covered. In this formulation, a cloud subensemble is defined to have the value of parameter λ between λ and $\lambda + d\lambda$. We can think as if there is a cloud ensemble present between the cloud base and the cloud top. Each cloud in this ensemble has its own entrainment rate and mass flux across the cloud base. It is assumed that each cloud in this ensemble has the same cloud base. However, their tops may vary. Furthermore, it is assumed that the cloud ensemble occupies a sufficiently large area. However, the area of the large-scale system (our cloud ensemble is a part of this) is assumed to be much larger than the area of the cloud ensemble itself.

It is assumed that every thermodynamic structure of the environment is associated with a cloud ensemble and that the fractional entrainment rate μ can be used to determine all the properties of the cloud subensemble represented by λ_k in the interval (λ, $\lambda + d\lambda$). These properties include the precipitation rate, the rate of destruction of the convective instability of the environment, the work done by the buoyancy force, the speed of the updraft within the cloud, the cloud-top level, the vertical mass flux at any vertical level, the buoyancy of air inside the cloud, and the total mass of cloud air that is detrained at the cloud top.

The *dynamic control* deals with how the convective clouds are influenced by the large-scale environment. Furthermore, the dynamic control determines the spectral distribution of the clouds. The *static control* determines cloud thermodynamic properties and is often linked to the dynamic control. The *feedback mechanism* determines the effect of convection on the environment. In other words, the static control is a way of communication between the feedback and the dynamic control, the dynamic control determines the effect of the environment on cumulus clouds, and the feedback determines the effect of cumulus clouds on the environment.

AS proposed that the amount of convection is related to the rate of destabilization of the environment. Furthermore, they assumed that the clouds would minimize this rate of destabilization of the large-scale environment. A single convective element was modeled as steady state jet by AS. They assumed a constant entrainment rate with height, and the mass flux in the cloud was solely determined by the entrainment rate and the net mass flux at the base of this cloud.

The AS scheme defines a cloud ensemble via the mass, moisture, and heat budgets of the cumulus-scale moist convection. Large-scale eddy flux convergence of these are related to the cloud ensemble properties.

The *fractional rate of entrainment* μ is defined by

$$\mu(z,\lambda) = \frac{1}{\eta(z,\lambda)}\frac{\partial \eta(z,\lambda)}{\partial z}. \tag{8.71}$$

Here η denotes the normalized vertical mass flux within the cloud. The *total cloud mass flux* M_c at a level z is defined by

$$M_c(z) = \int_0^{\lambda_D(z)} m(z,\lambda)\, d\lambda\,, \tag{8.72}$$

where $m(z,\lambda) = m_B(\lambda)e^{\lambda(z-z_B)}$. Here λ_D is the detrainment level, which is the top of the cloud. λ may be regarded as a vertical coordinate with $\lambda = 0$ residing at the top of the atmosphere. m_B is the cloud-base mass flux. $\lambda = \lambda_D$ is the top level of a cloud. It is assumed that all of the cloud detrainment occurs at the top. A normalized cloud mass function η can be defined as

$$\eta(z,\lambda) = \begin{cases} e^{\lambda(z-z_B)} & \text{if } z_B \ll z \ll z_D \\ 0 & \text{if } z > z_D(\lambda) \end{cases}.$$

The detrainment at the cloud top at level z is defined as

$$D(z) = -m\left[z, \lambda_D(z)\right]\frac{d\lambda_D(z)}{dz}. \tag{8.73}$$

In order to define a relationship between the large-scale and the cloud ensemble-scale heat and moisture budgets, we express their respective large-scale changes by

$$\rho\frac{\partial \overline{s}}{\partial t} + \rho\vec{V}\cdot\nabla\overline{s} + \rho\overline{w}\frac{\partial \overline{s}}{\partial z} = Q_R + L(\overline{c}-\overline{e}) - \frac{\partial}{\partial z}\rho\overline{w's'}\,, \tag{8.74}$$

$$\rho\frac{\partial \overline{q}}{\partial t} + \rho\vec{V}\cdot\nabla\overline{q} + \rho\overline{w}\frac{\partial \overline{q}}{\partial z} = -(\overline{c}-\overline{e}) - \frac{\partial}{\partial z}\rho\overline{w'q'}\,. \tag{8.75}$$

Here ρ denotes the density of air, $\overline{s}$ is the dry static energy, $\overline{q}$ is the specific humidity, Q_R denotes radiative heating, $\overline{c}$ denotes condensation rate, and $\overline{e}$ denotes evaporation rate. The last term in (8.74) and (8.75) denotes the eddy flux convergence of heat and moisture per unit volume of air, respectively. These eddy flux convergences can be related to the cloud mass flux. If w_c and $\tilde{w}$ are the vertical velocity of the clouds and their environment, respectively, we can write

$$M_c = \sigma w_c \qquad \text{and} \qquad \tilde{M} = (1-\sigma)\,\tilde{w}. \tag{8.76}$$

Furthermore, we can express other properties in the same way, for example

$$\overline{s} = \sigma s_c + (1-\sigma)\tilde{s}\,, \tag{8.77}$$

$$\overline{q} = \sigma q_c + (1-\sigma)\tilde{q}\,. \tag{8.78}$$

Next we derive the very useful relation $\overline{A'w'} = M_c(A_c - \tilde{A})$ where A is any property. We can write

$$\overline{A} = \sigma A_c + (1-\sigma)\tilde{A}\,, \tag{8.79}$$

$$\overline{w} = \sigma w_c + (1-\sigma)\tilde{w}, \tag{8.80}$$

$$\overline{Aw} = \sigma A_c w_c + (1-\sigma)\,\tilde{A}\tilde{w}. \tag{8.81}$$

To the order σ, we can write

$$\overline{A}\overline{w} = \tilde{A}\tilde{w} - 2\sigma\tilde{A}\tilde{w} + \sigma(A_c\tilde{w} + w_c\tilde{A}), \tag{8.82}$$

then $\overline{A'w'} = \overline{Aw} - \overline{A}\,\overline{w}$ can be approximated as

$$\sigma w_c(A_c - \tilde{A}) = M_c(A_c - \tilde{A}). \tag{8.83}$$

Here we assumed that $\sigma \ll 1$, $|w|_c = |\tilde{w}|$ and $A_c - \overline{A} \approx A_c - \tilde{A}$. Thus the eddy fluxes of s, q, and h can be written in the form

$$\overline{s'w'} = M_c(s_c - \overline{s}), \tag{8.84}$$

$$\overline{q'w'} = M_c(q_c - \overline{q}), \tag{8.85}$$

$$\overline{h'w'} = M_c(h_c - \overline{h}). \tag{8.86}$$

Then we can express the eddy fluxes by

$$\rho\overline{w's'} = \int_0^{\lambda_D(z)} m(z,\lambda)\left[s_c(z,\lambda) - \overline{s}(z)\right]d\lambda \tag{8.87}$$

and

$$\rho\overline{w'q'} = \int_0^{\lambda_D(z)} m(z,\lambda)\left[q_c(z,\lambda) - \overline{q}(z)\right]d\lambda, \tag{8.88}$$

where subscript c denotes a cloud variable and the overbar denotes a large-scale variable.

Using the notion of a one-dimensional steady-state entraining cloud model with the following definition for the cloud top, i.e.,

$$s_c\left[z_D, \lambda(z_D)\right] = \overline{s}(z_D)$$

and

$$q_c\left[z_D, \lambda(z_D)\right] = \overline{q}(q_D),$$

we can write eddy convergence of fluxes as

$$\frac{\partial}{\partial z}\rho\overline{w's'} = L\overline{c} - M_c\frac{\partial\overline{s}}{\partial z} \tag{8.89}$$

and

$$\frac{\partial}{\partial z}\rho\overline{w'q'} = D(\overline{q} - \overline{q}^*) - M_c\frac{\partial\overline{q}}{\partial z} - \overline{c}, \tag{8.90}$$

where $\overline{q}^*$ is the saturation specific humidity of the cloud environment. Substituting (8.89) and (8.90) into (8.74) and (8.75), we can express the large-scale equation by the relations

$$\rho\frac{\partial\overline{s}}{\partial t} + \rho\overline{\vec{V}}\cdot\nabla\overline{s} + \rho\overline{w}\frac{\partial\overline{s}}{\partial z} = -DL\hat{l} + M_c\frac{\partial\overline{s}}{\partial z} + Q_R \tag{8.91}$$

and

$$\rho\frac{\partial\overline{q}}{\partial t}+\rho\overline{\vec{V}}\cdot\nabla\overline{q}+\rho\overline{w}\frac{\partial\overline{q}}{\partial z}=D(\overline{q}^{*}-\overline{q}+\hat{l})+M_{c}\frac{\partial\overline{q}}{\partial z}. \qquad (8.92)$$

Here $\hat{l}$ denotes the liquid-water mixing ratio and D is the detrainment. We have replaced $\overline{e}$ by $D\hat{l}$; i.e., cloud evaporation occurs entirely at the detrainment level (at the cloud top).

A single equation for the dry static energy $(\overline{h}=\overline{s}+L\overline{q})$ can be obtained from the above two equations, i.e.,

$$\rho\frac{\partial\overline{h}}{\partial t}+\rho\overline{\vec{V}}\cdot\nabla\overline{h}+\rho\overline{w}\frac{\partial\overline{h}}{\partial z}=DL(\overline{q}^{*}-\overline{q})+M_{c}\frac{\partial\overline{h}}{\partial z}+Q_{R}. \qquad (8.93)$$

In (8.91) and (8.92), $M_c\partial\overline{s}/\partial z$ and $M_c\partial\overline{q}/\partial z$ are to be interpreted as the heating and drying effects from the cumulus-induced subsidence. It should be noted that $D\hat{l}$ is exactly equal to evaporation e if evaporation of falling rain is not considered. $D(\overline{q}^{*}-\overline{q})$ is interpreted as the detrainment of cloud water vapor into the environment. The following expressions for the apparent heat source Q_1 and the apparent moisture sink Q_2 are implicit in the above analysis

$$Q_{1}=\rho\left(\frac{\partial\overline{s}}{\partial t}+\overline{\vec{V}}\cdot\nabla\overline{s}+\overline{w}\frac{\partial\overline{s}}{\partial z}\right)=-DL\hat{l}+M_{c}\frac{\partial\overline{s}}{\partial z}+Q_{R}, \qquad (8.94)$$

$$Q_{2}=-L\left(\frac{\partial\overline{q}}{\partial t}+\overline{\vec{V}}\cdot\nabla\overline{q}+\overline{w}\frac{\partial\overline{q}}{\partial z}\right)$$

$$=-DL(\overline{q}^{*}-q+\hat{l})-LM_{c}\frac{\partial\overline{q}}{\partial z}. \qquad (8.95)$$

We next move on to the major issues of the AS cumulus parameterization scheme. This includes the quasi-equilibrium hypothesis and the definition of a cloud work function. In the dynamic control, a cloud work function, which is a measure of the buoyancy force associated with a subensemble, is described. The cloud work function is derived through the following equation:

$$\frac{dw_{c}}{dt}=Bu-F_{r}=\frac{1}{w_{c}}\frac{d}{dt}\frac{w_{c}^{2}}{2}, \qquad (8.96)$$

where w_c is the vertical velocity for a cloud subensemble, Bu is the acceleration due to buoyancy, and F_r represents the deceleration due to friction. It should be noted that w_c is a function of both λ and z.

If we multiply the above equation by $\rho_c w_c$ and integrate the resultant equation over the depth of the cloud, we obtain

$$\frac{d}{dt}\int_{z_B}^{z_T}\rho_{c}\frac{w_{c}^{2}}{2}dz=m_{B}(\lambda)\int_{z_B}^{z_T}\eta Bu\,dz-Dm_{B}(\lambda). \qquad (8.97)$$

We have made use of $m_z(\lambda,z)=m_B(\lambda)\eta(\lambda,z)$ and $m_z=\rho_c w_c$. Here D represents the cloud-scale kinetic energy dissipation per unit mass flux, z_B is the height of the cloud base, and z_T is the cloud-top level (level of zero buoyancy). One can write (8.97) as

$$\frac{d}{dt}\overline{KE}=A(\lambda)m_B(\lambda)-D(\lambda)m_B(\lambda). \tag{8.98}$$

The *work function* $A(\lambda)$ is defined by

$$A(\lambda)=\int_{z_B}^{z_D(\lambda)}\frac{g}{c_p\overline{T}(z)}\eta(z,\lambda)\left[s_c(z,\lambda)-\overline{s}(z)\right]dz, \tag{8.99}$$

where s_c is the cloud static energy. A is a measure of the kinetic energy generated by the buoyancy force for the subensemble λ. The quasi-equilibrium hypothesis assumes that the stabilization by the cumulus-scale forcing and the destabilization by the large-scale forcing are in an essential balance.

The time derivative of the work function is expressed by

$$\frac{\partial A(\lambda)}{\partial t}=\frac{\partial h_M}{\partial t}\int_{z_B}^{z_D(\lambda)}\rho(z)\beta(z)dz+\int_{z_B}^{z_D(\lambda)}\rho(z)b(z,\lambda)\lambda\eta(z,\lambda)\frac{\partial\overline{h}}{\partial t}dz$$
$$-\int_{z_B}^{z_D(\lambda)}\rho(z)\alpha(z)\eta(z,\lambda)\frac{\partial\overline{s}}{\partial t}dz, \tag{8.100}$$

where

$$\alpha(z)=\frac{g}{c_p\overline{T}(z)\rho(z)},$$

$$\beta(z)=\frac{g}{c_p\overline{T}(z)}\frac{1}{1+\dfrac{L}{c_p}\left.\dfrac{\partial\overline{q}^*}{\partial\overline{T}}\right|_{\overline{p}}}\frac{1}{\rho(z)},$$

and

$$b(z,\lambda)=\int_z^{z_D(\lambda)}\frac{\rho(z')\beta(z')}{\rho(z)}dz'.$$

Here h_M denotes the total moist static energy of the mixed layer. If we substitute for $\partial\overline{s}/\partial t$ and $\partial\overline{h}/\partial t$ from (8.91) and (8.93), we obtain

$$\frac{\partial A(\lambda)}{\partial t}=F_M(\lambda) \tag{8.101}$$
$$+\int_0^{\lambda_{\max}}[K_v(\lambda,\lambda')+K_D(\lambda,\lambda')]m_B(\lambda')d\lambda'+F_c(\lambda),$$

where

$$F_c(\lambda)=\int_{z_B}^{z_D(\lambda)}\eta(z,\lambda)\left[\lambda b(z,\lambda)\left(-\overline{w}\frac{\partial\overline{h}}{\partial z}-\overline{\vec{V}}\cdot\nabla\overline{h}+Q_R\right)\right.$$

$$-\alpha(z)\left(-\bar{w}\frac{\partial \bar{s}}{\partial z}-\bar{\vec{V}}\cdot\nabla\bar{s}+Q_R\right)\Bigg]\rho(z)dz\ ,$$

$$F_M(\lambda)=\frac{\partial h_M}{\partial t}\int_{z_B}^{z_D(\lambda)}\rho(z)\beta(z)dz\ ,$$

$$K_v(\lambda,\lambda')=\int_{z_B}^{z_D(\lambda)}\eta(z,\lambda)\eta(z,\lambda')\left(\lambda b(z,\lambda)\frac{\partial \bar{h}}{\partial z}-\alpha(z)\frac{\partial \bar{s}}{\partial z}\right)dz\ ,$$

$$K_D(\lambda,\lambda')=\eta(z'_D,\lambda)\eta(z'_D,\lambda')\{\lambda b(z'_D,\lambda)L\left[\bar{q}^*(z'_D)-\bar{q}(z'_D)\right]$$
$$+\alpha(z'_D)L\hat{l}(z'_D)\}\ .$$

The cloud work function $A(\lambda)$ is a measure of the generation of kinetic energy inside the cloud. When $A(\lambda)$ is positive, it implies that the environment has moist convective instability. For a given value of λ, it defines a property of the environment. Furthermore, we can obtain the rate of generation of kinetic energy by the vertical motion through $A(\lambda)$.

It is known that the convective instability of the environment is destroyed by cumulus convection through subsidence. Subsidence also reduces the buoyancy and kinetic energy of the cloud updraft. It should be noted that there are two types of processes (namely, cloud-cloud interaction and the influence of large-scale physics and dynamics) that can influence $A(\lambda)$. If we assume that there are two cloud types (i.e., cloud type λ and cloud type λ'), then cumulus convection tends to destroy the convective instability of the environment and subsidence reduces the buoyancy as well as the kinetic energy of the cloud updrafts.

8.3.3 *An Example of Cumulus Parameterization*

Thirteen research ships were deployed during the GATE experiment. Twelve of these ships formed two embedded hexagons near 10° N and 20° W, and one was located at the center of the hexagons. These ships provided upper-air data sets and some carried a shipboard precipitation radar as well. With this array of ships, it was possible to carry out tests of the aforementioned cumulus parameterization methods.

The Kuo and the Arakawa-Schubert hexagonal schemes were run in this hexagonal array of grid points. One-time step forecasts (called the semiprognostic method) were carried out. The result of a comparison between the observed and the model-based rain rates is shown in Fig. 8.1. The passage of African easterly waves gives rise to enhanced rain amounts on the order of 30 mm day^{-1}. Both models performed extremely well in predicting the observed rainfall rates. The observed rainfall over the hexagonal ship array comes from the area-averaged rainfall determined by the radar reflectively, which in turn had been calibrated against rain gauge estimates.

In these tests, the semiprognostic evaluation of rainfall from both schemes using GATE observation appears to be quite comparable. The vertical distribution of heating and moistening also appeared quite similar for these two methods. Details of these comparisons appear in Krishnamurti et al. (1980).

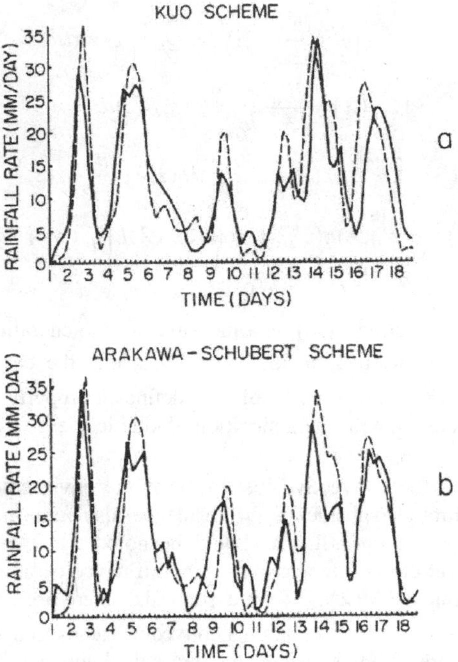

Figure 8.1. A comparison between observed and semiprognostic rainfall rates during GATE using (a) the Kuo (1974) and (b) the Arakawa-Schubert (after Lord 1978) cumulus parameterization schemes.

A further test of the sensitivity of model forecasts to cumulus parameterization is illustrated in Fig. 8.2. Here we illustrate the formation of a hurricane-like disturbance called the onset vortex of the monsoon. Figure 8.2a illustrates the initial flow field at 850 mb over the monsoon area, while the observed field six days later is shown in Fig. 8.2b. Figure 8.2c shows the six-day forecast based on a refined cumulus parameterization scheme (Krishnamurti and Bedi 1988). The forecast for the same period using classical Kuo cumulus parameterization is shown in Fig. 8.2b. The initial flow field over the Arabian Sea is anticyclonic. Six days later, the commencement of a moist current over the southern part of India resulted in the onset of the monsoon. This was brought on by the formation of a hurricane-like vortex that formed over the northern Arabian Sea.

The model using a classical Kuo parameterization scheme did not predict any of these features, whereas the model with the refined cumulus parameterization scheme captured the onset of the monsoon as well as the formation of the onset vortex. This forecast also successfully predicted the first seasonal monsoon rains over southwestern India. This sensitivity to cumulus parameterization is attributed mainly to a more robust heating from condensation in the improved scheme.

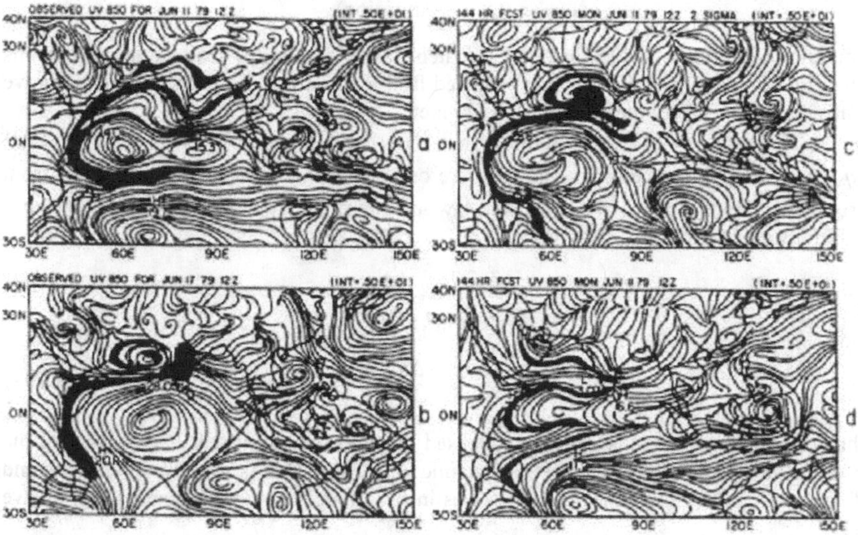

Figure 8.2. Sensitivity of the monsoon onset to cumulus parameterization. Shown at 850 mb is the observed flow (a) at the initial time and (b) on day 6. Also shown at 850 mb is the predicted flow on day 6 from (c) improved cumulus parameterization and (d) classical cumulus parameterization. Wind speed indicated in m s^{-1}.

8.4 Large-Scale Condensation

8.4.1 *Disposition of Supersaturation*

Large-scale condensation is usually invoked in a numerical weather prediction experiment if dynamic ascent of absolutely stable, near saturated air occurs at any level of the atmosphere. The ascent is usually a consequence of large-scale dynamics such as differential vorticity advection, thermal advection, slow orographic ascent, or even buoyancy-driven ascent from the lower troposphere. In the latter case, convective and nonconvective clouds coexist over the same region.

The stable air refers to absolute stability, where both potential and equivalent potential temperature increase with height. The saturation refers to the ratio of specific humidity to saturation specific humidity, which is close to unity. However, one uses a saturation ratio on the order of 0.8 to take into account possible sub-grid-scale saturation.

We may express these conditions via the relations

$$\omega < 0 \quad \text{(ascent)},$$

$$-\partial\theta / \partial p, \; -\partial\theta_e / \partial p > 0 \quad \text{(stable)},$$

and

$$q / q_s > 0.8 \quad \text{(saturation)}.$$

Saturation is defined at 80% in order to catch possible sub-grid-scale saturation at this lower threshold. This value has been verified from experimentation. In the following we consider regions where these conditions are met.

The removal of supersaturation will now be addressed. Let $\Delta q = q - q_s$. If $\Delta q > 0$, then at that level of the atmosphere one can incorporate the contributions from large-scale condensation into the first law of thermodynamics and the water-vapor continuity equation by the relations

$$c_p \frac{T}{\theta}\frac{\partial \theta}{\partial t} = +\frac{L\Delta q}{\Delta t} \tag{8.102}$$

and

$$\frac{\partial q}{\partial t} = -\frac{\Delta q}{\Delta t}. \tag{8.103}$$

Thus the supersaturation is simply condensed out with an equivalent heat release at the level of the atmosphere in the thermodynamic equation. This is usually done at the end of each time step Δt when other processes in the model have contributed to a positive Δq.

Calculation of the saturation specific humidity q_s is usually carried out with the use of various approximations, such as *Teten's formula* given below. In this formula, saturation vapor pressure is expressed by.

$$e_s = 6.11 \exp\left(\frac{a(T-273.16)}{T-b}\right), \tag{8.104}$$

and the saturation specific humidity is given by

$$q_s \cong \frac{0.622 e_s}{p}, \tag{8.105}$$

where the constants a and b are defined in terms of saturation over water ($a = 17.26,\ b = 35.86$) or over ice ($a = 21.87,\ b = 7.66$). Teten's formula has been tested and found to be a reasonable approximation for the construction of moist adiabats in the troposphere.

8.4.2 *Using a Local Moist Adiabat*

In the simplest formulation for nonconvective heating,

$$H_{NC} = -L\frac{dq_s}{dt}$$

is approximated by

$$H_{NC} = -L\omega\frac{\partial q_s}{\partial p}, \tag{8.106}$$

where $\partial q_s / \partial p$ is measured along a moist adiabat. From Teten's formula we can obtain $\partial q_s / \partial p$

$$q_s = 0.622 \times 6.11 \frac{\exp\left(\dfrac{a(T_s - 273.16)}{T_s - b}\right)}{p},$$

$$\frac{\partial q_s}{\partial p} = 0.622 \times 6.11 \frac{\exp\left(\dfrac{a(T_s - 273.16)}{T_s - b}\right) - p\dfrac{\partial}{\partial p}\exp\left(\dfrac{a(T_s - 273.16)}{T_s - b}\right)}{p^2},$$

where

$$\begin{aligned}\frac{\partial}{\partial p}\exp\left(\frac{a(T_s - 273.16)}{T_s - b}\right) &= \frac{a\dfrac{\partial T_s}{\partial p}(T_s - b) - a\dfrac{\partial T_s}{\partial p}(T_s - 273.16)}{(T_s - b)^2}\exp\left(\frac{a(T_s - 273.16)}{T_s - b}\right)\\ &= \frac{\dfrac{\partial T_s}{\partial p}a\left[(T_s - b) - (T_s - 273.16)\right]}{(T_s - b)^2}\exp\left(\frac{a(T_s - 273.16)}{T_s - b}\right)\\ &= \frac{a(273.16 - b)}{(T_s - b)^2}\frac{\partial T_s}{\partial p}\exp\left(\frac{a(T_s - 273.16)}{T_s - b}\right),\end{aligned}$$

so that

$$\begin{aligned}\frac{\partial q_s}{\partial p} &= -\frac{0.622 \times 6.11}{p}\exp\left(\frac{a(T_s - 273.16)}{T_s - b}\right)\\ &\times\left(\frac{1}{p} - \frac{a(273.16 - b)}{(T_s - b)^2}\frac{\partial T_s}{\partial p}\right).\end{aligned} \tag{8.107}$$

Conservation of moist static energy along a moist adiabat can be expressed by the conservation equation

$$gz_s + c_p T_s + Lq_s = E_s. \tag{8.108}$$

After differentiating with respect to pressure we obtain

$$g\frac{\partial z_s}{\partial p} + c_P\frac{\partial T_s}{\partial p} + L\frac{\partial q_s}{\partial p} = 0,$$

or by using the hydrostatic equation

$$-\frac{RT_s(1 + 0.61q_s)}{p} + c_P\frac{\partial T_s}{\partial p} + L\frac{\partial q_s}{\partial p} = 0. \tag{8.109}$$

Eliminating $\partial T_s / \partial p$ from (8.107) and (8.109), we obtain the relation

$$\frac{\partial q_s}{\partial p} = -\frac{0.622 \times 6.11}{p}\exp\left(\frac{a(T_s - 273.16)}{T_s - b}\right) \tag{8.11}$$

$$\times\left[\frac{1}{p}-\frac{a(273.16-b)}{(T_s-b)^2}\left(\frac{RT_s}{c_P p}(1+0.61q_s)-\frac{L}{c_P}\frac{\partial q_s}{\partial p}\right)\right].$$

We can solve for $\partial q_s/\partial p$ from this relation and obtain

$$\frac{\partial q_s}{\partial p}=-\frac{C_1C_2}{1+\frac{L}{c_p}C_1C_3}, \tag{8.111}$$

where

$$C_1=-\frac{0.622\times 6.11}{p}\exp\left(\frac{a(T_s-273.16)}{T_s-b}\right),$$

$$C_3=\left(-\frac{a(273.16-b)}{(T_s-b)^2}\right),$$

and

$$C_2=\frac{1}{p}+C_3\left(\frac{RT_s}{c_p p}(1+0.61q_s)\right).$$

The use of (8.111) within (8.106) provides the desired heating function. Stable rainfall rate is expressed by the integral

$$R_s=\frac{1}{g}\int_0^{p_s}\frac{H_{NC}}{L}dp. \tag{8.112}$$

8.4.3 *Scheme Based on Saturation Conditions*

An approach for the estimation of large-scale condensation can also be obtained from a complete form of the relative humidity equation. The heating is obtained by seeking a condition based on the local change of relative humidity set to zero. Saturation is maintained by a number of dynamic, thermodynamic, and moist processes. Here we assume first that the air is absolutely stable and is undergoing dynamic large-scale ascent.

Relative humidity is defined as $r=q/q_s$, or $q=rq_s$. Here q is the specific humidity, q_s is its saturation value, and r is the relative humidity. Using the horizontal advective operator

$$\frac{d}{dt}=\frac{\partial}{\partial t}+\vec{V}_H\cdot\nabla_p,$$

we obtain

$$\frac{dq}{dt}=r\frac{dq_s}{dt}+q_s\frac{dr}{dt}. \tag{8.113}$$

Since q_s is a function of potential temperature θ, we may write

$$\frac{dq}{dt}=r\left.\frac{dq_s}{d\theta}\right|_p\frac{d\theta}{dt}+q_s\frac{dr}{dt}. \tag{8.114}$$

We next define saturation specific humidity by

$$q_s \cong \frac{\in e_s}{p}, \tag{8.115}$$

where $\in = 0.622$, e_s denotes the saturation vapor pressure, and p stands for pressure. Hence, using the Clausius-Clapeyron equation, we obtain

$$\left.\frac{dq_s}{dT}\right|_p = \left.\frac{\in}{p}\frac{\partial e_s}{\partial T}\right|_p = \frac{\in Lq_s}{RT^2}, \tag{8.116}$$

where T denotes the temperature. Hence

$$\left.\frac{dq_s}{d\theta}\right|_p = \left.\frac{\in Lq_s}{RT^2}\frac{\partial T}{\partial \theta}\right|_p = \frac{\in Lq_s}{RT\theta}. \tag{8.117}$$

Now we can rewrite the relative humidity equation (8.114) as

$$\frac{dq}{dt} = \frac{\in Lrq_s}{RT\theta}\frac{d\theta}{dt} + q_s\frac{dr}{dt}. \tag{8.118}$$

We next substitute for dq/dt and $d\theta/dt$ from the moisture conservation equation and the first law of thermodynamics, respectively, that is,

$$\frac{dq}{dt} = -\omega\frac{\partial q}{\partial p} - P + g\frac{\partial F_q}{\partial p} + D_q \tag{8.119}$$

and

$$\frac{d\theta}{dt} = -\omega\frac{\partial \theta}{\partial p} + H + g\frac{\partial F_\theta}{\partial p} + D_\theta, \tag{8.120}$$

where P denotes the precipitation rate and F_q and F_θ denote vertical eddy fluxes of moisture and heat, respectively. H represents all forms of diabatic heating. D_q and D_θ are the horizontal diffusion of q and θ, respectively.

After substituting (8.119) and (8.120) into (8.118), we obtain

$$\begin{aligned}
\frac{dr}{dt} &= -\frac{1}{q_s}\left[\frac{\in Lrq_s}{RT\theta}\left(-\omega\frac{\partial \theta}{\partial p} + H + g\frac{\partial F_\theta}{\partial p} + D_\theta\right)\right. \\
&\quad \left. -\left(-\omega\frac{\partial q}{\partial p} - P + g\frac{\partial F_q}{\partial p} + D_q\right)\right] \\
&= -\frac{\omega}{q_s}\frac{\partial q}{\partial p} - \frac{\in Lr}{RT\theta}\left(-\omega\frac{\partial \theta}{\partial p} + H + g\frac{\partial F_\theta}{\partial p} + D_\theta\right) \\
&\quad + \frac{1}{q_s}\left(g\frac{\partial F_q}{\partial p} - P + D_q\right).
\end{aligned} \tag{8.121}$$

The first term on the right-hand side denotes relative humidity change from the vertical advection of moisture, the second term within the brackets denotes the effect of temperature change on relative humidity, and the last term denotes the effect from changes in specific humidity due to vertical fluxes, horizontal diffusion, and precipitation. If we do not permit supersaturation, then the saturation limit is governed by a balance condition (i.e., $\partial r/\partial t = 0$) which can be expressed as

$$\frac{\in Lr}{RT\theta}H_{NC} = -\frac{\omega}{q_s}\frac{\partial q_s}{\partial p} - \frac{\in Lr}{RT\theta}\left(-\omega\frac{\partial \theta}{\partial p} + H_x + g\frac{\partial F_\theta}{\partial p} + D_\theta\right)$$
$$+\frac{1}{q_s}\left(g\frac{\partial F_q}{\partial p} - P + D_q + q_s\vec{V}\cdot\nabla r\right). \tag{8.122}$$

Here we have set $\partial r/\partial t = 0$ and also divided the heating into two parts, H_{NC} and H_x, where H_{NC} denotes large-scale condensational heating and H_x denotes all other forms diabatic heating.

Noting that the large-scale condensation is

$$P = \frac{H_{NC}}{L}c_P\frac{T}{\theta} \tag{8.123}$$

and $r = 1$ for saturation, we obtain

$$H_{NC} = \frac{1}{c_P\dfrac{T}{\theta Lq_s} + \dfrac{\in L}{RT\theta}}\left[-\frac{\omega}{q_s}\frac{\partial q_s}{\partial p} + \frac{\omega \in L}{RT\theta}\frac{\partial \theta}{\partial p} - \frac{\in Lr}{RT\theta}\right.$$
$$\left.\left(H_x + g\frac{\partial F_\theta}{\partial p} + D_\theta\right) + \frac{1}{q_s}\left(g\frac{\partial F_q}{\partial p} + D_q + q_s\vec{V}\cdot\nabla r\right)\right]. \tag{8.124}$$

The first bracketed term on the right-hand side denotes the moist adiabatic heating, while the remaining terms are somewhat less important. Thus the computation of large-scale condensational heating H_{NC} and the associated precipitation P requires the estimation of all of these terms.

What do these other smaller terms imply? They state such things as the following: If any process, adiabatic or diabatic, in the first law of thermodynamics raises (or lowers) the temperature, then the saturation specific humidity accordingly increases (or decreases) and saturation takes a little different supply of moisture to achieve $\partial r/\partial t = 0$. This sort of interpretation is required for each of the terms.

In the monsoon region, the cumulonimbus anvils and the cloud debris from previously active deep convection can last for long periods in the form of active stable clouds in the middle and upper troposphere. As one proceeds from tropical waves to tropical depressions to hurricanes and to intense extratropical cyclones, the percent of nonconvective rain increases. The successive increases for these categories are generally close to 15%, 25%, 40%, and 75%, respectively.

Thus, stable heating H_{NC} can be very important. Although precipitation occurs from long-lasting clouds, not much of this precipitation reaches the ground since it undergoes considerable evaporation. That is especially true for cold air blowing from the north (in the Northern Hemisphere), which is usually relatively dry and contributes to a significant evaporation of the falling rain from overriding warmer moist air.

8.5 Parameterization of Radiative Processes

In numerical weather prediction and climate modeling, the following details of radiative parameterization need consideration: longwave and shortwave radiative transfer, effects of clouds, diurnal change, and surface energy balance. We first introduce the concepts of radiative transfer for longwave and shortwave radiation for clear and cloudy skies based on two current methods. The first of these is based on an emissivity-absorptivity method. The second utilizes a band model for radiative transfer. Next we address some of the current methods for the specification of clouds within the radiative transfer computations.

The surface energy balance is an integral part of the vertical column irradiances. Surface energy balance addresses the balance between the net radiation reaching the earth's surface and the radiant energy emitted from the earth's surface. The components of the surface energy balance include shortwave and longwave irradiances as well as the fluxes of latent and sensible heat from the earth's surface. The soil heat flux, the ground hydrology, and diurnal changes are important aspects of this problem. The surface energy balance needs to be closely coupled to the surface similarity theory discussed in Section 8.2.3.

8.5.1 *The Emissivity-Absorptivity Model*

Shortwave Radiation (*the Absorptivity Method*). The zenith angle ζ is given by the relation

$$\cos\zeta = \sin\phi\sin\delta + \cos\phi\cos\delta\cos hr\,, \tag{8.125}$$

where ϕ is the latitude, δ is the declination of the sun, and hr is the hour angle of the sun (measured from local solar noon, e.g., six hours = 90°). The declination of the sun is its angular distance north (+) or south (-) of the celestial equator.

The *optical depth* of the atmosphere is a function of the mixing ratio of the atmospheric constituents, the pressure, and temperature distribution. It is usually expressed by the relation

$$W(p) = \frac{1}{g}\int_0^p q\left(\frac{p}{p_0}\right)^{0.85}\left(\frac{T_0}{T}\right)^{0.5} dp\,, \tag{8.126}$$

where the path length is estimated from the top of the atmosphere ($p = 0$) to a reference level p. In radiative flux calculations, $W(p)$ is frequently regarded as a vertical coordinate increasing downwards. q is regarded here as the specific humidity of the absorbing constituent. In this simple formulation, only water vapor is being considered. The empirical coefficients are from the work of Kuhn (1963).

The solar radiation incident at the top of the atmosphere is broken into a scattered and an absorbed part following Katayama (1966) and Joseph (1966). Shortwave radiation is depleted due to absorption by water vapor and Rayleigh scattering by aerosols. The treatment of aerosols is poor in the present state of the art. We consider two parts (i.e., the scattered and the absorbed part) of shortwave radiation. Thus following Katayama we write

$$\text{scattered part} = S^s = 0.651 S_0 \cos\zeta ,$$
$$\text{absorbed part} = S^a = 0.349 S_0 \cos\zeta ,$$

where S_0 is the solar constant (the currently accepted value is approximately 1367 W m^{-2}). These two parts are handled somewhat differently in numerical calculations. It should be noted that we are primarily interested in illustrating a computational procedure for estimating the role of shortwave radiative warming of the atmosphere and the earth's surface.

In the following analysis, we first omit the attenuation of shortwave radiation by clouds. From empirical studies, Joseph (1966) has defined an *absorptivity function* $A(W)$. This function defines the depletion of solar radiation by the absorbing constituent, e.g., water vapor. Here W is the path length through which the radiation has to pass. He defines $A(W)$ by the empirical relation

$$A(W) = 0.271 \left(W \sec\zeta\right)^{0.303} . \tag{8.127}$$

The absorbed part of the direct solar radiation reaching a reference level i is written as

$$S^a \left[1 - A\left(W_i \sec\zeta\right)\right] . \tag{8.128}$$

In order to estimate the net downward flux of shortwave radiation at a reference level of the atmosphere, one should take into account the amount of diffuse radiation that comes up from the earth's surface. Here one should take into account the albedo of the earth's surface and also consider the absorptivity of the layer between the earth's surface and the reference level so that $S^a[1 - A(W_0)]\alpha_s$ is the amount of diffuse shortwave radiation that is reflected by the earth's surface. Note that the diffuse radiation is not a function of the zenith angle. Here α_s denotes the albedo of the earth's surface. The diffuse radiation, in general, experiences a longer path length as compared to direct solar radiation.

Following Joseph (1966), the absorptivity for diffuse radiation is expressed by $A(1.66W)$ instead of $A(W)$. The factor 1.66 was shown to account for the increased path length for diffuse radiation. Hence we can write an expression for the diffuse radiation that reaches a level i from the earth's surface, that is,

$$S^a \left[1 - A\left(W_0 \sec\zeta\right)\right] \alpha_s \left\{1 - A\left[1.66\left(W_0 - W_i\right)\right]\right\} . \tag{8.129}$$

The total downward flux of shortwave radiation at the level i is given by the relation

$$\begin{aligned} Si = {} & S^a \left[1 - A\left(W_i \sec\zeta\right)\right] \\ & - S^a \left[1 - A\left(W_0 \sec\zeta\right)\right] \alpha_s \left\{1 - A\left[1.66\left(W_0 - W_i\right)\right]\right\} . \end{aligned} \tag{8.130}$$

Next we outline the inclusion of clouds. We only illustrate a single-layer cloud configuration. S^a is the absorbed part of the shortwave radiation at the top of the atmosphere. $S^a[1-A(W_i \sec\zeta)]$ is the amount of shortwave radiation that reaches a reference level *i* just above the cloud level. We place this single cloud below the reference level *i*. The diffuse radiation that emanates upwards at the cloud level is determined by $S^a[1-A(W_{ct} \sec\zeta)]\alpha_c$, where W_{ct} is the path length at the cloud-top level and α_c is the albedo of the cloud. Part of this diffuse radiation would be absorbed before its arrival at reference level *i*. The upward diffuse radiation that reaches level *i* would be expressed as

$$S^a[1-A(W_{ct} \sec\zeta)]\alpha_c\left\{1-A\left[1.66(W_{ct}-W_i)\right]\right\}. \tag{8.131}$$

The net absorbed downward flux of shortwave radiation passing through a reference level *i* when there is a cloud layer present below is expressed as

$$\begin{aligned} Si^a &= S^a\left[1-A(W_i \sec\zeta)\right] \\ &-S^a\left[1-A(W_{ct} \sec\zeta)\right]\alpha_c\left\{1-A\left[1.66(W_{ct}-W_i)\right]\right\}. \end{aligned} \tag{8.132}$$

The next step is to examine the amount of shortwave radiation that passes through a cloud layer. To do this, one should define the absorptivity of the cloud. Since there are both liquid water and water vapor within clouds, Katayama (1966) defines the absorptivity of the clouds by a function $A(W_{ci}^*)$, where W_{ci}^* is an augmented path length which takes into account the equivalent amount of water vapor within the cloud. If S^a is the absorbed part of the shortwave radiation reaching the top of the atmosphere, then we write

$$S^a\left[1-A(W_{ct} \sec\zeta)\right](1-\alpha_c) \tag{8.133}$$

as the amount enters the cloud from above. The amount that reaches below the cloud is written as

$$S^a\left[1-A(W_{ct} \sec\zeta)\right](1-\alpha_c)\left[1-A(W_{ci}^*)\right]. \tag{8.134}$$

If we want to know the downward flux of net shortwave radiation below a single-cloud atmosphere, then we have to consider the upward flux of diffuse shortwave radiation that comes up from the earth's surface. This latter calculation should be performed similar to the cloud-free case. The total downward flux of absorbed shortwave radiation at a reference level *i* below a single-cloud atmosphere is thus given by

$$\begin{aligned} Si^a = S^a&\left[1-A(W_{ct} \sec\zeta)\right](1-\alpha_c)\Big\langle 1-A\left[W_{ci}^*+1.66(W_i-W_{cb})\right] \\ &-\left\{1-A\left[W_{ci}^*+1.66(W_0-W_{cb})\right]\right\}\alpha_s \\ &\times\left\{1-A\left[1.66(W_0-W_i)\right]\right\}\Big\rangle, \end{aligned} \tag{8.135}$$

where W_i is the path length at the reference level, W_0 is the path length at the ground, W_{cb} is the path length at the cloud base, W_{ci}^* is the equivalent path length of the cloud, α_c is the albedo of the cloud, and α_s is the albedo of the earth's surface. If there exists more than one cloud layer, then we carry out a simple logical extension of the above analysis.

Thus far we have not addressed the scattered part of the shortwave radiation. In general we can say that the rate of warming of the atmosphere by the scattered part of shortwave radiation is very small. This scattered part cannot, however, be neglected in the energy balance of the earth's surface. Following Chang (1978), we present two empirical formulas that are frequently used to define the scattered part of the shortwave radiation:

$$\alpha_0 = 0.085 - 0.245 \ln\left(\frac{p_s}{p_0}\cos\zeta\right), \tag{8.136}$$

where α_0 is the albedo of the atmosphere, p_s is the surface pressure, and p_0 is 1000 mb. The scattered part of the solar radiation reaching the earth's surface is given by

$$S_g^s = S^s(1-\alpha_0)(1-\alpha_0\alpha_s), \tag{8.137}$$

where S^s is the scattered part at the top of the atmosphere and α_s is the albedo of the earth's surface. It can be shown that, over periods of the order of several days, this is not a negligible effect.

Longwave Radiation (the Emissivity Method). All of the longwave radiation originates at the earth's surface or from the atmosphere (and clouds). The atmosphere absorbs longwave radiation much more strongly than solar radiation. Among ozone, water vapor, and carbon dioxide, the absorption by water vapor is considered here. A similar formulation is needed for the other constituents. The water vapor absorption is strong around 6 and 20μm (in the vibrational and the rotational bands, respectively). The atmosphere both absorbs and re-emits longwave radiation.

We start from *Schwartzchild's equation,*

$$dF_\lambda = k_\lambda\left[\phi(\lambda,T) - E_\lambda\right]du, \tag{8.138}$$

where k_λ is the absorption coefficient, dF_λ is the flux change in a layer of optical thickness du, $\phi(\lambda,T)$ is the blackbody emission as given by Planck's equation, and E_λ is the flux density at wavelength λ. By Kirchhoff's law, the emissivity of the layer is equal to the absorptivity, $k_\lambda du$. The above equation expresses the difference between absorption and emission in a layer. In principle, this equation can be used for a model atmosphere. However, it is not very well suited for line absorbers since k_λ varies greatly.

We describe some simple calculation procedures for the evaluation of longwave radiative flux divergence. The aim in the end is to evaluate the rate of longwave heating or cooling at the earth's surface and in the atmosphere. Calculations for clear and cloudy sky situations are illustrated.

First we consider the cloud-free case and describe the so-called emissivity method for estimating longwave radiative effects. We examine the upward flux of longwave radiation at a reference level *i*. We can divide this into two parts. The part that comes up to level *i* from the earth's surface (whose temperature is T_g) may be written as

$$F_g \uparrow = \sigma T_g^4 \left[1 - \in \left(W_0 - W_i \right) \right], \tag{8.139}$$

and the part which is emitted by the layer between the reference level *i* and the ground is given by

$$F_{gi} \uparrow = \int_{W_i}^{W_0} \sigma T^4 \frac{\partial}{\partial W} \in (W - W_i)\, dW, \tag{8.140}$$

where σ is the Stefan-Boltzmann constant, W_0 is the path length at the ground, W_i is the path length at the reference level, and $\in$ is the emissivity.

It is possible to make use of tables of emissivity as a function of path length to obtain reliable estimates of the longwave fluxes (Kuhn 1963). Rodgers (1967) has shown that emissivity tabulations yield results nearly as good as those one obtains from exact integration of the transfer equations. The error estimates are of the order of 0.1° C day^{-1} in the atmosphere. This is tolerable for most purposes.

The total upward flux of longwave radiation at a reference level *i* in the cloud-free case is given by the sum of the two terms, that is,

$$F_i \uparrow = \sigma T_g^4 \left[1 - \in \left(W_0 - W_i \right) \right] + \int_{W_i}^{W_0} \sigma T^4 \frac{\partial \in (W - W_i)}{\partial W} dW. \tag{8.141}$$

Here $\partial \in / \partial W$ is a measure of the change of emissivity with respect to optical depth. The downward flux in a cloud-free case is given by just one term, i.e.,

$$F_i \downarrow = -\int_0^{W_i} \sigma T^4 \frac{\partial \in (W_i - W)}{\partial W} dW + D. \tag{8.142}$$

Here *D* stands for the incoming longwave radiation at the top of the model.

If we have one cloud layer above the reference level, then the cloud will affect the downward flux at the reference level. In this case we write

$$F_i \downarrow = \sigma T_{cb}^4 \left[1 - \in \left(W_i - W_{cb} \right) \right] - \int_{W_{cb}}^{W_i} \sigma T^4 \frac{\partial \in (W_i - W)}{\partial W} dW, \tag{8.143}$$

where T_{cb} is the temperature at the cloud base and W_{cb} is the path length at that level. If there is one cloud layer below the reference level *i*, then the formula for the upward longwave radiative flux would be

$$F_i \uparrow = \sigma T_{ct}^4 \left[1 - \in \left(W_{ct} - W_i \right) \right] + \int_{W_i}^{W_{ct}} \sigma T^4 \frac{\partial \in (W - W_i)}{\partial W} dW. \tag{8.144}$$

Multiple cloud layers require a logical extension of the above illustrated principle.

The question of what one should do within a cloud remains an unsolved problem at this stage. For simplicity, one could set the net heating (or cooling) equal to zero if the reference level falls within a cloud layer. If we let $F = F_i \downarrow - F_i \uparrow$ represent the longwave radiative flux at any level, then warming (or cooling) would be determined by the divergence (or convergence) of flux, i.e.,

$$c_p \left. \frac{\partial T}{\partial t} \right|_{\text{longwave}} = -g \frac{\partial F}{\partial p}. \tag{8.145}$$

The negative sign is consistent with flux convergence, noting that F is positive for downward flux.

8.5.2 *The Band Model: Longwave Radiation*

The main absorber of solar radiation in the troposphere is water vapor, while the main absorbers of longwave radiation are water vapor, carbon dioxide, and ozone. The spectral line-by-line computation of shortwave and longwave fluxes is in practice difficult and computationally very expensive. Current radiative transfer models avoid this problem by resorting to certain assumptions regarding the distribution of spectral lines in the various absorption bands and constructing approximate band models.

We describe a band model based on the radiation scheme of the University of California, Los Angeles/Goddard Laboratory for Atmospheric Studies (UCLA/GLAS) general circulation model (GCM) (Harshvardhan and Corsetti 1984). The equations for the upward and downward fluxes in a clear sky can be expressed as

$$F_{clr}\uparrow(p)=\int_{\Delta\nu}\left(B_\nu(T_s)\tau_\nu(p,p_s)+\int_{p_s}^{p}B_\nu[T(p')]\frac{d\tau_\nu(p,p')}{dp'}dp'\right)d\nu, \tag{8.146}$$

$$F_{clr}\downarrow(p)=\int_{\Delta\nu}\left(\int_{p_t}^{p}B_\nu[T(p')]\frac{d\tau_\nu(p,p')}{dp'}dp'\right)d\nu. \tag{8.147}$$

These equations are for a clear sky case integrated over the spectral range $\Delta\nu$. These equations are quite analogous to (8.141) and (8.142). Here $B_\nu(T)$ is the blackbody flux at surface temperature T and wavelength ν, p_s is the surface pressure, p_t is the pressure at the top of the atmosphere, $T(p')$ is the air temperature at pressure p', $\tau_\nu(p,p')$ is the diffuse transmittance between levels p and p', and ν is the spectral wavenumber. Integration of (8.146) and (8.147) gives

$$F_{clr}\uparrow(p)=B\left[T(p)\right]+G(p,p_s,T_s)-G\left[p,p_s,T(p_s)\right]+\int_{T(p)}^{T(p_s)}\frac{\partial G[p,p',T(p')]}{\partial T}dT(p'), \tag{8.148}$$

$$F_{clr}\downarrow(p)=B\left[T(p)\right]+G\left[p,p_t,T(p_t)\right]+\int_{T(p)}^{T(p_t)}\frac{\partial G[p,p',T(p')]}{\partial T}dT(p'), \tag{8.149}$$

where

$$B\left[T(p)\right]=\int_\nu B_\nu(T)d\nu,$$

$$G(p,p',T)=\int_{\Delta\nu}\tau_\nu(p,p')B_\nu(T)d\nu,$$

$$\frac{\partial G(p,p',T)}{\partial T}=\int_{\Delta\nu}\tau_\nu(p,p')\frac{\partial B_\nu(T)}{\partial T}d\nu.$$

The spectral width of the 9.6μm and 15μm bands over which carbon dioxide and ozone, respectively, absorb the terrestrial radiation are narrow enough to use a mean value of the transmission function. The equation for G then becomes

$$G(p,p',T) = \int_{\Delta\nu} \tau_\nu(p,p') \frac{d\nu}{\Delta\nu} \int_{\Delta\nu} B_\nu(T)\, d\nu . \tag{8.150}$$

The radiative cooling rate or the divergence of net flux is the final output of the model. It is given by

$$-\frac{dT}{dt} = \frac{g}{c_p} \frac{d(F\downarrow - F\uparrow)}{dp} . \tag{8.151}$$

The model uses (8.148) and (8.149) to calculate the longwave flux of the atmosphere. Special considerations are given for overcast or partial cloud cover. The surface flux is also calculated.

The methods for solving the fluxes vary slightly for each type of radiatively active atmospheric constituent. The water vapor flux is solved by the methods of Chou and Arking (1980) and Chou (1984). The carbon dioxide flux is solved by the method of Chou and Peng (1983). The ozone flux is solved by the method of Rodgers (1968). The method used to compute each band is described in detail below.

Water Vapor Bands. The IR spectrum is divided into the water vapor bands, the 15-μm band and the 9.6-μm band. The spectral ranges for the water vapor band centers and wings are listed in Table 8.1. According to Chou and Arking (1980), the diffuse transmittance associated with a molecular line at wavenumber ν is

$$\tau_\nu(p_1,p_2) = 2\int_0^1 \exp\left(\frac{-k_\nu(p_r,T_r)\, w(p_1,p_2)}{\mu}\right) \mu\, d\mu , \tag{8.152}$$

where k_ν is the molecular line-absorption coefficient, T is the temperature, μ is the cosine of the zenith angle, p is the pressure, p_r is the reference pressure, T_r is the reference temperature (p_r and T_r are listed in Table 8.1), and w is the scaled water vapor amount given by

$$w(p_1,p_2) = \int_{p_1}^{p_2} \frac{p}{p_r} \frac{R[T(p)]q(p)}{g} dp , \tag{8.153}$$

where g is gravity, q is the water vapor mixing ratio, and $R(T)$ is the temperature scaling factor which is given by $R(T) = \exp[r(T - T_r)]$. Here r is a factor from Chou (1984) and is also listed in Table 8.1.

The diffuse transmittance associated with e-type absorption is

$$\tau_\nu(p_1,p_2) = 2\int_0^1 \exp\left(\frac{-\sigma_\nu(T_0)u(p_1,p_2)}{\mu}\right) \mu\, d\mu , \tag{8.154}$$

where σ_ν is the e-type absorption coefficient, $T_0 = 296$ K, and u is a scaled water vapor

Table 8.1. Water vapor absorption parameters from Chou (1984).

	H_2O Band Center	H_2O Band Wing	15-μm Band	9.6-μm Band
Spectral Range (cm^{-1})	0-340 1380-1900	340-540 800-980 1100-1380 1900-3000	540-800	980-1100
p_r (mb)	275	550	550	-
T_r (K)	225	256	256	-
r (K^{-1})	0.005	0.016	0.016	-

amount given by

$$u(p_1,p_2)=\int_{p_1}^{p_2} e(p)\exp\left[1800\left(\frac{1}{T(p)}-\frac{1}{T_0}\right)\right]\frac{q(p)}{g}dp. \tag{8.155}$$

In this case, $e(p)$ is the water vapor pressure (in units of atmospheres), $T_0 = 296\text{ K}$, and 1800 represents 1800 K, a temperature-dependent constant. Roberts et al. (1976) state that this is in accord with the fact that the hydrogen bond between two water molecules is in the neighborhood of 3-4 kcal, which leads to ≈ 1800 K.

Broad transmission functions can be derived by averaging (8.152) and (8.154) over wide spectral intervals. In the water vapor bands, the Planck weighted transmission function is

$$\tau(w,u,T)=\int\frac{B_v(T)\tau_v(w)\tau_v(u)}{B(T)}dv, \tag{8.156}$$

where $B(T)$ is the spectrally integrated Planck flux, $B_v(T)$ is the Planck flux, $\tau_v(w)$ is the transmittance given by (8.152), and $\tau_v(u)$ is the transmittance given by (8.154). Since $\tau(w,u,T)$ is a slowly varying function of temperature, it can be fitted by the quadratic function (Chou 1984)

$$\tau(w,u,T)=\tau(w,u,250)\left[1+\alpha(w,u)(T-250)+\beta(w,u)(T-250)^2\right],$$

where $\alpha(w,u)$ and $\beta(w,u)$ are the regression coefficients and $\tau(w,u,250)$ is the standardized transmission function given by Chou (1984).

The band center region absorption can be neglected because it is dominated by molecular lines and the effect due to e-type absorption is small. Therefore in the tables of Chou (1984), the last column shows that τ, α, and β are functions of w only. Equation (8.156) gives the Planck weighted transmission function to within an error of less than 0.002.

For the 15-μm region, Chou (1984) has fitted the diffuse transmission function averaged over the entire band as

$$\tau(w) = \exp\left(\frac{-6.7w}{1+16w^{0.6}}\right), \tag{8.157}$$

$$\tau(u) = \exp\left(-27u^{-0.83}\right). \tag{8.158}$$

Since the molecular line absorption is weak in the 9.6-μm region, only the e-type absorption is considered. Chou (1984) has also fitted the diffuse transmission associated with the e-type absorption in this region as

$$\tau(u) = \exp(-9.79u). \tag{8.159}$$

The difference between the transmission functions using (8.157), (8.158), and (8.159) and those derived from (8.152) and (8.154) is less than 0.015 ($<5\%$ of the mean absorption).

CO_2 Bands. The spectral thermal radiative flux in a nonscattering atmosphere can be found by integrating the Schwartzchild equation,

$$F_\nu(p)\downarrow = \int_0^p \frac{B_\nu[T(p')]d\tau_\nu(p,p')}{dp'}dp', \tag{8.160}$$

where $\tau_\nu(p,p')$ is the transmittance averaged over zenith angles θ and is given by

$$\tau_\nu(p,p') = 2\int_0^1 \exp\left(\frac{-u_\nu(p,p')}{\mu}\right)\mu d\mu. \tag{8.161}$$

Also,

$$u_\nu(p,p') = \int_{p'}^p \frac{c(p'')k_\nu\left[p'',T(p'')\right]}{g}dp'', \tag{8.162}$$

where $c(p'')$ is the CO_2 concentration and $k_\nu[p'',T(p'')]$ is the absorption coefficient. For a small enough spectral interval $\Delta\nu_i$, integration of (8.160) gives

$$F_{\nu_i}(p)\downarrow = \int_0^p \frac{B_i[T(p')]d\tau_i(p,p')}{dp'}dp', \tag{8.163}$$

where $B_i[T(p')]$ is the spectrally integrated Planck flux and $\tau(p,p')$ is the spectrally averaged diffuse transmittance given by

$$\tau_i(p,p') = \int_{\Delta\nu_i} \frac{\tau_\nu(p,p')}{\Delta\nu_i}d\nu. \tag{8.164}$$

With (8.164) we now have an equation where the spectrally averaged diffuse transmittance depends on the transmittance at a single wavenumber τ_ν. However, we see from (8.161) and (8.162) that τ_ν is dependent on the absorption coefficient, which is a function of wavenumber, temperature, and pressure. Therefore, τ_ν requires computations at numerous points in the spectral interval for each atmospheric situation in order to find the mean diffuse transmittance. Fortunately, the computations can be simplified by relating the absorption coefficient to a reference pressure p_r and a reference temperature T_r through a scaling function $f(p,T)$ such as

$$k_\nu(p,T) = k_\nu(p_r,T_r)f(p,T). \tag{8.165}$$

We may now combine and rewrite (8.161) and (8.162) as

$$\tau_\nu(p,p') = 2\int_0^1 \exp\left(\frac{-k_\nu(p_r,T_r)w(p,p')}{\mu}\right)\mu d\mu, \tag{8.166}$$

where $w(p,p')$ is the scaled CO_2 amount and is given by

$$w(p,p') = \int_{p'}^{p} \frac{c(p'')f\left[p'',T(p'')\right]}{g} dp''. \tag{8.167}$$

Here, (8.166) treats an inhomogeneous atmosphere as a homogeneous atmosphere with a constant pressure and temperature of p_r and T_r. This is possible because (8.167) scales the CO_2 concentration to simulate the absorption in an inhomogeneous atmosphere.

Equation (8.165) separates the absorption coefficient from the pressure and temperature variables. We may now write (8.164) to express the mean transmittance averaged over the spectral interval $\Delta\nu_i$ as a function of only the scaled absorber amount w as

$$\tau_i(w) = \int_{\Delta\nu_i} \frac{\tau_\nu(w)}{\Delta\nu_i} d\nu. \tag{8.168}$$

We can now accurately precompute τ_i as a function of w, with $k_\nu(p_r,T_r)$ obtained from accurate line-by-line calculations. τ_i may then be stored in a look-up table for quick and efficient use. Chou and Peng (1983) have shown that the spectrally averaged diffuse transmittance τ_i can be accurately precomputed from the analytical function

$$\tau_i(w) = \exp\left(\frac{-aw}{1+bw^n}\right), \tag{8.169}$$

where *a, b,* and *n* are chosen for individual spectral intervals such that the computed root-mean-square error in $\tau_i(w)$ is minimized. Chou and Peng (1983) have computed values for *a, b,* and *n*. They are shown in Table 8.2.

The absorption coefficients at the selected reference levels are accurately computed using line-by-line methods. Radiative transfer in the regions close to the reference levels will also be accurately computed. Therefore the values of p_r and T_r should be close to the regions where accurate computations of radiative transfer are important.

One important region is the stratosphere, where radiative cooling due to CO_2 emission is a dominant factor. Another important region is the lower troposphere, where the downward radiative flux at the surface is an important component affecting the

Table 8.2. 15 μm spectral band parameters from Chou and Peng (1983).

	Band Center	Band Wings	
		Narrow	Wide
$\Delta\nu$ (cm^{-1})	620-720	580-620 720-760	540-620 720-800
p_r (mb)	30	300	300
p_c (mb)	1	1	1
m	0.85	0.85	0.50
n	0.56	0.55	0.57
a	3.1	0.08	0.04
b	15.1	0.9	0.9
T_r (K)	240	240	240
$R(T,T_r)$	0.0089	0.025	0.025

surface temperature. Cooling of the stratosphere is mostly due to the absorption band center, and cooling in the lower troposphere is mostly due to the absorption band wings. We must choose a p_r representative of the lower troposphere. Chou and Peng (1983) state that the temperature is less critical to the equations and they use a value of 240 K, which they considered to be an intermediate value for both the stratosphere and the troposphere.

The scaling function used is

$$f(p,T)=\left(\frac{p}{p_r}\right)^m R(T,T_r), \tag{8.170}$$

which is from the work of Chow and Arking (1980) . Here m is a parameter for correcting the error arising from the assumption of linear dependence of the absorption coefficient on pressure. $R(T,T_r)$ is a temperature scaling factor and is given by

$$R(T,T_r)=\exp\left[r(T-T_r)\right],$$

where $r = 0.0089$ for the band center and 0.025 for the band wings. Also, $T_r = 240$ K for both the band center and band wings. The pressures scaling in (8.170) assumes that the absorption coefficient follows the Lorentz function. This assumption is not valid at low pressures, where broadening of absorption lines due to Doppler shift is important. In the 15-μm spectral region, the height where line broadening due to molecular collision and Doppler shift is equally important is approximately 10 mb. For a Doppler line function, the absorption is independent of pressure. To account for the Doppler effect, we define a critical pressure level p_c. At pressures higher than the critical pressure, the absorption coefficient is independent of pressure.

Table 8.3. Ozone coefficients from Rodgers (1968).

k (cm g^{-1})	a (cm^{-1})	α (cm^{-1})	Δ (cm^{-1})
208	81.21	0.28	0.1

We may write (8.170) as

$$f(p,T)=\left(\frac{p_c}{p_r}\right)^m R(T,T_r) \quad \text{for} \quad p<p_c, \tag{8.171}$$

where the values for p_r, T_r, p_c, and m were obtained empirically by Chou and Peng (1983) and are listed in Table 8.2.

Ozone Bands. Rodgers (1968) defines a transmission function for a Lorentz line shape as

$$T(k_i,m,p)=\exp\left\{-\frac{\pi\alpha p}{2\delta}\left[\left(1+\frac{4km}{\pi\alpha p}\right)^{0.5}-1\right]\right\}, \tag{8.172}$$

where α is the line width at one atmosphere, δ is the mean spectral interval, k is the line strength, and m is the ozone concentration. Rodgers (1968) used this transmission function to find the 9.6-μm ozone band absorption. The absorption equation is

$$A(m,p)\approx\sum_i a_i\left[1-T(k_i,m,p)\right], \tag{8.173}$$

where a_i is the spectral interval for the ith absorption band.

The values of the coefficients were chosen empirically by Rodgers (1968) and are listed in Table 8.3. Using the fact that transmission can be defined as one minus the absorption, Harshvardhan and Corsetti (1984) obtained the ozone transmission through the atmosphere simply by subtracting $A(m, p)$ from 1.0.

Treatment of Clouds. Consider a simple five-layer atmosphere with only one cloud layer. The fractional cloud cover in that layer is defined as *N*. Equation (8.149) for a level below the cloud may then be rewritten as

$$\begin{aligned} F_{cld}\downarrow(p) &= B\left[T(p)\right]-(1-N)G\left[p,p_t,T(p_t)\right] \\ &+(1-N)\int_{T(p)}^{T(p_t)}\frac{\partial G\left[p,p',T(p')\right]}{\partial T}dT(p') \\ &+N\int_{T(p)}^{T(p_{cb})}\frac{\partial G\left[p,p',T(p')\right]}{\partial T}dT(p'), \end{aligned} \tag{8.174}$$

where cb is the cloud base, t is the top of the atmosphere, N is the fractional cloud cover in the layer, and $1-N$ is the probability of a clear line of sight from p to p_t.

By splitting the limits of integration, (8.174) may be written as

$$F_{cld}\downarrow(p)=B\left[T(p)\right]-(1-N)G\left[p,p_t,T(p_t)\right]$$
$$+(1-N)\int_{T(p_{cb})}^{T(p_t)}\frac{\partial G\left[p,p',T(p')\right]}{\partial T}dT(p')$$
$$+\int_{T(p)}^{T(p_{cb})}\frac{\partial G\left[p,p',T(p')\right]}{\partial T}dT(p')\,. \tag{8.175}$$

The probability of a clear line of sight between any two levels will always be in the range from zero to one, as shown below.

Probability	Levels
$1-N$	p to p_t
$1-N$	p to p' (where p' is between p_{cb} and p_t)
1.0	p to p' (where p' is between p and p_{cb})

Equation (8.175) may now be written as

$$F_{cld}\downarrow(p)=B\left[T(p)\right]-C(p,p_t)G\left[p,p_t,T(p_t)\right]$$
$$+\int_{T(p)}^{T(p_t)}C(p,p')\frac{\partial G\left[p,p',T(p')\right]}{\partial T}dT(p')\,, \tag{8.176}$$

where $C(p,p')$ is the probability of a clear line of sight from p to p'. Equation (8.148) may be manipulated in the same way to yield

$$F_{cld}\uparrow(p)=B\left[T(p)\right]+C(p,p_s)\left\{G(p,p_s,T_s)-G\left[p,p_s,T(p_s)\right]\right\}$$
$$+\int_{T(p)}^{T(p_s)}C(p,p')\frac{\partial G\left[p,p',T(p')\right]}{\partial T}dT(p')\,. \tag{8.177}$$

Equations (8.176) and (8.177) are still valid even when there is more than one cloud layer. The longwave parameterization utilizes a random overlap of clouds. The random overlap of clouds is equal to the product of all fractional cloud amounts for all levels that have clouds.

8.5.3 *The Band Mode: Shortwave Radiation*

The solar radiation at the earth's surface and in the atmosphere is the initial source of energy causing atmospheric motions. The main absorbers of solar radiation in the earth's atmosphere are water vapor in the troposphere and ozone in the stratosphere. Water vapor absorbs primarily in the near-infrared region, $0.7\ \mu\text{m}<\lambda<4\ \mu\text{m}$. Ozone ($O_3$) is effective in the ultraviolet region, $\lambda<0.35\ \mu\text{m}$, and in the visible region, $0.5\ \mu\text{m}<\lambda<0.7\ \mu\text{m}$.

As an introduction to understanding the shortwave radiation parameterization scheme used, the transfer of shortwave radiation through a nonscattering clear atmosphere is described first. The transfer of shortwave radiation through the atmosphere with negligible scattering is given by

$$S\downarrow(z,\mu_0)=\mu_0\int_0^\infty S_\nu(\infty)\tau_\nu(z,\infty,\mu_0)d\nu, \tag{8.178}$$

where $S\downarrow$ is the downward radiation at height z having a solar radiance $S_\nu(\infty)$ at the top of the atmosphere and inclined at a zenith angle θ_0 (or $\mu_0=\cos\theta_0$) and $\tau_\nu(z,\infty,\mu_0)$ is the monochromatic transmittance given by

$$\tau_\nu(z,\infty,\mu_0)=\exp\left(-\frac{1}{\mu_0}\int_z^\infty k_\nu du\right). \tag{8.179}$$

Here k_ν is the monochromatic absorption coefficient and u is the optical path length for the particular absorber.

A *mean transmission function* can be defined as given in Stephens (1984) by

$$\tau(z,\infty,\mu_0)=\frac{1}{\Delta\nu}\int_{\Delta\nu}\exp\left(-m_r(\mu_0)\int_z^\infty k_\nu du\right)d\nu, \tag{8.180}$$

where the relative air mass factor $m_r(\mu_0)$ is used in place of $1/\mu_0$. This factor takes into account the earth's curvature and atmospheric refraction and is given by

$$m_r(\mu_0)=\frac{35}{(1224\mu_0^2+1)^{1/2}}. \tag{8.181}$$

It is easy to see that for small zenith angles ($\mu_0\approx 1$), we obtain $m_r\approx 1/\mu_0$.

Mean transmittance can also be defined as the convolution of the transmission function and $S_\nu(\infty)$ over the entire solar spectrum, that is,

$$\tau_\nu(z,\infty,\mu_0)=\frac{1}{S(\infty)}\int_0^\infty S_\nu\tau_\nu(z,\infty,\mu_0)\,d\nu, \tag{8.182}$$

where $S(\infty)$ is the net solar radiation at the top of the atmosphere. Knowledge of the absorption coefficient k_ν and $S_\nu(\infty)$ is enough to provide the mean transmission function for a particular optical path u. The downward solar flux at height z for a nonscattering atmosphere can now be written as

$$S\downarrow(z)=\mu_0 S(\infty)\tau(z,\infty,\mu_0). \tag{8.183}$$

The transmission function $\tau(z,\infty,\mu_0)$ can also written as $\tau(\mu)$. That is, it can be written as a function of optical path. With this notation, the upward solar radiative flux at level z by reflection from the ground is similarly defined as

$$S\uparrow(z)=\mu_0 S(\infty)R_g\tau(\mu^*), \tag{8.184}$$

where R_g is the surface albedo integrated over the entire spectral range and μ^* is the effective optical path traversed by diffusively reflected radiation and can be approximated by (Lacis and Hansen 1974)

$$\mu^*=m_r(\mu_0)\mu_0+(\mu_0-u)\overline{m}, \tag{8.185}$$

where μ_0 is the total thickness up from the ground and $\overline{m}$ is an effective magnification factor for diffuse radiation.

The heating rate at height z due to the shortwave radiative flux is given by

$$\frac{\partial T}{\partial t} = -\frac{1}{\rho c_P}\frac{d}{dz}(S\uparrow - S\downarrow) = \frac{g}{c_P}\frac{d}{dp}(S\uparrow - S\downarrow), \tag{8.186}$$

where

$$\begin{aligned}\frac{d}{dp}(S\uparrow - S\downarrow) &\approx \frac{S\uparrow(p+\Delta p) - S\downarrow(p+\Delta p) - S\uparrow(p) + S\downarrow(p)}{\Delta p}\\ &= \big\langle \mu_0\, S(\infty)R_g\left\{\tau\left[u^*(p+\Delta p)\right] - \tau\left[u^*(p)\right]\right\}\\ &\quad -\mu_0 S(\infty)\left\{\tau\left[u(p+\Delta p)\right] - \tau\left[u(p)\right]\right\}\big\rangle / \Delta p\\ &= \big\langle \mu_0\, S(\infty)R_g\left\{A\left[u^*(p)\right] - A\left[u^*(p+\Delta p)\right]\right\}\\ &\quad -\mu_0 S(\infty)\left\{A\left[u(p)\right] - A\left[u(p+\Delta p)\right]\right\}\big\rangle / \Delta p.\end{aligned} \tag{8.187}$$

Here the absorption function A is given by $A = 1 - \tau$. The heating rate by shortwave radiation is therefore proportional to dA/dp. Alternatively stated, to determine the heating rate, we must first determine the absorption functions.

The parameterization of the absorption and transmission functions in the model is based on the ULCA/GLAS GCM scheme and is described to some extent by Davies (1982). It includes a parameterization for the major absorption processes in the stratosphere, troposphere, and at the earth's surface. The parameterization is a function of the water vapor distribution, the cloud coverage, the zenith angle of the sun, the albedo of the earth's surface, and the ozone distribution. In this scheme, ozone absorption and water vapor absorption are assumed to be in the above-mentioned separable spectral regions. Multiple scattering is taken into account whenever it is significant. We next give a brief description of the specification of clouds used in the scheme.

8.5.4 *Specification of Clouds*

The specification of clouds is based on threshold values of relative humidity (Slingo 1985; Dickinson and Temperton 1985). Three types of clouds (low, medium, and high) are allowed. Low clouds are assumed to be present between 900 and 700 mb, medium clouds between 700 and 400 mb, and high clouds between 400 and 100 mb. Clouds are assumed to be present when the mean relative humidity $\overline{RH}$ in a layer exceeds the threshold value RH_c. The cloud amount N for each cloud type is defined by

$$C_{L,\,M,\text{or}\,H} = \left(\frac{\overline{RH} - RH_c}{1 - RH_c}\right)^2, \quad \overline{RH} \geq RH_c, \tag{8.188}$$

where RH_c is set to 0.66, 0.50, and 0.40 for low, medium, and high clouds, respectively. When $\overline{RH} < RH_c$, the cloud amount N is set to 0. The maximum possible value of N is 1. This definition of cloud cover allows eight categories of sky conditions to be defined:

$C_1 = (1-C_L)(1-C_M)(1-C_H)$	clear sky
$C_2 = C_L(1-C_M)(1-C_H)$	low clouds only
$C_3 = C_L C_M (1-C_H)$	low and middle clouds
$C_4 = C_L C_M C_H$	low, middle and high clouds
$C_5 = C_M(1-C_L)(1-C_H)$	middle clouds only
$C_6 = C_M C_H (1-C_L)$	middle and high clouds
$C_7 = C_H(1-C_L)(1-C_M)$	high clouds only
$C_8 = C_L C_H (1-C_M)$	low and high clouds

The relative humidity distribution determines the low (C_L), middle (C_M), and high (C_H) cloud amounts. The radiation calculations are first carried out eight times, assuming a weight of one for each of the above categories. Then the radiative fluxes are weighted by fractional cloud coverage under each of the above eight categories to give a total flux at any level *i* as

$$F_i \uparrow = \sum_{n=1}^{8} C_n F_{in} \uparrow . \tag{8.189}$$

8.5.5 *Surface Energy Balance*

The heat balance of the earth's surface will now be considered. For most tropical meteorological problems, the ocean is assumed to have an infinite heat capacity. Since the ocean's diurnal temperature changes are not as large as for land areas, one does not address the problems of heat balance of the ocean surface when small time scales of atmospheric changes are considered. When one is concerned with monthly or seasonal changes, the oceanic problem becomes very important. The land surface heat balance problem is very important for the present study, since the desert areas exhibit diurnal changes of the order of 30°C to 40°C. The elements of the heat balance at the earth's surface are described next.

First we discuss sensible and latent heat fluxes. The sensible heat flux from land areas, following Chang (1978), is expressed by the relations

$$H_S = \rho c_p C_H |V_a| (T_g - T_a), \tag{8.190}$$

$$H_L = \rho L C_q |V_a| (q_g - q_a) G_w . \tag{8.191}$$

Here we assumed an unsteady planetary boundary layer over homogenous terrain. The above expressions are consistent for a log-linear profile in the constant flux layer. C_H and C_q are the stability-dependent exchange coefficients based on similarity theory as discussed in Section 8.2.3. G_w is the ground wetness parameter with values between 0 and 1.

The following steps are carried out in the computation of ground temperature T_g. We first estimate the net downward flux of shortwave radiation $F_S \downarrow$ and longwave radiation $F_L \downarrow$ using emissivity or a band radiation model as described in Sections 8.5.1 and 8.5.2. The soil moisture is estimated from an empirical relation based on the surface albedo, that is,

$$G_w = 0.85\left(1 - \exp^{-200(0.25-\alpha_g)^2}\right), \qquad (8.192)$$

where α_g is the albedo of the surface.

The proper evaluation of ground wetness and the evaporative flux is important. Any excess in these measures usually results in excessive rainfall over desert areas. On the contrary, the occasional rainfall in the dry sub-Saharan belts is not predicted by the model if the ground wetness is significantly underestimated. The empirical formulas presented here are not guaranteed to represent the hydrology of the semiarid regions.

The heat balance condition may be expressed by

$$C\frac{\partial T_g}{\partial t} = (1-\alpha_g)F_S \downarrow + F_L \downarrow - (H_S + H_L) - \sigma T_g^{\,4}. \qquad (8.193)$$

T_g is the ground surface temperature, σ is the Stefan-Boltzmann constant, α_g is the surface albedo, and C is the heat capacity of the soil. The left-hand side of this equation is small since C is very small. However, $\partial T_g / \partial t$ is appreciable. If the heat capacity of the ground is assumed to be zero, then we may write

$$(1-\alpha_g)F_S \downarrow + F_L \downarrow - (H_S + H_L) - \sigma T_g^{\,4} = 0. \qquad (8.194)$$

This equation may be used to solve for the diurnally varying surface temperature. Surface temperature explicitly appears in the formulations of H_S and H_L. This is a transcendental equation in the surface temperature, and numerical methods such as the Newton-Raphson method are used for its determination.

The diurnal change arises via the inclusion of a varying zenith angle of the sun. In this balanced state, diurnal variation over warm land areas is accompanied by substantial diurnal changes in the fluxes of $F_S \downarrow$, $F_L \downarrow$, H_S, and H_L. This formula was also tested against ground-based measurements of surface temperature and was found to give a reasonable diurnal cycle of surface temperature.

8.5.6 *Surface Energy Balance and Similarity Theory Coupling*

The surface energy balance equation can be solved for the ground temperature T_g without any reference to the surface fluxes from the similarity theory. If that were done, then we would have to separate solutions for the surface fluxes of sensible and latent heat - that is, one from surface energy balance and the other from similarity theory. In order to avoid this inconsistency, we propose an iterative solution of the two problems that couples the two solutions.

The surface energy balance equation may be expressed by:

$$F_L - \sigma T_g^4 + (1-\alpha_g)F_S + \rho c_p u_* \theta_* + \rho L u_* q_* = 0. \qquad (8.195)$$

Here F_L denotes the downward flux of longwave radiation, F_S denotes the downward flux of shortwave radiation, α_g denotes the surface albedo, and θ_* and q_* denote the characteristic temperature and specific humidity of the surface layer, respectively. Furthermore, it should be noted that $\rho c_p u_* \theta_*$ denotes the flux of sensible heat from the earth's surface to the atmosphere, and $\rho L u_* q_*$ denotes the flux of latent heat.

In section 8.2.3 we have discussed a procedure for the solutions of u_*, θ_*, and q_* for stable and unstable conditions. Stability is determined from the sign of Monin-Obukhov length L or bulk Richardson number Ri_B. Given a first-guess (superscript 1) value of the surface temperature $T_g^{(1)}$, one can evaluate the corresponding bulk Richardson number by

$$Ri_g^{(1)} = \frac{\beta(T_a - T_g^{(1)})\Delta z}{(u_a - u_g)^2}, \tag{8.196}$$

where T_a is an interpolated air temperature at the top of the constant flux layer where the wind speed is u_a, Δz is the thickness of the constant flux layer, and $\beta = g/\theta_0$, θ_0 being the reference temperature. The ground temperature is T_g and the wind speed at the surface is $u_g = 0$. One next solves for $(\Delta z / L)^{(1)}$ for the stable ($Ri_B > 0$) and unstable ($Ri_B < 0$) situations from (8.29) and (8.23), respectively.

The stability exchange coefficients $C_D^{(1)}$, $C_H^{(1)}$, and $C_q^{(1)}$ are next evaluated by utilizing the first-guess value of the Monin-Obukhov length, $L^{(1)}$. It is now possible to update the values of the surface fluxes u_*^2, $-u_*\theta_*$, and $-u_*q_*$ as a function of the Monin-Obukhov length and stability. Here we use (8.21). At this point, u_*, θ_*, and q_* are substituted into the energy balance equation and one solves for an updated ground temperature $T_g^{(2)}$. This cycle is repeated to minimize the difference of $|T_g^{\nu+1} - T_g^{\nu}|$ as a function of the scan ν. This procedure converges very rapidly, yielding a coupling between the surface energy balance and the surface similarity theory.

8.5.7 *Column Model Results*

The UCLA/GLAS radiation algorithm was verified against line-by-line calculations using standard atmosphere profiles. The Florida State University Global Spectral Model (FSUGSM) algorithm was verified using the same standard atmosphere profiles. Figure 8.3a shows the column model results of the band model versus the emissivity model for a standard tropical atmosphere. The largest difference between the two models is stronger cooling of the troposphere in the band model.

Figure 8.3b shows the same atmospheric profile, except that the moisture in the lower layers has been modified to produce clouds. The maximum cooling in the emissivity model is several layers above the cloud tops. In addition, the band model shows stronger cooling than the emissivity model. This strong radiative cooling plays an

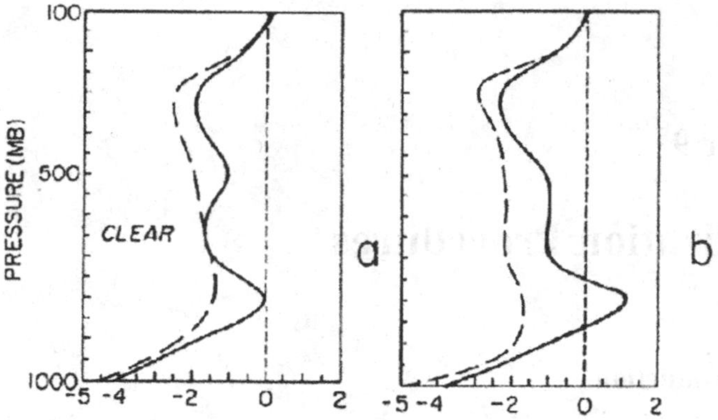

Figure 8.3. Vertical distribution of longwave radiative warming in units of °C day^{-1} for (a) clear and (b) cloudy conditions for the band model (dashed lines) and the emissivity model (solid lines).

important role in many tropical processes. Krishnamurti et al. (1991b) have demonstrated the effect of these two radiation schemes on the life cycle of typhoon Hope. The emissivity method, with smaller cooling at the cloud-top level, could not maintain an adequate radiative destabilization for the typhoon to continue development. On the other hand, the band model developed a stronger typhoon.

Chapter 9

Initialization Procedures

9.1 Introduction

In this chapter we describe two of the most commonly used initialization procedures. These are the *dynamic normal mode initialization* and the *physical initialization* methods. Historically, initialization for primitive equation models started from a hierarchy of static initialization methods. These include balancing the mass and the wind fields using a linear or nonlinear balance equation (Charney 1955; Phillips 1960), variational techniques for such adjustments satisfying the constraints of the model equations (Sasaki 1958), and dynamic initialization involving forward and backward integration of the model over a number of cycles to suppress high-frequency gravity oscillations before the start of the integration (Miyakoda and Moyer 1968; Nitta and Hovermale 1969; Temperton 1976). A description of these classical methods can be found in textbooks such as Haltiner and Williams (1980).

Basically, these methods invoke a balanced relationship between the mass and motion fields. However, it was soon realized that significant departures from the balance laws do occur over the tropics and the upper-level jet stream region. It was also noted that such departures can be functions of the heat sources and sinks and dynamic instabilities of the atmosphere. The procedure called nonlinear normal mode initialization with physics overcomes some of these difficulties. Physical initialization is a powerful method that permits the incorporation of realistic rainfall distribution in the model's initial state.

9.2 Normal Mode Initialization

This is an elegant and successful initialization procedure based on selective damping of the normal modes of the atmosphere, where the high-frequency gravity modes are suppressed while the slow-moving Rossby modes are left untouched. Williamson (1976) used the normal modes of a shallow-water model for initialization by setting the initial amplitudes of the high-frequency gravity modes equal to zero. Machenhauer (1977) and Baer (1977) developed the procedure for nonlinear normal mode initialization (NMI), which takes into account the nonlinearities in the model equations. Kitade (1983) incorporated the effect of physical processes in this initialization procedure.

We describe here the normal mode initialization procedure. Essentially following Kasahara and Puri (1981), we first derive the equations for vertical and horizontal modes of the linearized form of the model equations. Thereafter, the procedures for determining the normal modes of atmospheric models with a discrete number of vertical levels and a discrete horizontal resolution will be outlined.

9.2.1 *Basic Equations*

As discussed in Chapter 7, the basic equations for atmospheric motion on a spherical surface in the σ coordinate system are

$$\frac{\partial u}{\partial t} = -\left(\vec{V}\cdot\nabla u + \dot{\sigma}\frac{\partial u}{\partial \sigma}\right) + \left(f + u\frac{\tan\theta}{a}\right)v - \frac{1}{a\cos\theta}\left(\frac{\partial \phi}{\partial \lambda} + RT\frac{\partial q}{\partial \lambda}\right), \tag{9.1}$$

$$\frac{\partial v}{\partial t} = -\left(\vec{V}\cdot\nabla v + \dot{\sigma}\frac{\partial v}{\partial \sigma}\right) - \left(f + u\frac{\tan\theta}{a}\right)u - \frac{1}{a}\left(\frac{\partial \phi}{\partial \theta} + RT\frac{\partial q}{\partial \theta}\right), \tag{9.2}$$

$$\frac{\partial T}{\partial t} = -\left(\vec{V}\cdot\nabla T + \dot{\sigma}\frac{\partial T}{\partial \sigma}\right) + \kappa T\left(\frac{\dot{\sigma}}{\sigma} - \nabla\cdot\bar{\vec{V}} + \left(\vec{V} - \bar{\vec{V}}\right).\nabla q\right), \tag{9.3}$$

$$\frac{\partial q}{\partial t} = -\left(\bar{\vec{V}}\cdot\nabla q + \nabla\cdot\bar{\vec{V}}\right), \tag{9.4}$$

$$\frac{\partial \dot{\sigma}}{\partial \sigma} + \nabla\cdot\left(\vec{V} - \bar{\vec{V}}\right) + \left(\vec{V} - \bar{\vec{V}}\right)\cdot\nabla q = 0, \tag{9.5}$$

$$\frac{\partial \phi}{\partial \sigma} + \frac{RT}{\sigma} = 0. \tag{9.6}$$

Here $q = \ln p_s$, $\bar{\vec{V}} = \int_0^1 \vec{V}\, d\sigma$, and $k = R/c_p$. The vertical boundary conditions for this system of equations are $\dot{\sigma} = 0$ at $\sigma = 0$ and $\sigma = 1$.

We now introduce two new variables P and W defined by

$$P = \phi + RT_0 q, \tag{9.7}$$

$$W = \dot{\sigma} - \sigma\left(\nabla\cdot\bar{\vec{V}} + \bar{\vec{V}}\cdot\nabla q\right), \tag{9.8}$$

where T_0 is the mean horizontal temperature and is a function of σ only. Differentiating (9.7) with respect to σ and using the hydrostatic relation, we get,

$$\frac{\partial P}{\partial \sigma} = -\frac{RT}{\sigma} + R\frac{dT_0}{d\sigma}q. \tag{9.9}$$

The horizontal momentum equations (9.1) and (9.2) along with (9.5) may be written in terms of P and W as

$$\frac{\partial u}{\partial t} - fv + \frac{1}{a\cos\theta}\frac{\partial P}{\partial \lambda} = C_1, \tag{9.10}$$

$$\frac{\partial v}{\partial t} + fu + \frac{1}{a}\frac{\partial P}{\partial \theta} = C_2, \tag{9.11}$$

$$\frac{\partial W}{\partial \sigma} + \nabla \cdot \vec{V} = -\vec{V} \cdot \nabla q. \tag{9.12}$$

Differentiating (9.9) with respect to t and utilizing (9.3) and (9.4) gives

$$\frac{\sigma}{RT_0}\frac{\partial}{\partial t}\left(\frac{\partial P}{\partial \sigma}\right) + W = C_3. \tag{9.13}$$

Equations (9.10) to (9.13) are a new set of atmospheric equations which will be used for further analysis. The right-hand sides of these equations are the nonlinear terms and are given by

$$C_1 = -\left(\vec{V}\cdot\nabla u + \dot{\sigma}\frac{\partial u}{\partial \sigma}\right) - \frac{RT'}{a\cos\theta}\frac{\partial q}{\partial \lambda} + \frac{uv\tan\theta}{a},$$

$$C_2 = -\left(\vec{V}\cdot\nabla u + \dot{\sigma}\frac{\partial v}{\partial \sigma}\right) - \frac{RT'}{a}\frac{\partial q}{\partial \theta} - \frac{u^2\tan\theta}{a},$$

$$C_3 = \frac{1}{\Gamma_0}\left(\vec{V}\cdot\nabla T' - \kappa\frac{T'}{\sigma}W + \dot{\sigma}\frac{\partial T'}{\partial \sigma} - \kappa T\vec{V}\cdot\nabla q\right),$$

where $T' = T - T_0(\sigma)$ and $\Gamma_0 = (\kappa T_0)/\sigma - \partial T_0/\partial\sigma$ is a measure of the static stability of the basic (horizontally averaged) atmosphere. The boundary conditions for these equations are

$$W = 0 \qquad \text{at } \sigma = 0 \tag{9.14}$$

and

$$\frac{\partial P_H}{\partial t} - RT_0\big|_H = 0 \qquad \text{at } \sigma = 1, \tag{9.15}$$

where P_H and W_H are the surface values ($\sigma = 1$) of P and W given by

$$P_H = \Phi_H + RT_0\big|_H\, q,$$

$$W_H = -\left(\nabla\cdot\bar{\vec{V}} + \bar{\vec{V}}\cdot\nabla q\right).$$

Φ_H and $(T_0)_H$ are the geopotential and mean temperature at $\sigma = 1$, respectively.

9.2.2 *Linearized Equations*

The linearized forms of (9.10) to (9.13) with the basic state at rest and the horizontally averaged temperature T_0 as a function of σ are

$$\frac{\partial u'}{\partial t} - 2\Omega v' \sin\theta = -\frac{\partial P'}{a\cos\theta\partial\lambda}, \tag{9.16}$$

$$\frac{\partial v'}{\partial t} + 2\Omega u' \sin\theta = -\frac{\partial P'}{a\partial\theta}, \tag{9.17}$$

$$\frac{\partial W'}{\partial \sigma} + \nabla\cdot\vec{V}' = 0, \tag{9.18}$$

$$\frac{\sigma}{R\Gamma_0}\frac{\partial}{\partial t}\left(\frac{\partial P'}{\partial\sigma}\right) + W' = 0. \tag{9.19}$$

Elimination of W' between (9.18) and (9.19) yields

$$\frac{\partial}{\partial t}\left[\frac{\partial}{\partial\sigma}\left(\frac{\sigma}{R\Gamma_0}\frac{\partial P'}{\partial\sigma}\right)\right] - \nabla\cdot\vec{V}' = 0. \tag{9.20}$$

Equations (9.16), (9.17), and (9.20) constitute a system of equations for the perturbation quantities u', v', and P'.

The upper-boundary condition is now

$$\frac{\partial P'}{\partial\sigma} = \text{a finite value at } \sigma = 0. \tag{9.21}$$

At the lower boundary, that is, at $\sigma = 1$, (9.19) becomes

$$\frac{1}{R\Gamma_0}\frac{\partial}{\partial t}\left(\frac{\partial P'}{\partial\sigma}\right) + W_H' = 0. \tag{9.22}$$

On elimination of W_H' between the linear form of (9.15) and (9.22) we obtain

$$\left(\frac{\partial P'}{\partial\sigma}\right)_H + \frac{\Gamma_0}{T_0}P_H' = 0 \text{ at } \sigma = 1. \tag{9.23}$$

We now consider the separation of the vertical and horizontal dependence of the variables by assuming

$$u' = \overline{u}(\lambda,\theta)\psi(\sigma), \tag{9.24}$$

$$v' = \overline{v}(\lambda,\theta)\psi(\sigma), \tag{9.25}$$

and

$$P' = \overline{\phi}(\lambda,\theta)\psi(\sigma). \tag{9.26}$$

Substituting the above values of u', v', and P' into (9.16), (9.17), and (9.20) and separating the variables, we get

$$\frac{\partial\overline{u}}{\partial t} - 2\Omega\overline{v}\sin\theta = -\frac{1}{a\cos\theta}\frac{\partial\overline{\phi}}{\partial\lambda}, \tag{9.27}$$

$$\frac{\partial\overline{v}}{\partial t} + 2\Omega\overline{u}\sin\theta = -\frac{1}{a}\frac{\partial\overline{\phi}}{\partial\theta}, \tag{9.28}$$

and

$$\frac{\partial \overline{\phi}}{\partial t} + gD\nabla \cdot \overline{\vec{V}} = 0\,, \tag{9.29}$$

where D is the separation constant with the dimensions of height, and it satisfies the differential equation of vertical dependency, that is,

$$\frac{d}{d\sigma}\left(\frac{\sigma}{R\Gamma_0}\frac{d\psi}{d\sigma}\right) + \frac{1}{gD}\psi = 0\,. \tag{9.30}$$

The boundary conditions (9.21) and (9.23) now become

$$\frac{\partial \psi}{\partial \sigma} = \text{a finite constant at } \sigma = 0, \tag{9.31}$$

which for practical purposes can be taken as 0, and

$$\frac{\partial \psi}{\partial \sigma} + \frac{\Gamma_0}{T_0}\psi = 0 \text{ at } \sigma = 1\,. \tag{9.32}$$

Equation (9.30) is the vertical structure equation with D as an eigenvalue. Equations (9.27) to (9.29) are the equations for horizontal structure. We note that this set of equations is identical to the linearized shallow-water equations with a mean height of the fluid D, which is commonly referred to as the *equivalent height*.

9.2.3 *Vertical Structure Functions*

The vertical structure equation (9.30) with the boundary conditions (9.31) and (9.32) can be solved using vertical discretization and finite differencing in the model. If the atmospheric models has L vertical levels with vertical discretization shown in Fig. 9.1, then the finite-difference form of (9.30) for levels $l = 1$ to L may be written as

$$\frac{2}{\Delta\sigma_1 + \Delta\sigma_2}\left(\beta_{1\frac{1}{2}}\frac{\psi(2) - \psi(1)}{\Delta\sigma_2} - 0\right) + \frac{\psi(1)}{D} = 0\,, \tag{9.33}$$

$$\frac{2}{\Delta\sigma_2 + \Delta\sigma_3}\left(\beta_{2\frac{1}{2}}\frac{\psi(3) - \psi(2)}{\Delta\sigma_3} - \beta_{1\frac{1}{2}}\frac{\psi(2) - \psi(1)}{\Delta\sigma_2}\right) + \frac{\psi(2)}{D} = 0\,, \tag{9.34}$$

$$\frac{2}{\Delta\sigma_l + \Delta\sigma_{l+1}}\left(\beta_{l-\frac{1}{2}}\frac{\psi(l+1) - \psi(l)}{\Delta\sigma_{l+1}} - \beta_{l-1\frac{1}{2}}\frac{\psi(l) - \psi(l-1)}{\Delta\sigma_l}\right) + \frac{\psi(l)}{D} = 0\,, \tag{9.35}$$

and

$$\frac{2}{\Delta\sigma_L + \Delta\sigma_{L+1}}\left(-\beta_H\frac{\frac{\Gamma}{T_0}\psi(L)}{1 + \frac{\Gamma}{T_0}\Delta\sigma_{L+\frac{1}{2}}} - \beta_{L-\frac{1}{2}}\frac{\psi(L) - \psi(L-1)}{\Delta\sigma_L}\right) + \frac{\psi(L)}{D} = 0\,, \tag{9.36}$$

where $\beta_l = [(\sigma g)/(R\Gamma_0)]_l$ and $\beta_H = g/(R\Gamma_0|_H)$. The subscript H indicates the surface ($\sigma = 1$) value. While writing (9.33) and (9.36), use has been made of boundary

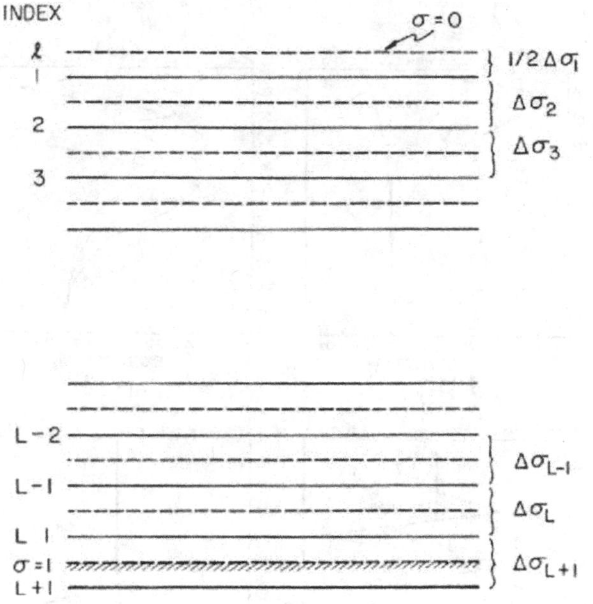

Figure 9.1. Vertical discretization of the FSU model.

conditions (9.31) and (9.32). Also, at the lower boundary it is assumed that $\psi(L+1/2)=[\psi(L)+\psi(L+1)]/2$.

The above set of equations may be written as an algebraic eigenvalue problem of the type $A\overline{\psi}+D^{-1}\overline{\psi}=0$, where A is matrix of finite coefficients of dimension L and $\overline{\psi}$ is a column vector. The solution is L eigenvalues (equivalent depth D) with L eigenvectors associated with each equivalent depth.

Figure 9.2 shows the equivalent depth and eigenvectors corresponding to each equivalent depth for the 12-level FSUGSM. The gravest mode (first mode with the largest equivalent depth) is the external mode and has very little vertical structure. The other modes are the internal modes. With decreasing equivalent depth, the modes have more and more vertical structure.

9.2.4 *Horizontal Structure Functions*

To solve the horizontal equations (9.27) to (9.29), we assume

$$\begin{pmatrix}\overline{u}\\ \overline{v}\\ \overline{\phi}\end{pmatrix}=\begin{pmatrix}u_m\\ iv_m\\ 2\Omega\phi_m\end{pmatrix}e^{i(m\lambda-2\Omega\sigma t)}, \tag{9.37}$$

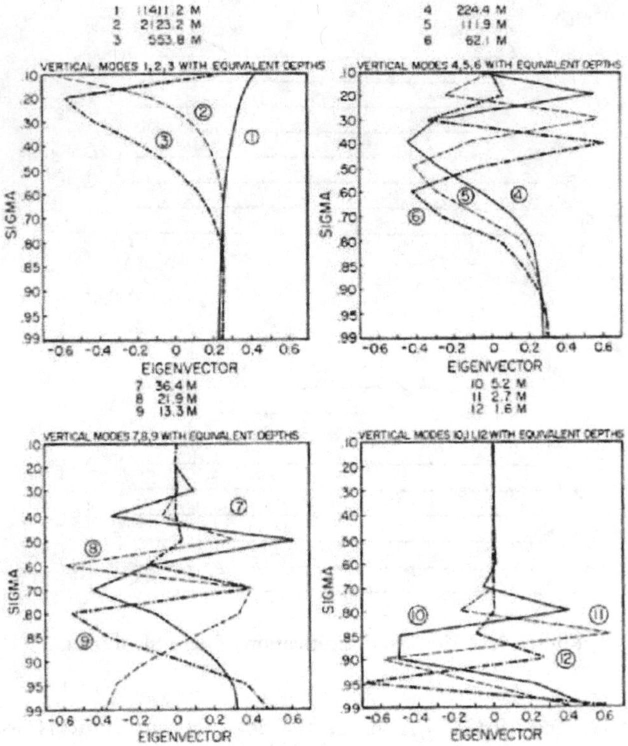

Figure 9.2. Vertical modes (eigenfunctions) and the corresponding equivalent depths (eigenvalues) for the 12-layer FSU model.

where m is the zonal wavenumber, σ is the nondimensional frequency and u_m, v_m, and ϕ_m have dimensions LT^{-1}. With this, (9.27) to (9.29) reduce to

$$\sigma u_m + v_m \sin\theta - \frac{m\phi_m}{a\cos\theta} = 0, \tag{9.38}$$

$$\sigma v_m + u_m \sin\theta + \frac{1}{a}\frac{\partial\phi_m}{\partial\theta} = 0, \tag{9.39}$$

and

$$\sigma\phi_m - \frac{gD}{4\Omega^2 a\cos\theta}\left(mu_m + \frac{\partial}{\partial\theta}(v_m\cos\theta)\right) = 0. \tag{9.40}$$

From (9.38) and (9.39), we get

$$u_m = \frac{1}{\sigma^2 - \sin^2\theta}\left(\frac{\sin\theta}{a}\frac{\partial}{\partial\theta} + \frac{m\sigma}{a\cos\theta}\right)\phi_m \tag{9.41}$$

and

$$v_m = -\frac{1}{\sigma^2 - \sin^2\theta}\left(\frac{\sigma}{a}\frac{\partial}{\partial\theta} + \frac{m\sin\theta}{a\cos\theta}\right)\phi_m . \tag{9.42}$$

Substituting the values of u_m and v_m into (9.40), and after some simplification, we get the horizontal structure equation in ϕ_m as

$$\left[\frac{1}{\cos\theta}\frac{\partial}{\partial\theta}\left(\frac{\cos\theta}{\sigma^2 - \sin^2\theta}\frac{\partial}{\partial\theta}\right) + \frac{1}{\sigma^2 - \sin^2\theta}\left(\frac{m}{\sigma}\frac{\sigma^2 + \sin^2\theta}{\sigma^2 - \sin^2\theta} - \frac{m^2}{\cos^2\theta}\right)\right]\phi_m$$
$$+\frac{2\Omega^2 a^2}{gD}\phi_m = 0. \tag{9.43}$$

The eigenvalues for this equation are the nondimensional frequency σ, and the eigenvectors are ϕ_m. Substituting ϕ_m into (9.38) and (9.39), we obtain the eigenvectors u_m and v_m.

However, (9.43) is not convenient for solving. Instead, it is easy to solve (9.38) to (9.40) without combining them. In spectral modeling, if the horizontal equations are written in terms of vorticity and divergence equations, the determination of the horizontal structure is more convenient.

Equations (9.27) to (9.29) may be written in vorticity and divergence form as

$$\frac{\partial\zeta'}{\partial t} + 2\Omega\sin\theta D' + 2\Omega\frac{\cos\theta}{a}\left(\frac{\partial\chi'}{a\partial\theta} + \frac{1}{a\cos\theta}\frac{\partial\psi'}{\partial\lambda}\right) = 0, \tag{9.44}$$

$$\frac{\partial D'}{\partial t} - 2\Omega\sin\theta\zeta + 2\Omega\frac{\cos\theta}{a}\left(\frac{1}{a\cos\theta}\frac{\partial\chi'}{\partial\lambda} - \frac{\partial\psi'}{a\partial\theta}\right) = -\frac{g}{a^2}\hat{\nabla}^2 h', \tag{9.45}$$

and

$$\frac{\partial h'}{\partial t} + \frac{H}{a^2}\hat{\nabla}^2\chi' = 0, \tag{9.46}$$

where

$$\hat{\nabla}^2 = \frac{1}{\cos^2\theta}\left[\frac{\partial^2}{\partial\lambda^2} + \cos\theta\frac{\partial}{\partial\theta}\left(\cos\theta\frac{\partial}{\partial\theta}\right)\right],$$

$\zeta' = 1/a^2\hat{\nabla}^2\psi'$, and $D' = 1/a^2\hat{\nabla}^2\chi'$. Note that ψ' and χ' are the streamfunction and velocity potential, respectively. Writing

$$\begin{pmatrix}\psi' \\ \chi' \\ h'\end{pmatrix} = \begin{pmatrix}\hat{\psi}(gH/2\Omega) \\ \hat{\chi}(gH/2\Omega) \\ \hat{h}H\end{pmatrix} e^{i(m\lambda - 2\Omega\sigma t)}, \tag{9.47}$$

we have

$$\frac{\partial}{\partial\lambda} = im \qquad \text{and} \qquad \frac{\partial}{\partial t} = -\, i2\Omega\sigma\,,$$

where $\hat{\psi}$, $\hat{\chi}$, $\hat{h}$ and σ are the nondimensional values of the streamfunction ψ, velocity potential χ, height h, and frequency ν, respectively.

The above set of equations reduces to

$$\left(\sigma\hat{\nabla}^2 - m\right)\hat{\psi} + i\left(\mu\hat{\nabla}^2 + (1-\mu^2)\frac{\partial}{\partial\mu}\right)\hat{\chi} = 0\,, \tag{9.48}$$

$$i\left(\sigma\hat{\nabla}^2 - m\right)\hat{\chi} + \left(\mu\hat{\nabla}^2 + (1-\mu^2)\frac{\partial}{\partial\mu}\right)\hat{\psi} = \hat{\nabla}^2\hat{h}\,, \tag{9.49}$$

and

$$\sigma\hat{h} = -\frac{i}{\in}\hat{\nabla}^2\hat{\chi}\,, \tag{9.50}$$

where $\in = 4a^2\Omega^2/(gH)$, $\sigma = \nu/(2\Omega)$, $\mu = \sin\theta$, and

$$\hat{\nabla}^2 = \frac{d}{d\mu}\left((1-\mu^2)\frac{d}{d\mu}\right) - \frac{m^2}{1-\mu^2}\,.$$

$\hat{\psi}, \hat{\chi}$, and $\hat{h}$ are functions of latitude and may be represented as

$$\begin{pmatrix} \hat{\chi} \\ \hat{\psi} \\ \hat{h} \end{pmatrix} = \sum_{m=-M}^{M} \sum_{n=|m|}^{N} \begin{pmatrix} iA_n^m \\ B_n^m \\ C_n^m \end{pmatrix} P_n^m(\mu)\,, \tag{9.51}$$

where $P_n^m(\mu)$ are normalized Legendre functions. As shown in Chapter 6, $P_n^m(\mu)$ satisfy the following relations:

$$\hat{\nabla}^2 P_n^m = -n(n+1)P_n^m\,,$$

$$\mu P_n^m = \in_{n+1}^m P_{n+1}^m + \in_n^m P_{n-1}^m\,,$$

and

$$(1-\mu^2)\frac{dP_n^m}{d\mu} = -n \in_{n+1}^m P_{n+1}^m + (n+1) \in_n^m P_{n-1}^m\,,$$

where

$$\in_n^m = \left(\frac{n^2 - m^2}{4n^2 - 1}\right)^{1/2}\,,$$

so that

$$\left(\sigma\hat{\nabla}^2 - m\right)P_n^m = \left[-n(n+1)\sigma - m\right]P_n^m \tag{9.52}$$

and

$$\left(\mu\hat{\nabla}^2 + (1-\mu^2)\frac{\partial}{\partial\mu}\right)P_n^m = -n(n+2)\in_{n+1}^m P_{n+1}^m$$
$$-(n-1)(n+1)\in_n^m P_{n-1}^m. \tag{9.53}$$

Substituting the expansions of $\hat{\psi}$, $\hat{\chi}$, and $\hat{h}$ from (9.51) into (9.48) to (9.50) and making use of (9.52) and (9.53) and collecting coefficients of particular P_n^m gives

$$\frac{n-1}{n}\in_n^m A_{n-1}^m + \frac{n+2}{n+1}\in_{n+1}^m A_{n+1}^m - \left(\frac{m}{n(n+1)}+\sigma\right)B_n^m = 0, \tag{9.54}$$

$$-\left(\frac{m}{n(n+1)}+\sigma\right)A_n^m + \frac{n-1}{n}\in_n^m B_{n-1}^m$$
$$+\frac{n+2}{n+1}\in_{n+1}^m B_{n+1}^m - C_n^m = 0, \tag{9.55}$$

and

$$-\frac{n(n+1)}{\in}A_n^m - \sigma C_n^m = 0. \tag{9.56}$$

Assuming a rhomboidal truncation at wavenumber N, we must then have $n = |m|,\ |m|+1,\ |m|+2, \ldots, N$. This gives a set of $3(N+1)$ equations for each zonal wavenumber m, which may be written as $(\underline{A} - \sigma I)X = 0$, where $\underline{A}$ is a $3(N+1)\times 3(N+1)$ coefficient matrix and X is a column vector given by

$$X = \begin{pmatrix} A_m^m \\ B_m^m \\ C_m^m \\ A_{m+1}^m \\ B_{m+1}^m \\ C_{m+1}^m \\ \vdots \\ A_N^m \\ B_N^m \\ C_N^m \end{pmatrix}.$$

This eigenvalue problem has $3(N+1)$ eigenvalues σ (nondimensional frequencies) corresponding to each vertical mode H [which come through $\in = 4a^2\Omega^2/(gH)$]. Out of these, $N+1$ eigenvalues correspond to Rossby modes and $2(N+1)$ eigenvalues correspond to gravity modes. Each eigenvalue has $3(N+1)$ eigenvectors corresponding to the spectra of variables ψ, χ, and h (or u, v, and ϕ).

The normal mode initialization involves selective damping of the amplitudes of the gravity modes. The first (gravest) mode has a phase speed very close to the speed of sound. For the higher modes, the phase speed decreases as the equivalent height decreases, and the phase speed of the gravity modes becomes comparable to that of the Rossby modes. It is desirable to suppress the amplitudes of the higher-frequency gravity modes only. In a numerical model with a vertical resolution of 10-15 levels, the gravity modes higher than about the first eight modes have phase speeds comparable to that of the Rossby modes. In practice, it is therefore sufficient to suppress the amplitudes of the first four to six modes only.

9.3 Physical Initialization

Krishnamurti et al. (1991a) developed a procedure of physical initialization that assimilates observed measures of rain rates into an atmospheric model. During this process, the surface fluxes of moisture, the vertical distribution of the humidity variable, the mass divergence, the convective heating, the apparent moisture sink [following Yanai et al. (1973)], and the surface pressure experience a spin-up consistent with the model physics and the imposed (observed) rain rates.

This is accomplished through a number of reverse physical algorithms within the assimilation mode. These include a reverse similarity algorithm, a reverse cumulus parameterization algorithm, and an algorithm that restructures the vertical distribution of relative humidity to provide a match between the model-calculated outgoing longwave radiation (OLR) and its satellite-based observations.

The reverse similarity algorithm is structured from the vertically integrated equations for the apparent moisture sink $\hat{Q}_2$ and the apparent heat source $\hat{Q}_1$. Using the observed rain rates, the surface evaporative flux can be obtained from the sum of the apparent moisture sink $\hat{Q}_2$ and the observed rain rate *P*. The surface sensible heat flux can also be obtained from a knowledge of the apparent heat source $\hat{Q}_1$ and the net radiative heating $\hat{Q}_R$.

During the assimilation, $\hat{Q}_1$, $\hat{Q}_2$, and $\hat{Q}_R$ continually evolve from the insertion of the observed rain rates. The resulting surface fluxes (called Yanai fluxes) tend to exhibit a consistency with the observed rain rates, which is an important component of the reverse similarity theory (Krishnamurti et al. 1991a, 1993, 1994a). These fluxes are then used within the similarity theory, where one solves for potential temperature and the humidity variable (assumed to be unknowns) at the top of the constant flux layer. The assimilation of these data provides a consistency among the observed rain rates and the surface fluxes. A robust coupling of the ocean and the atmosphere is also seen to result from this approach (Krishnamurti et al. 1993).

We have shown (Krishnamurti et al. 1994a) that this procedure results in a very high skill for the nowcasting of rainfall. The observed rain rates are obtained from a mix of Special Sensor Microwave/Imager (SSM/I) and OLR based algorithm and rain gauge data sets. Skill is measured from correlations of accumulated rain over six-hour periods and over transform grid squares of the very high-resolution global model.

9.3.1 *Reverse Similarity Theory*

Given the Yanai fluxes of sensible and latent heat as input to the similarity theory, one can in principle solve for a potential temperature and the moisture variable on top of the constant flux layer. In the conventional problem, one solves the similarity equations for the variables L, u^*, θ^*, and q^*. Here L is the Monin-Obukhov length and the remaining variables represent the momentum, heat, and moisture fluxes, respectively.

Different sets of similarity equations describe stable ($L > 0$) and unstable ($L < 0$) surface layers. In this problem, the basic variables such as wind, temperature, and moisture at the bottom and top of the constant flux layer and the surface roughness are prescribed. We will not discuss these equations in detail here, since they were already provided in Chapter 8. We express the similarity fluxes of momentum, heat, and moisture, respectively, by the relations

$$F_M = \rho C_M (U_2 - U_1)^2, \tag{9.57}$$

$$F_H = \rho C_H c_p (U_2 - U_1)(\theta_2 - \theta_1), \tag{9.58}$$

$$F_Q = \rho C_q c_p (U_2 - U_1)(q_2 - q_1) g_w, \tag{9.59}$$

where g_w is the ground wetness parameter and the similarity exchange coefficients are given as follows:

For the stable and neutral cases, the bulk Richardson number Ri_B is greater than 0, and we have

$$C_M = \frac{U_*^2}{(U_2 - U_1)^2} = \frac{k^2}{[\ln(Z_2/Z_1)]^2} \frac{1}{(1+4.7\mathrm{Ri}_B)^2}, \tag{9.60}$$

$$\begin{aligned} C_H &= \frac{U_*\theta_*}{(U_2 - U_1)(\theta_2 - \theta_1)} \\ &= \frac{-1}{0.74} \frac{k^2}{[\ln(Z_2/Z_1)]^2} \frac{1}{(1+4.7\mathrm{Ri}_B)^2}, \end{aligned} \tag{9.61}$$

and

$$C_q = 1.7 C_H. \tag{9.62}$$

For the unstable case, $\mathrm{Ri}_B < 0$ and

$$C_M = \frac{U_*^2}{(U_2 - U_1)^2} = \frac{k^2}{[\ln(Z_2/Z_1)]^2} \left(1 - \frac{9.4\mathrm{Ri}_B}{1 + C|\mathrm{Ri}_B|^{1/2}}\right), \tag{9.63}$$

$$\begin{aligned} C_H &= \frac{U_*\theta_*}{(U_2 - U_1)(\theta_2 - \theta_1)} \\ &= \frac{-1}{0.74} \frac{k^2}{[\ln(Z_2/Z_1)]^2} \left(1 - \frac{9.4\mathrm{Ri}_B}{1 + C|\mathrm{Ri}_B|^{1/2}}\right), \end{aligned} \tag{9.64}$$

and

$$C_q = 1.7 C_H. \tag{9.65}$$

Here Z_1 and Z_2 denote, respectively, the height of the bottom and top of the constant flux layer. Z_1 is identified with a roughness length Z_0. (U_1, U_2) and (θ_1, θ_2) denote the wind and potential temperature at the bottom and top of the surface layer. The bulk Richardson number Ri_B is a measure of stability, and k denotes the von Karman constant.

Over land areas, the roughness length Z_0 is defined as a function of elevation h following Manobianco (1989), that is,

$$Z_0 = 0.15 + 0.2 \times 10^{-8} (2368 + 18.42h)^2, \tag{9.66}$$

while over oceans Charnock's formula is used, i.e.,

$$Z_0 = \frac{0.04U^{*2}}{g}. \tag{9.67}$$

The expression for C follows the analysis of Louis (1979), where

$$C = \frac{7.4 \times 9.4k^2}{[\ln(Z_2/Z_1)]^2} \left(\frac{Z_2}{Z_1} \right)^{1/2}$$

for momentum and

$$C = \frac{5.3 \times 9.4k^2}{[\ln(Z_2/Z_1)]^2} \left(\frac{Z_2}{Z_1} \right)^{1/2}$$

for heat and moisture. Here the following definition of the bulk Richardson number is used:

$$\mathrm{Ri}_B \equiv \frac{g(\theta_2 - \theta_1)(Z_2 - Z_1)}{\bar{\theta}(U_2 - U_1)^2}. \tag{9.68}$$

We next address the reverse similarity theory, where one solves for θ_2 and q_2 given the surface fluxes of heat and moisture. For a closure of the reverse similarity equation, we assume that the wind at the lowest model level U_2 is known. Since the heat flux is defined by the stability, the only unknown in (9.61) and (9.64) is θ_2, which can be solved for directly or iteratively.

For the unstable case, where $U_* \theta_* < 0$, we define an objective function

$$F = \frac{U_* \theta_*}{(U_2 - U_1)(\theta_2 - \theta_1)} + \frac{k^2}{0.74[\ln(Z_2/Z_1)]^2} \left(1 - \frac{9.4\mathrm{Ri}_B}{1 + C|\mathrm{Ri}_B|^{1/2}} \right). \tag{9.69}$$

Next we search for the value of θ_2 which minimizes F. Given a guess of θ_2^n and a small increment of θ_2^n say $\delta\theta_2$, $F(\theta_2^n + \delta\theta_2)$ and $F(\theta_2^n - \delta\theta_2)$ can be computed using (9.69).

One can also compute an approximation for the derivative of F with respect to θ_2, that is,

$$\left(\frac{\Delta F}{\Delta \theta_2}\right)^n = \frac{F\left(\theta_2^n + \delta\theta_2\right) - F\left(\theta_2^n - \delta\theta_2\right)}{2\delta\theta}. \tag{9.70}$$

Then an updated value of θ_2 can be computed from the Newton-Raphson approach:

$$\theta_2^{n+1} = \theta_2^n + \frac{F\left(\theta_2^n\right)}{-\left(\Delta F / \Delta \theta_2\right)^n}. \tag{9.71}$$

This iterative procedure is continued until we have an accepted threshold value of $\left|\theta_2^{n+1} - \theta_2^n\right|$. Experiments show that the above scheme indeed converges extremely fast.

For stable conditions, (9.61) may be written as

$$4.7^2 \mathrm{Ri}_B^2 + (9.4 - A)\mathrm{Ri}_B + 1 = 0, \tag{9.72}$$

where

$$A = \frac{1}{0.74} \frac{\overline{\theta}}{g\Delta z} \frac{(U_2 - U_1)^3}{U_* \theta_*} \frac{k^2}{[\ln(Z_2 / Z_1)]^2}.$$

It is easy to prove that the sign of both roots of (9.72) is positive and that the root with the larger value is beyond the physical upper limit of Ri_B. In other words, the solution should be the smaller root. With Ri_B known, θ_2 can be obtained from (9.68). Here, the influence of the change of $\overline{\theta}$ is disregarded.

Once θ_2 is computed, C_H and C_q can be obtained using (9.61), (9.62), (9.64), and (9.65), while q_2 is obtained from (9.59). For the stable case, the exact solutions for θ_2 and q_2 are given by (9.61) and (9.62) and have no convergence problem.

For the unstable case, the Newton-Raphson method provides a convergence to the prescribed Yanai fluxes within three to four iterations. The convergence is very rapid and accurate to roughly 1 W m^{-2}. Thus the Yanai fluxes are nearly exactly reproduced from the proposed reverse similarity theory. Figure 9.3 illustrates the typical convergence for the bulk Richardson number Ri_B. Only a few examples of convergence over land and ocean are illustrated here. The global convergence over all of the Gaussian grid points is easily obtained for the unstable surface layers. The exact solution is, as stated above, possible for all the stable surface layer points. In the global model over all of the Gaussian grid points, the convergence is met within three to four scans and the Yanai fluxes are recovered.

9.3.2 *Reverse Cumulus Parameterization*

The vertical distribution of specific humidity is reanalyzed such that the rainfall implied by the cumulus parameterization algorithm closely matches the prescribed observed rainfall rates. The procedure for the construction of a reverse cumulus parameterization algorithm is discussed in Krishnamurti et al. (1984, 1988). This particular procedure is designed for a modified Kuo's scheme that is described in Krishnamurti et al. (1983). Similar reverse cumulus parameterization algorithms have been proposed by Donner (1988).

As discussed in Chapter 8, our modified Kuo's scheme invokes two parameters, b and η, where b is a moistening parameter and η is a mesoscale convergence parameter. The large-scale moisture convergence is defined by

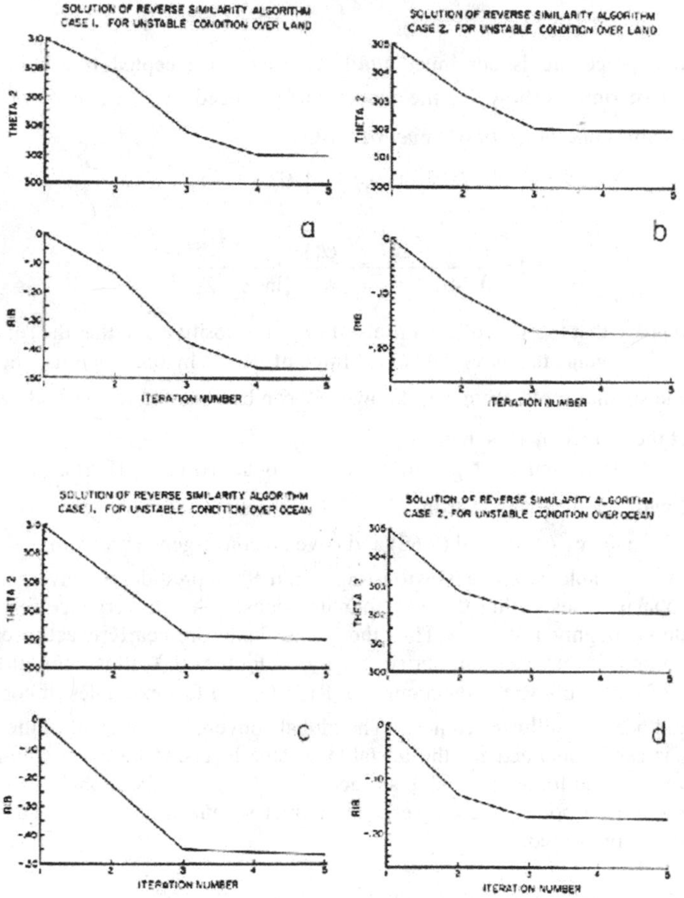

Figure 9.3. The convergence of $\theta\,(K)$ on top of the constant flux layer and of Ri_B as a function of iterations for oceanic and land grid points.

$$I_L = -\frac{p_s}{g}\int_{\sigma_T}^{\sigma_B} \dot{\sigma}\frac{\partial q}{\partial \sigma} d\sigma, \tag{9.73}$$

where σ_B and σ_T denote a sigma surface at the cloud base and cloud top, respectively. The total moisture supply is denoted by

$$I = I_L(1+\eta). \tag{9.74}$$

The total moistening and rainfall rates are expressed by the relations

$$M = I_L(1+\eta)b \tag{9.75}$$

and

$$R = I_L(1+\eta)(1-b). \tag{9.76}$$

The modified Kuo's scheme makes use of multiple regression to optimize the heating, moistening, and rainfall rates. Here, *M* and *R* are expressed as functions of several large-scale variables that have a strong control on deep convection. Thus the evolving large-scale variables of a forecast determine the local values of *M* and *R*, which in turn determine *b* and η. Thus a closure of the parameterization is accomplished.

In the reverse Kuo, the specific humidity *q* is modified using the relation

$$q_m = \frac{qR}{-\frac{p_s}{g}\int_{\sigma_T}^{\sigma_B} \dot{\sigma}\frac{\partial q}{\partial \sigma} d\sigma} + \frac{\frac{1}{g}\int_{\sigma_T}^{\sigma_B} q\, d\sigma}{\frac{1}{g}\int_{\sigma_T}^{\sigma_B} d\sigma}\left(1 - \frac{R}{-\frac{p_s}{g}\int_{\sigma_T}^{\sigma_B} \dot{\sigma}\frac{\partial q}{\partial \sigma} d\sigma}\right). \tag{9.77}$$

Here q_m is the modified specific humidity, *q* is the specific humidity prior to modification, and *R* is the observed rainfall rate. The moisture convergence corresponding to the modified specific humidity matches the observed rainfall, which is given by the relation

$$-\frac{p_s}{g}\int_{\sigma_T}^{\sigma_B} \dot{\sigma}\frac{\partial q_m}{\partial \sigma} d\sigma = R. \tag{9.78}$$

The total precipitable water remains an invariant, that is,

$$\frac{1}{g}\int_{\sigma_T}^{\sigma_B} q_m\, d\sigma = \frac{1}{g}\int_{\sigma_T}^{\sigma B} q\, d\sigma. \tag{9.79}$$

The limitations of the matching are obvious. In a region where $R > 0$, if the supply I_L is zero or negative, this method would not work. Furthermore, saturation is imposed as a limit; in other words, q_m cannot exceed q_s (the saturation value). Thus an exact match is not possible by this procedure in regions of excessive rainfall.

In Fig. 9.4 we illustrate an example of tropical rainfall distribution from (a) observed, (b) use of cumulus parameterization from the control analysis of humidity, and

(c) use of cumulus parameterization from the modified analysis of humidity. Figure 9.4 illustrates a typical distribution of these three fields for 12 UTC 26-27 July 1979.

The reverse cumulus parameterization is only applied to the tropical latitudes between 30° S and 30° N. Elsewhere, a smooth transition to the control values (i.e., unadjusted) is retained beyond 25° S and 25° N. The calculations shown in Fig. 9.4b, c illustrate a major improvement from the use of the reverse cumulus parameterization algorithm in the data-sparse tropics. The arrows in Fig. 9.4c identify regions where the control experiment failed to specify the initial rainfall, and therefore where improvements were possible from the use of this procedure. This illustration is a snapshot view of the reverse cumulus parameterization and matching of rainfall.

It should be noted that this procedure is not sufficient for the improvement of numerical weather prediction. This was also noted by Puri and Miller (1990). It is necessary to assimilate the related condensational heating and the spin-up of the divergent wind to the model's initial state. That is accomplished by a Newtonian relaxation of the humidity field during a preintegration phase prior to the day of the forecast. The interpolated rainfall is subject to the reverse cumulus parameterization at each time step. The moisture and the related heating field exert their influence on the divergence field, which is initialized in a consistent manner.

In Fig 9.5a, b we show the precipitation forecast skill. These are correlations among the observed and model rainfall totals over 24-hour periods. These data sets are

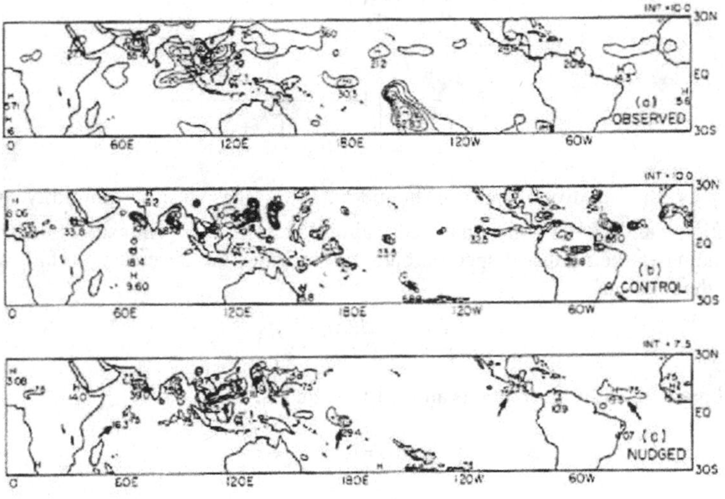

Figure 9.4. An example of precipitation initialization from 12 UTC 26-27 July 1979. (a) Based on satellite rain gauge observations. (b) Based on a control experiment's rainfall rates from the first time-step. (c) Based on a Newtonian relaxation experiment from day –1 to day 0.

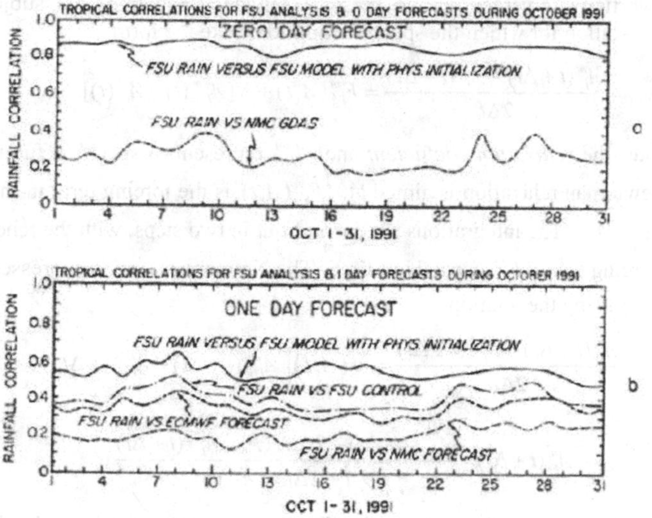

Figure 9.5. Correlations between the FSU analysis and physically initialized experiments, NMC, and ECMWF operational centers at day 0 and day 1. (a) FSU rainfall analysis correlated with postphysical initialization rainfall (solid line) and NMC global data assimilation system rainfall (dashed line). (b) FSU rainfall analysis correlated with the day 1 forecast of FSU postphysical initialization (solid line), FSU control (dot-dash line), ECMWF model (thin dashed line), and NMC model (thick dashed line).

averaged over six hours in time and over a transform grid square in space. The nowcasting of rainfall, shown in the top panel, illustrates a very high skill compared to the National Meteorological Center (NMC) operational forecasts. Also shown in the lower panel of this illustration is the one-day precipitation forecast skill. Here again, it is clearly noted that the forecasts with the physical initialization maintain a very high skill. These are the results from 31 experiments that were carried out for October 1991.

9.3.3 *Newtonian Relaxation*

Also termed *nudging* (Hoke and Anthes 1976), this is a powerful technique for the initialization of physical processes. It is possible to introduce the notions of physical initialization as proposed in the earlier sections. This will be carried out using a *Newtonian nudging* of the basic variables of the model during a preintegration phase. Some of the variables will be strongly relaxed in comparison to the others. This has to do with their overall distribution in the tropics, as well as the need to improve the regional and global spin-up.

The vorticity, divergence, and pressure tendency equations are subjected to a Newtonian relaxation for which the spectral equations take the form

$$\frac{A_l^m(t+\Delta t)-A_l^m(t-\Delta t)}{2\Delta t}=F_l^m(A,t)+N[A_l^{0m}(t)-A_l^m(t)]. \tag{9.80}$$

Here N denotes the *relaxation coefficient* and A_l^{0m} represents a specified future value to which the Newtonian relaxation is aimed at. $F_l^m(A,t)$ is the forcing term in the equation for the variable A_l^m. The integrations are carried out in two steps, with the tendencies for the normal forcing terms F_l^m carried out first. The Newtonian term is expressed in finite-difference form using the relation

$$\frac{A_l^m(t+\Delta t)-A_l^{*m}(t+\Delta t)}{2\Delta t}=N(A,t)\left[A_l^{0m}(t+\Delta t)-A_l^m(t+\Delta t)\right],$$

or

$$A_l^m(t+\Delta t)=\frac{A_l^{*m}(t+\Delta t)+2\Delta tN(A,t)A_l^{0m}(t+\Delta t)}{1+2\Delta tN(A,t)}. \tag{9.81}$$

Here $A_l^{*m}(t+\Delta t)$ denotes the value of A_l^m at time $t+\Delta t$ prior to the Newtonian relaxation.

It is clear that the value of A_l^m in the relaxation process is a weighted average of the model-predicted and the observed value, and thus it falls between these values as the relaxation proceeds. Following Krishnamurti et al. (1988), the relaxation coefficients were kept time-invariant. Their values were simply determined from numerical experimentation. The following values were used:

$$N=1\times10^{-4}\ \text{s}^{-1} \text{ for vorticity,}$$
$$N=5\times10^{-5}\ \text{s}^{-1} \text{ for divergence,}$$

and

$$N=1\times10^{-4}\ \text{s}^{-1} \text{ for surface pressure.}$$

A lower value of N for the divergence is used to permit the impact of physical initialization. The divergence field evolves strongly from the imposed heating (the prescribed rainfall rates) and is weakly relaxed to the analysis.

9.4 Initialization of the Earth's Radiation Budget

It is possible to produce a close match between OLR as inferred by the polar orbiting satellite and as determined from the model's radiation algorithm. The humidity measurements above 500 mb can be defined using a single parameter $\in$. The local difference between the two estimates of OLR (satellite versus model) can be minimized, thus determining an optimal value of $\in$. This procedure tends to improve the high and middle clouds and the planetary albedo, thus resulting in an overall improvement of the

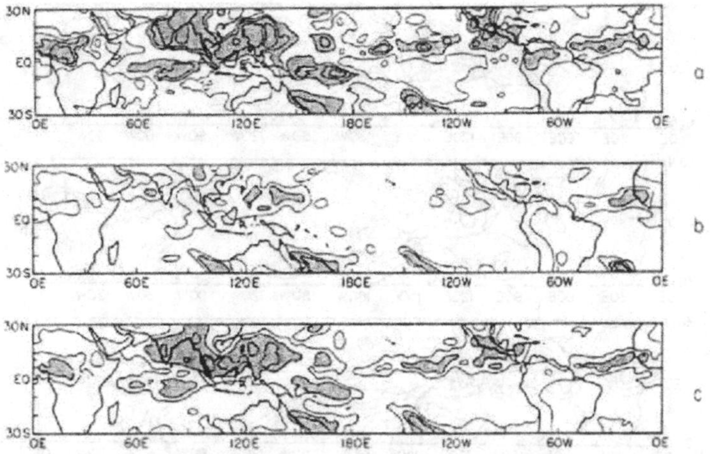

Figure 9.6. Field of the initial OLR from 27 July 1979. Analysis interval is 40 W m^{-2} (<240 W m^{-2} shaded): based on (a) polar orbiting satellite data sets; (b) a control experiment forecast at resolution *T*–42; (c) a Newtonian relaxation experiment at resolution *T*–42.

earth's radiation budget. This is not a unique method, since a matching of OLR can in principle be accomplished by altering the cloud fractions. In principle, infinitely many possible combinations of low, middle, and high clouds can provide this matching.

The humidity analysis entails an iterative procedure. Thus for l iterations we write

$$q_l = q_{l-1}(1+\in), \tag{9.82}$$

where for $l = 0$, $q_0 = q$ (analysis). Let $\delta = \text{OLR}_M - \text{OLR}_{SAT}$, where the subscript M denotes the model-based value and the subscript SAT denotes the satellite-based value. A tolerance value of 10 W m^{-2} for $|\delta|$ was assigned; that is, if $|\delta| < 10$ W m^{-2}, the iteration is discontinued.

If $\delta_{l-1}\delta_{l-2} > 0$ (i.e., when either $\text{OLR}_{M_{l-1}} > \text{OLR}_{SAT}$ and $\text{OLR}_{M_{l-2}} > \text{OLR}_{SAT}$ or $\text{OLR}_{M_{l-1}} < \text{OLR}_{SAT}$ and $\text{OLR}_{M_{l-2}} < \text{OLR}_{SAT}$), then we define $F_{l-1} = F_{l-2}$ and $\in_l = R_{l-1}(0.01F_{l-1})$, and we set

$$q_l = q_{l-1}(1+\in_l). \tag{9.83}$$

If $\delta_{l-1}\delta_{l-2} < 0$ (i.e., when either $\text{OLR}_{M_{l-1}} > \text{OLR}_{SAT}$ and $\text{OLR}_{M_{l-2}} < \text{OLR}_{SAT}$ or $\text{OLR}_{M_{l-1}} < \text{OLR}_{SAT}$ and $\text{OLR}_{M_{l-2}} > \text{OLR}_{SAT}$), then we define $F_{l-1} = 0.5F_{l-2}$ and $\in_l = R_{l-1}(0.01F_{l-1})$, and set

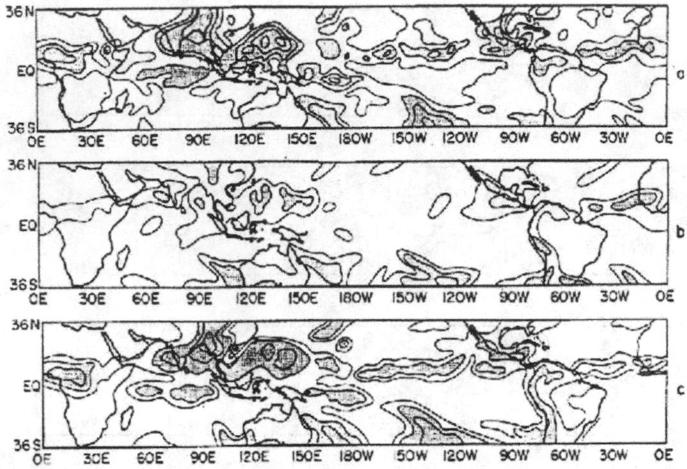

Figure 9.7. Field of the initial OLR and the satellite from 27 July 1979. Analysis interval is 40 W m^{-2} (<240 W m^{-2} shaded): based on (a) polar orbiting satellite data sets; (b) a control experiment forecast at resolution *T*–106; (c) a Newtonian relaxation experiment at resolution *T*–106.

$$q_l = q_{l-1}\left(1+\in_l\right). \tag{9.84}$$

Here $R_{l-1} = \mathrm{OLR}_{SAT} - \mathrm{OLR}_{M_{l-1}}$ is the residual at the end of iteration $l-1$. An initial value of the convergence factor F is set to 1.0. The process (9.82) to (9.84) exhibits a rapid convergence in roughly four scans. A match between the model-based OLR and the satellite-based values to within 10 W m^{-2} is realized by this bisection procedure.

Here we illustrate two examples of the initialization of OLR. The tropical distribution of OLR is illustrated in Fig 9.6. The three respective panels show the OLR values as determined from the satellite, the control model, and the model after initialization. These were calculated for a model resolution of *T*–42.

Figure 9.7 illustrates a second example which was initialized for the resolution *T*–106. Basically, a very close matching of OLR is indeed realized. The correlation coefficient between the satellite-based and initialized OLR is on the order 0.95, while that for the satellite-based and the control OLR is on the order of 0.4. The initial humidity analysis in the upper troposphere is thus constrained to the matching of OLR. The limitations of this method were stated earlier.

Chapter 10

Spectral Energetics

10.1 Introduction

In this chapter we present spectral energetics. This is a useful tool for the interpretation of model output. It can be used to interpret both short-term weather evolution and climate time scales. These same procedures can be used with both real atmospheric data and model output. A comparison of energetics histories can be very useful for the assessment of model performance.

In the first section of the chapter we derive the equations for atmospheric energetics. In the following section, a method for the representation of these equations in the Fourier domain is introduced, and the equations are derived in the one-dimensional (zonal) wavenumber domain. In the last section, we view the problem in two-dimensional wavenumber domain and derive expressions for barotropic energy exchanges and baroclinic energy conversions in this framework. Some sample results of the energetics in two-dimensional wavenumber domain are also presented in this section.

10.2 Energy Equations on a Sphere

In this section we consider a system of basic equations in spherical coordinates and derive the relevant energy equations for the zonally averaged flow and the eddy flow. These derivations are essentially based on the work of Saltzman (1957).

10.2.1 *Kinetic Energy*

We assume that the large-scale atmosphere is in a state of hydrostatic equilibrium, so that in spherical coordinates with pressure as the vertical coordinate we can write:

$$\frac{\partial u}{\partial t}+\vec{V}_H\cdot\nabla u+\omega\frac{\partial u}{\partial p}=v\left(f+\frac{u\tan\phi}{a}\right)-\frac{g}{a\cos\phi}\frac{\partial z}{\partial\lambda}-\mathrm{A}_1, \tag{10.1}$$

$$\frac{\partial v}{\partial t}+\vec{V}_H\cdot\nabla v+\omega\frac{\partial v}{\partial p}=-u\left(f+\frac{u\tan\phi}{a}\right)-\frac{g}{a}\frac{\partial z}{\partial\lambda}-\mathrm{B}_1, \tag{10.2}$$

$$g\frac{\partial z}{\partial p}=-\alpha, \tag{10.3}$$

$$\frac{\partial \omega}{\partial p} = -\nabla \cdot \vec{V}_H = -\left(\frac{1}{a\cos\phi}\frac{\partial u}{\partial \lambda} + \frac{1}{a}\frac{\partial v}{\partial \phi} - \frac{v\tan\phi}{a} \right), \tag{10.4}$$

$$c_p \frac{dT}{dt} - \omega\alpha = h, \tag{10.5}$$

$$\alpha = \frac{RT}{p}. \tag{10.6}$$

The horizontal momentum equations are given by (10.1) and (10.2), (10.3) is the hydrostatic equation, (10.4) is the continuity equation, (10.5) is the thermodynamic equation, and (10.6) is the equation of state. In these equations λ represents longitude, ϕ represents latitude, u is the eastward component of the wind, v is the northward component of the wind, $\vec{V}_H = u\hat{i} + v\hat{j}$, where $\hat{i}$ is a unit vector in the eastward direction and $\hat{j}$ is a unit vector in the northward direction, $\omega = dp/dt$, $\nabla = 1/(a\cos\phi)\hat{i}\partial/\partial\lambda + 1/a\hat{j}\partial/\partial\phi$, a is the radius of the earth, z is the height of an isobaric surface, $\alpha = 1/\rho$ is the specific volume, A_1 and B_1 are the frictional forces per unit mass, $f = 2\Omega\sin\phi$ is the Coriolis parameter, t is time, h is the heat sources and sinks, c_p is the specific heat of air at constant pressure, T is the temperature, and R is the gas constant. We also assume that the earth is a perfect sphere, so that orographic effects can be neglected.

Next we obtain the equations for total, mean, and eddy kinetic energy. Multiplying (10.1) by u and (10.2) by v, we obtain

$$\frac{\partial\left(\frac{u^2}{2}\right)}{\partial t} = -\vec{V}_H \cdot \nabla \frac{u^2}{2} - \omega \frac{\partial\left(\frac{u^2}{2}\right)}{\partial p} + uv\left(f + \frac{u\tan\phi}{a} \right) - \frac{gu}{a\cos\phi}\frac{\partial z}{\partial \lambda} - uA_1, \tag{10.7}$$

$$\frac{\partial\left(\frac{v^2}{2}\right)}{\partial t} = -\vec{V}_H \cdot \nabla \frac{v^2}{2} - \omega \frac{\partial\left(\frac{v^2}{2}\right)}{\partial p} - uv\left(f + \frac{u\tan\phi}{a} \right) - \frac{gv}{a}\frac{\partial z}{\partial \phi} - vB_1. \tag{10.8}$$

Notice that the term $uv[f + (u\tan\phi)/a]$ in (10.7) and (10.8) has opposite signs. This term therefore represents the transfer of kinetic energy between the zonal and meridional components of the wind, but it does not contribute towards change of total kinetic energy.

If we add (10.7) and (10.8), we obtain

$$\frac{\partial k}{\partial t} = -\vec{V}_H \cdot \nabla k - \omega \frac{\partial k}{\partial p} - g\vec{V}_H \cdot \nabla z - \vec{V}_H \cdot \vec{F}, \tag{10.9}$$

where $k = (u^2 + v^2)/2$ and $\vec{F} = A_1\hat{i} + B_1\hat{j}$ is the frictional force per unit mass of air. If we integrate (10.7), (10.8), and (10.9) along a latitude circle, we obtain

$$\frac{\partial\overline{\left(\frac{u^2}{2}\right)}}{\partial t}=-\overline{\vec{V}_H\cdot\nabla\left(\frac{u^2}{2}\right)}-\overline{\omega\frac{\partial\left(\frac{u^2}{2}\right)}{\partial p}}+\overline{u^2v}\frac{\tan\phi}{a}+f\overline{uv}-\frac{g}{a\cos\phi}\overline{u\frac{\partial z}{\partial\lambda}}-\overline{u\mathrm{A}_1}, \qquad (10.10)$$

$$\frac{\partial\overline{\left(\frac{v^2}{2}\right)}}{\partial t}=-\overline{\vec{V}_H\cdot\nabla\left(\frac{v^2}{2}\right)}-\overline{\omega\frac{\partial\left(\frac{v^2}{2}\right)}{\partial p}}-\overline{u^2v}\frac{\tan\phi}{a}-f\overline{uv}-\frac{g}{a}\overline{v\frac{\partial z}{\partial\phi}}-\overline{v\mathrm{B}_1}, \qquad (10.11)$$

$$\frac{\partial\overline{k}}{\partial t}=-\overline{\vec{V}_H\cdot\nabla k}-\overline{\omega\frac{\partial k}{\partial p}}-g\overline{\vec{V}_H\cdot\nabla z}-\overline{\vec{V}_H\cdot\vec{F}}, \qquad (10.12)$$

where $\overline{(\;)}=1/(2\pi)\int_0^{2\pi}(\;)d\lambda$ denotes a zonal average (average around a latitude circle). We can further resolve the kinetic energy averaged around a latitude circle into the kinetic energy of the zonally averaged flow and that of the eddy motions:

$$\overline{u^2}=\overline{u}^2+\overline{u'^2}, \qquad (10.13)$$

$$\overline{v^2}=\overline{v}^2+\overline{v'^2}, \qquad (10.14)$$

$$\overline{k}=\frac{1}{2}\left(\overline{u}^2+\overline{v}^2\right)+\frac{1}{2}\left(\overline{u'^2}+\overline{v'^2}\right)=\frac{1}{2}\left(\overline{\vec{V}_H}^2+\overline{\vec{V}'^2_H}\right), \qquad (10.15)$$

where the prime denotes deviations from the zonal mean, in other words, $u=\overline{u}+u'$ and $v=\overline{v}+v'$.

We now write the equations for the time rate of change of the kinetic energy of the zonally averaged flow. With the help of the continuity equation (10.4), we may write the momentum equations (10.1) and (10.2) in flux form as

$$\frac{\partial u}{\partial t}=-\left(\frac{\partial(uu)}{a\cos\phi\partial\lambda}+\frac{\partial(uv\cos\phi)}{a\cos\phi\partial\phi}+\frac{\partial(u\omega)}{\partial p}\right)$$
$$+v\left(f+\frac{u\tan\phi}{a}\right)-g\frac{\partial z}{a\cos\phi\partial\lambda}-\mathrm{A}_1, \qquad (10.16)$$

$$\frac{\partial v}{\partial t}=-\left(\frac{\partial(uv)}{a\cos\phi\partial\lambda}+\frac{\partial(vv\cos\phi)}{a\cos\phi\partial\phi}+\frac{\partial(v\omega)}{\partial p}\right)$$
$$-u\left(f+\frac{u\tan\phi}{a}\right)-g\frac{\partial z}{a\partial\phi}-\mathrm{B}_1. \qquad (10.17)$$

Multiplying (10.16) by $\overline{u}$ and integrating over $\lambda=0,2\pi$ gives

$$\frac{1}{2\pi}\int_0^{2\pi}\overline{u}\frac{\partial u}{\partial t}=-\frac{1}{2\pi}\int_0^{2\pi}\left(\overline{u}\frac{\partial(uu)}{a\cos\phi\partial\lambda}+\overline{u}\frac{\partial(uv\cos\phi)}{a\cos\phi\partial\phi}+\overline{u}\frac{\partial(u\omega)}{\partial p}\right)d\lambda$$
$$+\frac{1}{2\pi}\int_0^{2\pi}\overline{u}v\left(f+\frac{u\tan\phi}{a}\right)d\lambda$$
$$-\frac{1}{2\pi}g\int_0^{2\pi}\overline{u}\frac{\partial z}{a\cos\phi\partial\lambda}d\lambda-\frac{1}{2\pi}\int_0^{2\pi}\overline{u}\mathrm{A}_1d\lambda. \qquad (10.18)$$

Using the notation $1/(2\pi)\int_0^{2\pi}(\ )d\lambda=\overline{(\ )}$ and noting that

$$\int_0^{2\pi}\bar{u}\frac{\partial(uu)}{a\cos\phi\partial\lambda}d\lambda=0$$

and

$$\int_0^{2\pi}\bar{u}\frac{\partial z}{a\cos\phi\partial\lambda}d\lambda=0,$$

we get from (10.18)

$$\frac{\partial\left(\frac{\bar{u}^2}{2}\right)}{\partial t}=-\frac{\bar{u}}{a\cos\phi}\frac{\partial\left(\overline{uv}\cos\phi\right)}{\partial\phi}-\bar{u}\frac{\partial\left(\overline{u\omega}\right)}{\partial p}$$
$$+\bar{u}\left(f\bar{v}+\frac{\overline{uv}\tan\phi}{a}\right)-\bar{u}\,\overline{A_1}. \tag{10.19}$$

Similarly, multiplying (10.17) by $\bar{v}$ and integrating over $\lambda=0,2\pi$ gives

$$\frac{\partial\left(\frac{\bar{v}^2}{2}\right)}{\partial t}=-\frac{\bar{v}}{a\cos\phi}\frac{\partial\left(\overline{v^2}\cos\phi\right)}{\partial\phi}-\bar{v}\frac{\partial\left(\overline{v\omega}\right)}{\partial p}$$
$$-\bar{v}\left(f\bar{u}+\frac{\overline{u^2}\tan\phi}{a}\right)-\frac{g\bar{v}}{a}\frac{\partial\bar{z}}{\partial\phi}-\bar{v}\,\overline{B_1}. \tag{10.20}$$

Writing $u=\bar{u}+u'$ and $v=\bar{v}+v'$, we can represent the various terms on the right-hand side of (10.19) as follows:

$$-\frac{\bar{u}}{a\cos\phi}\frac{\partial\left(\overline{uv}\cos\phi\right)}{\partial\phi}=-\frac{1}{a\cos\phi}\frac{\partial\left(\bar{u}\overline{uv}\cos\phi\right)}{\partial\phi}+\frac{\left(\bar{u}\bar{v}+\overline{u'v'}\right)}{a\cos\phi}\cos\phi\frac{\partial\bar{u}}{\partial\phi}$$
$$=-\frac{1}{a\cos\phi}\frac{\partial\left(\bar{u}\overline{uv}\cos\phi\right)}{\partial\phi}+\bar{v}\cos\phi\frac{\partial\left(\frac{\bar{u}^2}{2}\right)}{a\cos\phi\partial\phi}$$
$$+\frac{\overline{u'v'}}{a\cos\phi}\cos\phi\frac{\partial\bar{u}}{\partial\phi}$$
$$=\frac{1}{a\cos\phi}\frac{\partial}{\partial\phi}\left(\bar{v}\frac{\bar{u}^2}{2}-\bar{u}\overline{uv}\right)\cos\phi-\frac{\bar{u}^2}{2}\frac{\partial\left(\bar{v}\cos\phi\right)}{a\cos\phi\partial\phi}$$
$$+\frac{\overline{u'v'}}{a}\cos\phi\frac{\partial}{\partial\phi}\left(\frac{\bar{u}}{\cos\phi}\right)-\overline{u'v'}\bar{u}\frac{\tan\phi}{a}, \tag{10.21}$$

$$-\bar{u}\frac{\partial\overline{u\omega}}{\partial p} = -\frac{\partial(\bar{u}\overline{u\omega})}{\partial p} + (\bar{u}\bar{\omega} + \overline{u'\omega'})\frac{\partial\bar{u}}{\partial p}$$

$$= -\frac{\partial(\bar{u}\overline{u\omega})}{\partial p} + \frac{\partial\frac{\bar{u}^2\bar{\omega}}{2}}{\partial p} - \frac{\bar{u}^2}{2}\frac{\partial\bar{\omega}}{\partial p} + \overline{u'\omega'}\frac{\partial\bar{u}}{\partial p}$$

$$= \frac{\partial}{\partial p}\left(\bar{\omega}\frac{\bar{u}^2}{2} - \bar{u}\overline{u\omega}\right) + \overline{u'\omega'}\frac{\partial\bar{u}}{\partial p} - \frac{\bar{u}^2}{2}\frac{\partial\bar{\omega}}{\partial p}, \tag{10.22}$$

$$\bar{u}\left(f\bar{v} + \overline{uv}\frac{\tan\phi}{a}\right) = \bar{u}\left[f\bar{v} + (\bar{u}\bar{v} + \overline{u'v'})\frac{\tan\phi}{a}\right]$$

$$= \bar{u}\bar{v}\left(f + \bar{u}\frac{\tan\phi}{a}\right) + \overline{u'v'}\bar{u}\frac{\tan\phi}{a}. \tag{10.23}$$

From (10.19) and (10.21) to (10.23) and noting that

$$\frac{\partial\bar{v}\cos\phi}{a\cos\phi\partial\phi} + \frac{\partial\bar{\omega}}{\partial p} = 0,$$

we get

$$\frac{\partial}{\partial t}\left(\frac{\bar{u}^2}{2}\right) = \left[\frac{1}{a\cos\phi}\frac{\partial}{\partial\phi}\left(\bar{v}\frac{\bar{u}^2}{2} - \bar{u}\overline{uv}\right)\cos\phi + \frac{\partial}{\partial p}\left(\bar{\omega}\frac{\bar{u}^2}{2} - \bar{u}\overline{u\omega}\right)\right]$$

$$+\overline{u'v'}\frac{\cos\phi}{a}\frac{\partial}{\partial\phi}\left(\frac{\bar{u}}{\cos\phi}\right) + \overline{u'\omega'}\frac{\partial\bar{u}}{\partial p}$$

$$+\bar{u}\bar{v}\left(f + \frac{\bar{u}\tan\phi}{a}\right) - \bar{u}\overline{A_1}. \tag{10.24}$$

Similarly, (10.20) can be written as

$$\frac{\partial}{\partial t}\left(\frac{\bar{v}^2}{2}\right) = \left[\frac{1}{a\cos\phi}\frac{\partial}{\partial\phi}\left(\bar{v}\frac{\bar{v}^2}{2} - \bar{v}\overline{vv}\right)\cos\phi + \frac{\partial}{\partial p}\left(\bar{\omega}\frac{\bar{v}^2}{2} - \bar{v}\overline{v\omega}\right)\right]$$

$$+\frac{\overline{v'v'}}{a}\frac{\partial\bar{v}}{\partial\phi} + \overline{v'\omega'}\frac{\partial\bar{v}}{\partial p} - \overline{u'u'}\bar{v}\frac{\tan\phi}{a}$$

$$-\bar{u}\bar{v}\left(f + \bar{u}\frac{\tan\phi}{a}\right) - g\frac{\bar{v}}{a}\frac{\partial\bar{z}}{\partial\phi} - \bar{v}\overline{B_1}. \tag{10.25}$$

Adding (10.24) and (10.25) gives the time rate of change of the kinetic energy of the zonally averaged flow as

$$\frac{\partial}{\partial t}\left(\frac{\overline{\vec{V}_H}^2}{2}\right) = \left[\frac{1}{a\cos\phi}\frac{\partial}{\partial\phi}\left(\bar{v}\frac{\overline{\vec{V}_H}^2}{2} - \bar{u}\overline{uv} - \bar{v}\overline{vv}\right)\cos\phi\right. \tag{10.26}$$

$$+\frac{\partial}{\partial p}\left(\overline{\omega}\frac{\overline{\vec{V}}_H^{2}}{2}-\overline{u}\,\overline{u\omega}-\overline{v}\,\overline{v\omega}\right)\Bigg]-\overline{u'u'}\overline{v}\frac{\tan\phi}{a}+\overline{v'v'}\frac{1}{a}\frac{\partial\overline{v}}{\partial\phi}$$

$$+\overline{u'v'}\frac{\cos\phi}{a}\frac{\partial}{\partial\phi}\left(\frac{\overline{u}}{\cos\phi}\right)+\overline{u'\omega'}\frac{\partial\overline{u}}{\partial p}+\overline{v'\omega'}\frac{\partial\overline{v}}{\partial p}-g\frac{\overline{v}}{a}\frac{\partial\overline{z}}{\partial\phi}-\overline{C},$$

where

$$\overline{C}=\overline{\vec{V}}_H\cdot\overline{\vec{F}}=\left(\overline{u}\,\overline{\mathrm{A}}_1+\overline{v}\,\overline{\mathrm{B}}_1\right)$$

is the rate of frictional dissipation of the kinetic energy of the mean flow.

One can obtain an equation for the rate of change of eddy kinetic energy by subtracting (10.24) from (10.10) and (10.25) from (10.11). Subtracting (10.24) from (10.10) and making use of (10.13), (10.14), and (10.15) leads to

$$\frac{\partial}{\partial t}\left(\frac{\overline{u^2}}{2}\right)-\frac{\partial}{\partial t}\left(\frac{\overline{u}^2}{2}\right)=\frac{\partial}{\partial t}\left(\frac{\overline{u'^2}}{2}\right)$$

$$=-\left[\frac{1}{a\cos\phi}\frac{\partial}{\partial\phi}\left(\overline{v\frac{u'^2}{2}}\right)\cos\phi+\frac{\partial}{\partial p}\left(\overline{\omega\frac{u'^2}{2}}\right)\right]$$

$$-\overline{u'v'}\frac{\cos\phi}{a}\frac{\partial}{\partial\phi}\left(\frac{\overline{u}}{\cos\phi}\right)-\overline{u'\omega'}\frac{\partial\overline{u}}{\partial p}$$

$$+\overline{u'v'}\left(f+\frac{\overline{u}}{a}\tan\phi\right)+\overline{v}\frac{\overline{u'u'}}{a}\tan\phi-\frac{g}{a\cos\phi}\overline{u'\frac{\partial z'}{\partial\lambda}}$$

$$+\overline{uu'\frac{v'}{a}}\tan\phi-\overline{u'\mathrm{A}'_1}. \tag{10.27}$$

Similarly, subtracting (10.25) from (10.11) and using (10.13), (10.14), and (10.15), we obtain

$$\frac{\partial}{\partial t}\left(\frac{\overline{v^2}}{2}\right)-\frac{\partial}{\partial t}\left(\frac{\overline{v}^2}{2}\right)=\frac{\partial}{\partial t}\left(\frac{\overline{v'^2}}{2}\right)$$

$$=-\left[\frac{1}{a\cos\phi}\frac{\partial}{\partial\phi}\left(\overline{v\frac{v'^2}{2}}\right)\cos\phi+\frac{\partial}{\partial p}\left(\overline{\omega\frac{v'^2}{2}}\right)\right]$$

$$-\overline{v'v'}\frac{1}{a}\frac{\partial\overline{v}}{\partial\phi}-\overline{v'\omega'}\frac{\partial\overline{v}}{\partial p}-\overline{u'v'}\left(f+\frac{\overline{u}}{a}\tan\phi\right)$$

$$-\overline{uu'\frac{v'}{a}}\tan\phi-\frac{g}{a}\overline{v'\frac{\partial z'}{\partial\phi}}-\overline{v'\mathrm{B}'_1}. \tag{10.28}$$

Finally, by adding (10.27) and (10.28) we obtain

$$\frac{\partial}{\partial t}\left(\frac{\overline{\vec{V}_H'^2}}{2}\right)=-\left[\frac{1}{a\cos\phi}\frac{\partial}{\partial\phi}\left(\overline{v\frac{\vec{V}_H'^2}{2}}\right)\cos\phi+\frac{\partial}{\partial p}\left(\overline{\omega\frac{\vec{V}_H'^2}{2}}\right)\right]$$
$$-\overline{u'v'}\frac{\cos\phi}{a}\frac{\partial}{\partial\phi}\left(\frac{\overline{u}}{\cos\phi}\right)-\overline{v'v'}\frac{1}{a}\frac{\partial\overline{v}}{\partial\phi}-\overline{u'\omega'}\frac{\partial\overline{u}}{\partial p}$$
$$-\overline{v'\omega'}\frac{\partial\overline{v}}{\partial p}+\overline{v}\frac{\overline{u'u'}}{a}\tan\phi-g\overline{\vec{V}_H'\cdot\nabla z'}-\overline{C'}, \tag{10.29}$$

where

$$\overline{C'}=\overline{\vec{V}_H'\cdot\vec{F}}=\overline{u'A_1'}+\overline{v'B_1'}.$$

Integrating (10.26) over a closed mass of air, which in our case is the entire globe, we get

$$\int_M\frac{\partial}{\partial t}\left(\frac{\overline{\vec{V}_H}^2}{2}\right)=\int_M\left[\frac{1}{a\cos\phi}\frac{\partial}{\partial\phi}\left(\overline{v}\frac{\overline{\vec{V}_H}^2}{2}-\overline{u}\overline{uv}-\overline{v}\overline{vv}\right)\cos\phi\right.$$
$$\left.+\frac{\partial}{\partial p}\left(\overline{\omega}\frac{\overline{\vec{V}_H}^2}{2}-\overline{u}\overline{u\omega}-\overline{v}\overline{v\omega}\right)\right]dm+\int_M\left[\overline{v'v'}\frac{1}{a}\frac{\partial\overline{v}}{\partial\phi}\right.$$
$$+\overline{u'v'}\frac{\cos\phi}{a}\frac{\partial}{\partial\phi}\left(\frac{\overline{u}}{\cos\phi}\right)+\overline{u'\omega'}\frac{\partial\overline{u}}{\partial p}+\overline{v'\omega'}\frac{\partial\overline{v}}{\partial p}$$
$$\left.-\overline{v}\frac{\overline{u'u'}}{a}\tan\phi-g\frac{\overline{v}}{a}\frac{\partial\overline{z}}{\partial\phi}-\overline{C}\right]dm, \tag{10.30}$$

where

$$dm=\frac{a^2\cos\phi}{g}d\lambda d\phi dp$$

and M denotes integration over the entire atmosphere, that is,

$$0\le\lambda\le 2\pi;\qquad -\frac{\pi}{2}\le\phi\le\frac{\pi}{2};\qquad 0\le p\le p_s.$$

Since M is considered to be independent of time,

$$\int_M\frac{\partial}{\partial t}\left(\frac{\overline{\vec{V}_H}^2}{2}\right)dm=\frac{\partial}{\partial t}\int_M\left(\frac{\overline{\vec{V}_H}^2}{2}\right)dm.$$

We also assume that $\omega=0$ at the top and bottom of the atmosphere. With this, the first integral on the right-hand side of (10.30) vanishes, hence

$$\frac{\partial}{\partial t}\int_M\left(\frac{\overline{\vec{V}_H}^2}{2}\right)dm=\int_M\left[\overline{u'v'}\frac{\cos\phi}{a}\frac{\partial}{\partial\phi}\left(\frac{\overline{u}}{\cos\phi}\right)\right.$$

$$+\overline{v'v'}\frac{1}{a}\frac{\partial \overline{v}}{\partial \phi}-\overline{v}\frac{\overline{u'u'}}{a}\tan\phi+\overline{u'\omega'}\frac{\partial \overline{u}}{\partial p}+\overline{v'\omega'}\frac{\partial \overline{v}}{\partial p}\Bigg]dm$$

$$-\int_M g\frac{\overline{v}}{a}\frac{\partial \overline{z}}{\partial \phi}dm-\int_M \overline{C}\,dm. \tag{10.31}$$

Similarly, integrating (10.29) over the entire globe gives

$$\frac{\partial}{\partial t}\int_M\left(\frac{\overline{\vec{V}'^2_H}}{2}\right)dm=-\int_M\Bigg[\overline{u'v'}\frac{\cos\phi}{a}\frac{\partial}{\partial\phi}\left(\frac{\overline{u}}{\cos\phi}\right)$$

$$+\overline{v'v'}\frac{1}{a}\frac{\partial \overline{v}}{\partial \phi}-\overline{v}\frac{\overline{u'u'}}{a}\tan\phi+\overline{u'\omega'}\frac{\partial \overline{u}}{\partial p}+\overline{v'\omega'}\frac{\partial \overline{v}}{\partial p}\Bigg]dm$$

$$-\int_M g\overline{\vec{V}_H'\cdot\nabla z'}\,dm-\int_M\overline{C'}\,dm. \tag{10.32}$$

The first integral on the right-hand side of (10.31) is equal to but with opposite sign of the first integral on the right-hand side of (10.32). This term therefore represents the transformation of kinetic energy between the zonal flow and the eddies. By adding (10.31) and (10.32) we obtain

$$\frac{\partial}{\partial t}\int_M\overline{k}\,dm=-\int_M g\overline{\vec{V}_H\cdot\nabla z}\,dm-\int_M\overline{\vec{V}_H\cdot\vec{F}}\,dm, \tag{10.33}$$

which is an equation for total kinetic energy. It can be shown that over a closed domain

$$\int_M g\overline{V_H\cdot\nabla z}\,dm=\int_M\overline{\alpha\omega}\,dm,$$

where α is the specific volume. With this, (10.33) can be written as

$$\frac{\partial}{\partial t}\int_M\overline{k}\,dm=-\int_M\overline{\alpha\omega}\,dm-\int_M\overline{\vec{V}_H\cdot\vec{F}}\,dm. \tag{10.34}$$

As will be shown in the next section, the first term on the right-hand side of (10.34) (i.e., $-\int_M\overline{\alpha\omega}\,dm$) represents conversion between available potential energy and kinetic energy. The last term in (10.34) is a measure of dissipation of kinetic energy by frictional forces.

10.2.2 *Available Potential Energy*

The available potential energy is that portion of the total potential energy which is available for conversion into kinetic energy. Lorenz (1955) defined the available potential energy as the difference between the total potential energy and the minimum potential energy achieved by an adiabatic rearrangement of the temperature field which yields a stable, horizontal stratification of the potential temperature field. Following this definition by Lorenz (1955), the total available potential energy of the atmosphere over the globe can be shown to be

$$A \approx \frac{1}{2} \frac{c_p}{g p_{00}^{\kappa} (1+\kappa)} \int_0^{\infty} \int_S \overline{p}^{1+\kappa} \kappa (1+\kappa) \left(\frac{p'}{\overline{p}} \right)^2 ds\, d\theta , \qquad (10.35)$$

where p_{00} is 1000 mb, $\overline{p}$ is the mean pressure on an isentropic surface, and p' is the departure of the actual pressure from its mean value. Integration with respect to s is over the whole global horizontal area S. Vertical integration is with respect to potential temperature θ, with lower and upper limits $\theta = 0$ and $\theta = \infty$, respectively.

We may write

$$p' = \left(\frac{\partial \overline{p}}{\partial \overline{\theta}} \right) \theta' \qquad (10.36)$$

and

$$d\theta = \frac{\partial \theta}{\partial p} dp \approx \frac{\partial \overline{\theta}}{\partial \overline{p}} d\overline{p} , \qquad (10.37)$$

where $\overline{p}$ is the mean pressure on an isentropic surface and $\overline{\theta}$ is the mean potential temperature on a pressure surface. Using (10.36) and (10.37), we can transfer the vertical integration in (10.35) from the θ to the p coordinate to get

$$A = \frac{1}{2} \frac{R}{g} \frac{1}{p_{00}^{\kappa}} \int_0^{p_0} \int_S \overline{p}^{\kappa - 1} \frac{\theta'^2}{\left(\frac{\partial \overline{\theta}}{\partial \overline{p}} \right)^2} \left(- \frac{\partial \overline{\theta}}{\partial \overline{p}} \right) ds\, d\overline{p} . \qquad (10.38)$$

From the equation of state and the definition of potential temperature on a constant pressure surface, we can write

$$\frac{\theta'}{\overline{\theta}} = \frac{T'}{\overline{T}} = \frac{\alpha'}{\overline{\alpha}} ,$$

or

$$\theta' = \frac{\overline{\theta}}{\overline{\alpha}} \alpha' , \qquad (10.39)$$

so that (10.38) may be written as

$$A = \frac{1}{2} \frac{R}{g} \frac{1}{p_{00}^{\kappa}} \int_0^{p_0} \int_S \overline{p}^{\kappa - 1} \frac{\overline{\theta}^2}{\overline{\alpha}^2} \frac{\alpha'^2}{\left(- \frac{\partial \overline{\theta}}{\partial \overline{p}} \right)} ds\, d\overline{p} . \qquad (10.40)$$

Noting that

$$\overline{\theta} = \overline{T} \left(\frac{p_{oo}}{p} \right)^{\kappa} = \frac{\overline{\alpha}}{R} \frac{p_{00}^{\kappa}}{\overline{p}^{\kappa - 1}} ,$$

we get

$$A = \frac{1}{g} \int_0^{p_0} \int_S \frac{1}{2} \frac{\overline{\theta}}{\overline{\alpha}} \frac{\alpha'^2}{\left(- \frac{\partial \overline{\theta}}{\partial p} \right)} ds\, dp ,$$

or

$$A = \frac{1}{g}\int_o^{p_0}\int_s \frac{\alpha'^2}{2\bar{\sigma}} ds\, dp, \tag{10.41}$$

or

$$A = \int_M \frac{\alpha'^2}{2\bar{\sigma}} dm. \tag{10.42}$$

Here

$$\bar{\sigma} = -\frac{\bar{\alpha}}{\bar{\theta}}\frac{\partial\bar{\theta}}{\partial p}$$

is the mean static stability and M is the total mass of the atmosphere.

To obtain an equation for available potential energy, we write the thermodynamic equation as

$$\frac{\partial \alpha}{\partial t} = -\vec{V}_\psi \cdot \nabla\alpha + \bar{\sigma}\omega + \frac{R}{c_p p}H = -\nabla\cdot\vec{V}_\psi\alpha + \bar{\sigma}\omega + \frac{R}{c_p p}H. \tag{10.43}$$

Taking the global average and noting that for a nondivergent flow the global average of ω is zero, we get

$$\frac{\partial\bar{\alpha}}{\partial t} = \frac{R}{c_p p}\bar{H}. \tag{10.44}$$

Subtracting (10.44) from (10.43) gives

$$\frac{\partial \alpha'}{\partial t} = -\vec{V}_\psi \cdot \nabla\alpha' + \bar{\sigma}\omega + \frac{R}{c_p p}H' = -\nabla\cdot\vec{V}_\psi\alpha' + \bar{\sigma}\omega + \frac{R}{c_p p}H'. \tag{10.45}$$

Multiplying by $\alpha'/\bar{\sigma}$, integrating over the entire global mass, and applying the $\overline{(\,)}$ operator, we get

$$\frac{\partial\bar{A}}{\partial t} = \int_M \overline{\frac{\alpha'}{\bar{\sigma}}\frac{\partial\alpha'}{\partial t}} dm = \int_M \overline{\omega\alpha'} dm + \int_M \frac{R}{c_p\bar{\sigma}p}\overline{H'\alpha'} dm. \tag{10.46}$$

Since $\bar{\omega} = 0$, the first term on the right-hand side of (10.46) can be written as $\int_M \overline{\alpha\omega}\, dm$. This term is equal and opposite in sign to the first term on the right-hand side of (10.34), and therefore represents conversion between available potential energy (APE) and kinetic energy (KE). If this term is negative, a conversion of APE to KE takes place. This is possible if the $\alpha\omega$ covariance is negative, which results from rising (negative ω) of warm air (positive α') and sinking (positive ω) of cold air (negative α'). The second term in (10.46) represents the generation of available potential energy, which can also be written as

$$\int_M \frac{R^2}{c_p\bar{\sigma}p^2} H'T' dm.$$

A net generation of APE results from a positive covariance between T' and H', which can result from warming of warm regions (positive $H'T'$) and cooling of cold regions (negative $H'T'$).

We now derive equations for zonal available potential energy and eddy available potential energy. For this let $\alpha' = \alpha'_Z + \alpha'_E$, where α'_Z is the zonal average of α' and α'_E is the departure of α' from α'_Z. Likewise we may write

$$\omega = \omega_Z + \omega_E,$$
$$u_\psi = u_{\psi Z} + u_{\psi E},$$
$$v_\psi = v_{\psi Z} + v_{\psi E} = v_{\psi Z}, \qquad \text{since } v_{\psi Z} = 0,$$
$$H' = H'_Z + H'_E,$$
$$T' = T'_Z + T'_E.$$

Taking the zonal average of the flux form of (10.45), we get

$$\frac{\partial \alpha'_Z}{\partial t} = -\frac{1}{a\cos\phi}\frac{\partial}{\partial\phi}(\alpha'_E v_{\psi E})_Z \cos\phi + \bar{\sigma}\omega_Z + \frac{R}{c_p p}H'_Z. \tag{10.47}$$

After multiplying by $\alpha' z/\bar{\sigma}$ and integrating over the whole global atmosphere, we get

$$\frac{\partial A_Z}{\partial t} = \int_M \frac{\partial}{\partial t}\left(\frac{\alpha'^2_Z}{2\bar{\sigma}}\right)dm = -\int_M \frac{\alpha'_Z}{\bar{\sigma} a\cos\phi}\frac{\partial}{\partial\phi}\left(\alpha'_E v_{\psi E}\right)_Z \cos\phi\, dm$$
$$+\int_M (\alpha'_Z \omega_Z)dm + \int_M \frac{R}{c_p p\bar{\sigma}}(\alpha'_Z H'_Z)dm \tag{10.48}$$

as the equation for zonal available potential energy. Also subtracting (10.47) from (10.45) and noting that $v_\psi = v_{\psi E}$, we obtain

$$\frac{\partial \alpha'_E}{\partial t} = -\frac{1}{a\cos\phi}\frac{\partial}{\partial\phi}(\alpha'_E v_{\psi E})_Z \cos\phi + \bar{\sigma}\omega_E + \frac{R}{c_p p}H'_E. \tag{10.49}$$

Multiplying by $\alpha'/\bar{\sigma}$ and integrating over the global domain gives

$$\frac{\partial A'_E}{\partial t} = \int_M \frac{\partial}{\partial t}\left(\frac{\alpha'^2_E}{2\bar{\sigma}}\right)dm = \int_M \frac{\alpha'_Z}{\bar{\sigma} a\cos\phi}\frac{\partial}{\partial\phi}\left(\alpha'_E v_{\psi E}\right)_Z \cos\phi\, dm$$
$$+\int_M (\alpha'_E \omega_E)dm + \int_M \frac{R}{c_p p\bar{\sigma}}(\alpha'_E H'_E)dm \tag{10.50}$$

as the equation for eddy available potential energy. The first term on the right-hand side of (10.50) is equal but opposite in sign to the first term on the right-hand side of (10.48). This term therefore represents the exchange of available potential energy between zonally averaged and eddy flows.

Using the hydrostatic equation and the continuity equation for the zonally averaged and eddy flow, it can be shown that

$$\int_M \alpha'_Z\, \omega_Z dm = \int_M g \frac{\bar{u}}{a} \frac{\partial \bar{Z}}{\partial \phi} dm \tag{10.51}$$

and

$$\int_M \alpha'_E\, \omega_E\, dm = \int_M g \vec{V}' \cdot \nabla Z' dm . \tag{10.52}$$

The second integral on the right-hand side in (10.48) is therefore equal to the second integral in (10.31), but has opposite sign. It therefore represents the conversion of zonal available potential energy to zonal kinetic energy. Similarly, the second integral in (10.32) and (10.50) gives the conversion of eddy available potential energy to eddy kinetic energy. The last integral in (10.48) and (10.50) represents the generation of zonal available potential energy and eddy available potential energy, respectively. The last term in (10.31) and (10.32) represents the dissipation of zonal kinetic energy and eddy kinetic energy, respectively, by frictional forces.

As already mentioned, the term

$$\int_M \left[\overline{u'v'} \frac{\cos\phi}{a} \frac{\partial}{\partial \phi}\left(\frac{\bar{u}}{\cos\phi} \right) + \overline{v'v'} \frac{1}{a} \frac{\partial \bar{v}}{\partial \phi} - \overline{u'u'} \bar{v} \frac{\tan\phi}{a} + \overline{u'\omega'} \frac{\partial \bar{u}}{\partial p} + \overline{v'\omega'} \frac{\partial \bar{v}}{\partial p} \right] dm \tag{10.53}$$

in (10.31) and (10.32) represents the exchange of kinetic energy between the zonal mean flow and the eddies. The term

$$\int_M \frac{\alpha'_Z}{\bar{\sigma}\, a \cos\phi} \frac{\partial}{\partial \phi} \left(\alpha'_E\, v_{\psi E} \right)_Z \cos\phi\, dm \tag{10.54}$$

in (10.48) and (10.50) appears with opposite signs. It therefore represents the exchange of available potential energy between the zonal flow and the eddies.

10.2.3 *Energy Budget of Nondivergent and Divergent Flow (ψ-χ Interactions)*

We can study the energy budget of the atmosphere by decomposing the wind field into its nondivergent (rotational) and divergent parts. As we shall see, this provides some interesting insight into the energy exchanges between the divergent and nondivergent parts of the wind.

We can write

$$\vec{V} = \vec{k} \times \nabla \psi - \nabla \chi , \tag{10.55}$$

$$\zeta = \vec{k} \cdot \nabla \times \vec{V} = \nabla^2 \psi , \tag{10.56}$$

and

$$D = \nabla \cdot \vec{V} = -\nabla^2 \chi . \tag{10.57}$$

Note that here we defined $u = -\partial\chi/\partial x$ and $v = -\partial\chi/\partial y$. The area-averaged kinetic energy is then given by

$$\overline{K} = \frac{1}{2}\overline{\left(\vec{V}_\psi + \vec{V}_\chi\right)^2} = \frac{1}{2}\overline{\left(\vec{V}_\psi \cdot \vec{V}_\psi\right)} + \frac{1}{2}\overline{\left(\vec{V}_\chi \cdot \vec{V}_\chi\right)} + \overline{\vec{V}_\psi \cdot \vec{V}_\chi}$$

$$= \frac{1}{2}\overline{(\nabla\psi \cdot \nabla\psi)} + \frac{1}{2}\overline{(\nabla\chi \cdot \nabla\chi)} - \overline{J(\psi,\chi)}.$$

Since $\overline{J(\psi,\chi)} = 0$, $\overline{K} = \overline{K}_\psi + \overline{K}_\chi$. Now

$$\overline{\nabla\psi \cdot \nabla\psi} = \overline{\nabla \cdot (\psi\nabla\psi)} - \overline{\psi\nabla^2\psi}.$$

Therefore since $\overline{\nabla \cdot (\psi\nabla\psi)} = 0$, $\overline{K}_\psi = -\overline{\psi\nabla^2\psi}$, and similarly $\overline{K}_\chi = -\overline{\chi\nabla^2\chi}$. Also,

$$\frac{\partial \overline{K}_\psi}{\partial t} = \frac{\partial}{\partial t}\overline{\frac{1}{2}(\nabla\psi \cdot \nabla\psi)} = \overline{\nabla\psi \cdot \frac{\partial}{\partial t}\nabla\psi} = \overline{\nabla \cdot \left(\psi \frac{\partial}{\partial t}\nabla\psi\right)} - \overline{\psi \frac{\partial}{\partial t}\nabla^2\psi}.$$

As

$$\overline{\nabla \cdot \left(\psi \frac{\partial}{\partial t}\nabla\psi\right)} = 0,$$

we get

$$\frac{\partial \overline{K}_\psi}{\partial t} = -\overline{\psi \frac{\partial}{\partial t}\nabla^2\psi}. \tag{10.58}$$

Similarly,

$$\frac{\partial \overline{K}_\chi}{\partial t} = -\overline{\chi \frac{\partial}{\partial t}\nabla^2\chi}. \tag{10.59}$$

The vorticity and divergence equations are

$$\frac{\partial \zeta}{\partial t} = -\vec{V}_H \cdot \nabla(\zeta + f) - \omega\frac{\partial \zeta}{\partial p} - (\zeta + f)\nabla \cdot \vec{V}_H$$

$$-\vec{k} \cdot \nabla\omega \times \frac{\partial \vec{V}_H}{\partial p} + F_\zeta \tag{10.60}$$

and

$$\frac{\partial D}{\partial t} = -\nabla \cdot (\vec{V}_H \cdot \vec{V}_H) - \nabla \cdot \left(\omega \frac{\partial \vec{V}}{\partial p}\right)$$

$$-\nabla \cdot (f\vec{k} \times \vec{V}_H) - \nabla^2\phi + F_D. \tag{10.61}$$

The thermodynamic equation is

$$\frac{\partial}{\partial t}c_p T = -\vec{V}_H \cdot \nabla c_p T - \omega\left(c_p \frac{\partial T}{\partial p} + \frac{\partial \phi}{\partial p}\right) + H. \tag{10.62}$$

Here F_ζ and F_D are frictional forces and H represents heat sources and sinks. Making use of (10.55), (10.56), and (10.57), we may write (10.60) to (10.62) in terms of ψ and χ as

$$\frac{\partial}{\partial t}\nabla^2{}_\psi = -J\left(\psi, \nabla^2{}_\psi + f\right) + \nabla\chi \cdot \nabla\left(\nabla^2\psi + f\right)$$

$$-\omega\frac{\partial}{\partial p}\nabla^2\psi+\left(\nabla^2\psi+f\right)\nabla^2\chi-\nabla\omega\cdot\nabla\frac{\partial\psi}{\partial p}+J\left(\omega,\frac{\partial\chi}{\partial p}\right)+F_\psi, \tag{10.63}$$

$$\begin{aligned}\frac{\partial}{\partial t}\nabla^2\chi &= \nabla^2\left[\frac{1}{2}(\nabla\psi)^2+\frac{1}{2}(\nabla\chi)^2-J(\psi,\chi)\right]\\ &-(\nabla^2\psi)^2-\nabla\psi\cdot\nabla(\nabla^2\psi)-\omega\frac{\partial}{\partial p}\nabla^2\chi-J\left(\omega,\frac{\partial\psi}{\partial p}\right)\\ &-\nabla\omega\cdot\nabla\frac{\partial\chi}{\partial p}-\nabla f\cdot\nabla\psi+J(f,\chi)+J(\nabla^2\psi,\chi)\\ &-f\nabla^2\psi+\nabla^2\phi+F_\chi,\end{aligned} \tag{10.64}$$

and

$$\frac{\partial}{\partial t}c_pT=J\left(c_pT,\psi\right)+\nabla\chi\cdot\nabla c_pT+(c_pT+\phi)\nabla^2\chi-\frac{\partial}{\partial t}\left[\omega(c_pT+\phi)\right]+G. \tag{10.65}$$

Note that here $F_\psi=F_\zeta$, $F_\chi=F_D$, and $G=H$.

After multiplying (10.63) by ψ, (10.64) by χ, and integrating over a closed domain, we get after some simplifications

$$\begin{aligned}\frac{\partial\overline{\overline{K_\psi}}}{\partial t} &= -\overline{\overline{\psi\frac{\partial\nabla^2\psi}{\partial t}}}\\ &= \left[\overline{\overline{f\nabla\psi\cdot\nabla\chi}}+\overline{\overline{\nabla^2\psi\nabla\psi\cdot\nabla\chi}}+\overline{\overline{\nabla^2\chi\frac{(\nabla\psi)^2}{2}}}+\overline{\overline{\omega J\left(\psi,\frac{\partial\chi}{\partial p}\right)}}\right]+\overline{\overline{F_\psi}}\end{aligned} \tag{10.66}$$

and

$$\begin{aligned}\frac{\partial\overline{\overline{K_\chi}}}{\partial t} &= -\overline{\overline{\chi\frac{\partial\nabla^2\chi}{\partial t}}}\\ &= -\left[\overline{\overline{f\nabla\psi\cdot\nabla\chi}}+\overline{\overline{\nabla^2\psi\nabla\psi\cdot\nabla\chi}}+\overline{\overline{\nabla^2\chi\frac{(\nabla\psi)^2}{2}}}\right.\\ &\left.+\overline{\overline{\omega J\left(\psi,\frac{\partial\chi}{\partial p}\right)}}\right]-\overline{\overline{\chi\nabla^2\phi}}+\overline{\overline{F_\chi}}.\end{aligned} \tag{10.67}$$

Also after integrating (10.65) over a closed domain, we get after some simplifications

$$\frac{\partial}{\partial t}\overline{\overline{(P+I)}}=\overline{\overline{\chi\nabla^2\psi}}+\overline{\overline{G}}. \tag{10.68}$$

Here $\overline{\overline{(\;)}}$ indicates integration over a three-dimensional (x, y, p) closed domain. $P=1/g\int_0^{p_0}\phi\,dp$ and $I=1/g\int_0^{p_0}c_vT\,dp$ are the potential energy and internal energy, respectively. Also note that $P+I=1/g\int_0^{p_0}c_pT\,dp$.

The terms within the brackets on the right-hand side of (10.66) and (10.67) involve both ψ and χ, and provide the interaction between the rotational and divergent parts of the wind (i.e., ψ-χ interactions). These terms in (10.66) and (10.67) are equal but have opposite sign. They therefore represent the exchange of kinetic energy between the rotational and divergent parts of the wind through ψ-χ interactions. The term $\left(\overline{\overline{\chi\nabla^2\phi}}\right)$ in (10.67) and (10.68) appears with opposite sign and represents the conversion between potential energy and kinetic energy of the divergent part of the wind.

Using the hydrostatic relation it can be shown that

$$\overline{\overline{\chi\nabla^2\phi}} = \overline{\overline{\omega\alpha}} = \frac{R}{p}\overline{\overline{\omega T}}.$$

Thus $\overline{\overline{\chi\nabla^2\phi}}$ is equal to the covariance of vertical velocity and specific volume $\overline{\overline{\omega\alpha}}$ or of vertical velocity and temperature $\overline{\overline{\omega T}}$, and is a measure of the conversion of potential energy to kinetic energy of the divergent wind. The terms $\overline{\overline{F_\psi}}$ and $\overline{\overline{F_\chi}}$ in (10.66) and (10.67) are the dissipation of kinetic energy of the rotational and divergent parts of the wind due to frictional forces. The second term on the right-hand side of (10.68), that is $\overline{\overline{H}}$, is the generation of potential energy.

If $F_\psi = 0$, $F_\chi = 0$, and $H = 0$, then on adding (10.66), (10.67), and (10.68) we get

$$\frac{\partial}{\partial t}\left(\overline{\overline{K_\psi}} + \overline{\overline{K_\chi}} + \overline{\overline{P+I}}\right) = 0. \tag{10.69}$$

Thus in the absence of energy generation and dissipation forces, the sum of the kinetic energy of the rotational and divergent flows and the potential and internal energies are conserved over a closed domain.

In the presence of dissipation and generation of energy, the energy conversion scenario is as follows: The potential energy is generated due to the covariance of temperature and heat sources and sinks $\overline{\overline{G}}$. Most of the potential energy generated thus is converted into kinetic energy of the divergent flow (through the $\chi\nabla^2\psi$ term). A small part of it may be used to change the total potential energy of the atmosphere (i.e., $\overline{\overline{\partial(P+I)/\partial t}}$). A larger part of the kinetic energy of the divergent flow converted from potential energy is converted to the kinetic energy of the nondivergent flow through ψ-χ interactions. A part of it ($\overline{\overline{F_\chi}}$) is also dissipated by frictional forces and a part is used in changing the total energy of the divergent flow (i.e., $\overline{\overline{\partial K_\chi / \partial t}}$). A larger part of the kinetic energy of the nondivergent flow that is converted from the kinetic energy of the divergent flow is dissipated by frictional forces $\overline{\overline{F_\psi}}$ while a small part of it may be used to change the total energy of the nondivergent flow (i.e., $\overline{\overline{\partial K_\psi / \partial t}}$).

10.3 Energy Equations in Wavenumber Domain

10.3.1 *Basic Concepts of Fourier Analysis*

Any real, single-valued function $f(\lambda)$ which is also piecewise differentiable in the interval $0 \le \lambda \le 2\pi$ may be represented in terms of a Fourier series as

$$f(\lambda) = \sum_{n=-\infty}^{\infty} F(n) e^{in\lambda}, \qquad (10.70)$$

where $F(n)$ is a complex coefficient given by

$$F(n) = \frac{1}{2\pi} \int_0^{2\pi} f(\lambda) e^{-in\lambda} d\lambda. \qquad (10.71)$$

Here $f(\lambda)$ may be any meteorological field along a given latitude circle. The above representation can then be used for writing the atmospheric equations in their spectral form. Also, λ is the longitude and n is the wavenumber (i.e., the number of waves around a latitude circle). $F(n)$, the spectral amplitude of wavenumber n for the space function $f(\lambda)$, can be obtained using (10.71). The function $f(\lambda)$ can be obtained from the spectral amplitudes $F(n)$ using (10.70). Therefore (10.70) and (10.71) are often referred to as a *Fourier transform pair.*

Given a function $f(\lambda, \phi, p, t)$, we may have a Fourier representation of its derivative as follows:

$$\frac{\partial f}{\partial \lambda} = \sum_{n=-\infty}^{\infty} inF(n) e^{in\lambda}; \qquad (10.72)$$

hence

$$inF(n) = \frac{1}{2\pi} \int_0^{2\pi} \frac{\partial f}{\partial \lambda} e^{-in\lambda} d\lambda. \qquad (10.73)$$

The Fourier representation of the derivative with respect to ϕ, p, or t, for which $f(\lambda, \phi, p, t)$ is not periodic may be written as

$$\frac{\partial f}{\partial \zeta} = \sum_{n=-\infty}^{\infty} F_\zeta(n) e^{in\lambda}, \qquad (10.74)$$

$$F_\zeta(n) = \frac{1}{2\pi} \int_0^{2\pi} \frac{\partial f}{\partial \zeta} e^{-in\lambda} d\lambda, \qquad (10.75)$$

where ζ can be ϕ, p, or t. The subscript here denotes partial differentiation with respect to ζ.

If we have the functions $f(\lambda)$ and $g(\lambda)$, then we can find the Fourier representation of their product. If $F(n)$ and $G(n)$ are Fourier amplitudes of $f(\lambda)$ and $g(\lambda)$, respectively, then we can write

$$\frac{1}{2\pi}\int_0^{2\pi} f(\lambda)g(\lambda)e^{-in\lambda}d\lambda = \frac{1}{2\pi}\int_0^{2\pi} f(\lambda)\left(\sum_{k=-\infty}^{\infty} G(k)e^{ik\lambda}\right)e^{-in\lambda}d\lambda . \tag{10.76}$$

Interchanging the summation and the integral sign, we may write

$$\frac{1}{2\pi}\int_0^{2\pi} f(\lambda)g(\lambda)e^{-in\lambda}d\lambda = \sum_{k=-\infty}^{\infty} G(k)\frac{1}{2\pi}\int_0^{2\pi} f(\lambda)e^{-i(n-k)\lambda}d\lambda . \tag{10.77}$$

From (10.71) we obtain

$$F(n-k) = \frac{1}{2\pi}\int_0^{2\pi} f(\lambda)e^{-i(n-k)\lambda}d\lambda .$$

After substituting this into (10.77), we finally obtain

$$\frac{1}{2\pi}\int_0^{2\pi} f(\lambda)g(\lambda)e^{-in\lambda}d\lambda = \sum_{k=-\infty}^{\infty} G(k)F(n-k) . \tag{10.78}$$

A special case of (10.78) is when $n = 0$. In this case we are calculating the area mean of the product $f(\lambda)\cdot g(\lambda)$ in the spectral domain, that is,

$$\frac{1}{2\pi}\int_0^{2\pi} f(\lambda)g(\lambda)d\lambda = \sum_{k=-\infty}^{\infty} G(k)F(-k) . \tag{10.79}$$

A second special case is if $f = g$. Then we obtain

$$\frac{1}{2\pi}\int_0^{2\pi} f^2(\lambda)d\lambda = \sum_{k=-\infty}^{\infty} \left|F(k)\right|^2 . \tag{10.80}$$

Note that $F(-k)$ is the complex conjugate of $F(k)$, hence

$$F(k)F(-k) = \left|F(k)\right|^2 .$$

Equations (10.79) and (10.80) are both known as Parseval's Theorem.

We make use of the above basic concepts of Fourier analysis in transforming the basic equations into their spectral from. This treatment essentially follows Saltzman (1957). The representation of atmospheric variables in the space domain and the spectral domain will also be similar to Saltzman (1957), and is given in Table 10.1.

As a simple example, consider the wave equation

$$\frac{\partial u}{\partial t} = -u\frac{\partial u}{\partial x} \tag{10.81}$$

in the domain $0 \le x \le 2\pi$. To transfer it into spectral form, let

$$u(x,t) = \sum_{m=-\infty}^{\infty} U(m,t)e^{imx} \tag{10.82}$$

and

$$\frac{\partial u(x,t)}{\partial x} = \sum_{m=-\infty}^{\infty} imU(m,t)e^{imx} , \tag{10.83}$$

Table 10.1. Fourier transform pairs of common meteorological variables.

$f(\lambda)$:	u	v	ω	z	T	h	A_1	B_1	α'
$F(n)$:	U	V	Ω	Z	B	H	P	Q	Λ

so that

$$u\frac{\partial u}{\partial x}=\sum_{m=-\infty}^{\infty}U(m,t)e^{imx}\sum_{m_1=-\infty}^{\infty}im_1U(m_1,t)e^{im_1x} \tag{10.84}$$

and

$$\frac{\partial u}{\partial t}=\sum_{n=-\infty}^{\infty}\frac{dU(n,t)}{dt}e^{inx}. \tag{10.85}$$

In (10.84) and (10.85) we have used the different wavenumber indices m, m_1, and n to permit various wave components of u and $\partial u/\partial x$ to interact with each other to produce different wave components of $\partial u/\partial t$.

With (10.84) and (10.85), we can write (10.81) as

$$\sum_{n=-\infty}^{\infty}\frac{dU(n)}{dt}e^{inx}=-\sum_{m=-\infty}^{\infty}\sum_{m_1=-\infty}^{\infty}im_1U(m)U(m_1)e^{i(m+m_1)x}. \tag{10.86}$$

Multiplying both sides of (10.86) by e^{-inx} and integrating over the domain $(0,\ 2\pi)$, we get

$$\frac{dU(n)}{dt}=-\frac{1}{2\pi}\sum_{m=-\infty}^{\infty}\sum_{m_1=-\infty}^{\infty}\int_0^{2\pi}im_1U(m)U(m_1)e^{i(m+m_1-n)x}dx. \tag{10.87}$$

The integral on the right-hand side of (10.87) vanishes except for those terms for which $m+m_1-n=0$ or $m_1=n-m$. As a result, (10.87) reduces to

$$\frac{dU(n)}{dt}=-\sum_{m=-\infty}^{\infty}i(n-m)U(m)U(n-m). \tag{10.88}$$

Thus each wavenumber m of u interacts with wavenumber $n-m$ of $\partial u/\partial x$ to contribute towards wavenumber n of $\partial u/\partial t$. Since m and m_1 could be positive or negative, it is clear that those Fourier components of u and $\partial u/\partial x$ for which the sum or difference is equal to n interact to produce wavenumber n of the product $u\partial u/\partial x$.

We can also write (10.88) as

$$\frac{dU(n)}{dt}=-\sum_{m=-\infty}^{\infty}imU(m)U(n-m), \tag{10.89}$$

which can be interpreted as the interaction of each wavenumber m of $\partial u/\partial x$ with wavenumber $n-m$ of u to produce $dU(n)/dt$. Equations (10.88) and (10.89) are both spectral forms of the wave equation (10.81).

10.3.2 *Governing Equations in Wavenumber Domain*

One can transform the basic equations (10.1) to (10.5) from space domain to wavenumber domain using the Fourier transform described in Section 10.3.1. Space and spectral representations of the various variables will be used as in Table 10.1. Space to spectral domain transformations can be carried out by multiplying the basic equations by $(1/2\pi)e^{-in\lambda}$ and integrating around a latitude circle. Multiplication of the zonal momentum equation (10.1) by $(1/2\pi)e^{-in\lambda}$ and integration around a latitude circle results in

$$\frac{1}{2\pi}\int_0^{2\pi}\frac{\partial u}{\partial t}e^{-in\lambda}d\lambda=-\frac{1}{2\pi}\int_0^{2\pi}\left(\vec{V}\cdot\nabla u+\omega\frac{\partial u}{\partial p}\right)e^{-in\lambda}d\lambda$$
$$+\frac{1}{2\pi}\int_0^{2\pi}\left[v\left(f+\frac{u\tan\phi}{a}\right)-\frac{g}{a\cos\phi}\frac{\partial z}{\partial\lambda}-A_1\right]e^{-in\lambda}d\lambda. \qquad (10.90)$$

Noting that

$$U(n)=\frac{1}{2\pi}\int_0^{2\pi}ue^{-in\lambda}d\lambda$$

and

$$\frac{\partial U(n)}{\partial t}=\frac{1}{2\pi}\int_0^{2\pi}\frac{\partial u}{\partial t}e^{-in\lambda}d\lambda$$

and transforming the various quadratic product terms into their spectral form, we obtain the equation for the zonal wind component in wavenumber domain as

$$\frac{\partial}{\partial t}U(n)=-\sum_{m=-\infty}^{\infty}\left(\frac{im}{a\cos\phi}U(m)U(n-m)+\frac{1}{a}U_\phi(m)V(n-m)\right.$$
$$\left.+U_p(m)\Omega(n-m)-\frac{\tan\phi}{a}U(m)V(n-m)\right)$$
$$-\frac{ing}{a\cos\phi}Z(n)+fV(n)-P(n). \qquad (10.91)$$

Similarly, the equation for the meridional wind component (10.2) is written in wavenumber domain as

$$\frac{\partial}{\partial t}V(n)=-\sum_{m=-\infty}^{\infty}\left(\frac{im}{a\cos\phi}V(m)U(n-m)+\frac{1}{a}V_\phi(m)V(n-m)\right.$$
$$\left.+V_p(m)\Omega(n-m)+\frac{\tan\phi}{a}U(m)U(n-m)\right)$$
$$-\frac{g}{a}Z_\phi(n)-fU(n)-Q(n). \qquad (10.92)$$

The hydrostatic equation (10.3) takes the form

$$Z_p(n)=-\frac{R}{pg}B(n) \qquad (10.93)$$

and the continuity equation (10.4) takes the following form:

$$\Omega_p(n) = -\left(\frac{in}{a\cos\phi}U(n) + \frac{1}{a}V_\phi(n) - \frac{\tan\phi}{a}V(n)\right). \quad (10.94)$$

Lastly, the thermodynamic energy equation (10.5) has the following form:

$$\frac{\partial}{\partial t}B(n) = -\sum_{m=-\infty}^{\infty}\left(\frac{im}{a\cos\phi}B(m)U(n-m) + \frac{1}{a}B_\phi(m)V(n-m) + B_p(m)\Omega(n-m) - \frac{R}{pc_p}B(m)\Omega(n-m)\right) + \frac{1}{c_p}H(n). \quad (10.95)$$

Given the frictional forcings P and Q and the heating distribution H, the set of equations (10.91) to (10.95) represents a closed system of equations with the five dependent variables as U, V, Ω, Z, and B; each a function of n, ϕ, p, and t.

We next derive the equation for the rate of change of kinetic energy for a given scale of motion. Wavenumber zero corresponds to the zonal mean flow, while wavenumber n corresponds to a wavelength of $2\pi / n$. The kinetic energy is given by $k = 1/2(u^2 + v^2)$. Therefore, the zonally averaged kinetic energy is given by

$$\bar{k} = \frac{1}{2\pi}\int_0^{2\pi}\frac{1}{2}\left(u^2 + v^2\right)d\lambda .$$

Using Parseval's Theorem,

$$\bar{k} = \frac{1}{2}\sum_{n=-\infty}^{\infty}\left(|U(n)|^2 + |V(n)|^2\right). \quad (10.96)$$

Since $|U(n)| = |U(-n)|$, (10.96) may be written as

$$\bar{k} = \frac{1}{2}\left(|U(0)|^2 + |V(0)|^2\right) + \sum_{n=1}^{\infty}\left(|U(n)|^2 + |V(n)|^2\right), \quad (10.97)$$

where

$$k(0) = \frac{1}{2}\bar{\bar{V}}^2 = \frac{1}{2}\left(|U(0)|^2 + |V(0)|^2\right)$$

is the mean kinetic energy of the zonally averaged flow, while

$$k(n) = \sum_{n=1}^{\infty}\left(|U(n)|^2 + |V(n)|^2\right)$$

is the mean kinetic energy of the zonal eddy flow in wavenumber domain.

We now present an equation for kinetic energy in wavenumber domain. If we multiply (10.91) by $U(-n)$, we obtain

$$U(-n)\frac{\partial}{\partial t}U(n) = -U(-n)\sum_{m=-\infty}^{\infty}\left(\frac{im}{a\cos\phi}U(m)U(n-m)\right.$$

$$+\frac{1}{a}U_\phi(m)V(n-m)+U_p(m)\Omega(n-m)$$
$$-\frac{\tan\phi}{a}U(m)V(n-m)\Bigg)-U(-n)\frac{ing}{a\cos\phi}Z(n)$$
$$+fU(-n)V(n)-U(-n)P(n). \tag{10.98}$$

Similarly, we also get

$$U(n)\frac{\partial}{\partial t}U(-n)=-U(n)\sum_{m=-\infty}^{\infty}\Bigg(\frac{im}{a\cos\phi}U(m)U(-n-m)$$
$$+\frac{1}{a}U_\phi(m)V(-n-m)+U_p(m)\Omega(-n-m)$$
$$-\frac{\tan\phi}{a}U(m)V(-n-m)\Bigg)+U(n)\frac{ing}{a\cos\phi}Z(-n)$$
$$+fU(n)V(-n)-U(n)P(-n), \tag{10.99}$$

as

$$U(-n)\frac{\partial}{\partial t}U(n)+U(n)\frac{\partial}{\partial t}U(-n)=\frac{\partial}{\partial t}[U(n)U(-n)]=\frac{\partial}{\partial t}|U(n)|^2.$$

Therefore, adding (10.98) and (10.99) gives

$$\frac{\partial}{\partial t}|U(n)|^2=-\sum_{m=-\infty}^{\infty}\Bigg(\frac{im}{a\cos\phi}U(m)[U(-n)U(n-m)+U(n)U(-n-m)]$$
$$+\frac{1}{a}U_\phi(m)[U(-n)V(n-m)+U(n)V(-n-m)]$$
$$+U_p(m)[U(-n)\Omega(n-m)+U(n)\Omega(-n-m)]\Bigg)$$
$$+\frac{\tan\phi}{a}\sum_{m=-\infty}^{\infty}U(m)[U(-n)V(n-m)+U(n)V(-n-m)]$$
$$-\frac{ing}{a\cos\phi}[U(-n)Z(n)-U(n)Z(-n)]$$
$$+f[U(-n)V(n)+U(n)V(-n)]$$
$$-[U(-n)P(n)+U(n)P(-n)]. \tag{10.100}$$

Separating the term for $m=0$ from the terms under the summation $\sum_{m=-\infty}^{\infty}$, writing $U(0)=\overline{u}$ and $V(0)=\overline{v}$, and making use of the continuity equation (10.94), (10.100) can be simplified to

$$\frac{\partial}{\partial t}|U(n)|^2=-[U(-n)V(n)+U(n)V(-n)]\frac{\cos\phi}{a}\frac{\partial}{\partial\phi}\left(\frac{\overline{u}}{\cos\phi}\right)$$
$$-[U(-n)\Omega(n)+U(n)\Omega(-n)]\frac{\partial\overline{u}}{\partial p}$$

$$+\frac{\tan\phi}{a}\overline{v}\left[U(-n)U(n)+U(n)U(-n)\right]$$

$$-\sum_{\substack{m=-\infty\\m\neq 0}}^{\infty}\left[\left(\frac{in}{a\cos\phi}U(m)\left[U(-n)U(n-m)\right.\right.\right.$$

$$\left.-U(n)U(-n-m)\right]\Big)$$

$$+\frac{1}{a\cos\phi}\left\{U(-n)\left[U(m)V(n-m)\cos\phi\right]_{\phi}\right.$$

$$\left.+U(n)\left[U(m)V(-n-m)\cos\phi\right]_{\phi}\right\}$$

$$+U(-n)\left[U(m)\Omega(n-m)\right]_{p}+U(n)\left[U(m)\Omega(-n-m)\right]_{p}$$

$$\left.-\frac{\tan\phi}{a}V(m)\left[U(n-m)U(-n)+U(-n-m)U(n)\right]\right]$$

$$-\frac{ing}{a\cos\phi}\left[Z(n)U(-n)-Z(-n)U(n)\right]$$

$$+\left(f+\frac{\overline{u}}{a}\tan\phi\right)\left[U(-n)V(n)+U(n)V(-n)\right]$$

$$-\left[U(-n)P(n)+U(n)P(-n)\right]. \qquad (10.101)$$

Similarly,

$$\frac{\partial}{\partial t}\left|V(n)\right|^2=V(-n)\frac{\partial}{\partial t}V(n)+V(n)\frac{\partial}{\partial t}V(-n)$$

$$=-\sum_{m=-\infty}^{\infty}\left(\frac{im}{a\cos\phi}V(m)\left[V(-n)U(n-m)+V(n)U(-n-m)\right]\right.$$

$$+\frac{1}{a}V_{\phi}(m)\left[V(-n)V(n-m)+V(n)V(-n-m)\right]$$

$$\left.+V_{P}(m)\left[V(-n)\Omega(n-m)+V(n)\Omega(-n-m)\right]\right)$$

$$-\frac{\tan\phi}{a}U(m)\left[V(-n)U(n-m)+V(n)U(-n-m)\right]$$

$$-\frac{g}{a}\left[V(-n)Z_{\phi}(n)+V(n)Z_{\phi}(-n)\right]$$

$$-f\left[V(-n)U(n)+V(n)U(-n)\right]$$

$$-\left[V(-n)Q(n)+V(n)Q(-n)\right], \qquad (10.102)$$

which can be further be simplified to give the final form as

$$\frac{\partial}{\partial t}\left|V(n)\right|^2=-\left[V(-n)V(n)+V(n)V(-n)\right]\frac{1}{a}\frac{\partial\overline{v}}{\partial\phi}$$

$$-\left[V(-n)\Omega(n)+V(n)\Omega(-n)\right]\frac{\partial\overline{v}}{\partial p}$$

$$-\sum_{\substack{m=-\infty\\m\neq 0}}^{\infty}\left(\frac{in}{a\cos\phi}V(m[V(-n)U(n-m)-V(n)U(-n-m)]\right.$$

$$+\frac{1}{a\cos\phi}\{V(-n)[V(m)V(n-m)\cos\phi]_{\phi}$$

$$+V(n)[V(m)V(-n-m)\cos\phi]_{\phi}\}$$

$$+\{V(-n)[V(m)\Omega(n-m)]_{p}+V(n)[V(m)\Omega(-n-m)]_{p}\}\}$$

$$+\frac{\tan\phi}{a}U(m)[V(-n)U(n-m)$$

$$\left.+V(n)U(-n-m)]\right)-\frac{g}{a}\left[Z_{\phi}(n)V(-n)+Z_{\phi}(-n)V(n)\right]$$

$$-\left(f+\bar{u}\frac{\tan\phi}{a}\right)[U(n)V(-n)+U(-n)V(n)]$$

$$-[V(-n)Q(n)+V(n)Q(-n)]. \tag{10.103}$$

By adding (10.101) and (10.103) and integrating the resulting equation over the entire mass of the atmosphere, one can obtain the time rate of change of the total kinetic energy for a given wavenumber as

$$\frac{\partial}{\partial t}\int_{M}K(n)\,dm=-\int_{M}\left[[U(n)V(-n)+U(-n)V(n)]\frac{\cos\phi}{a}\frac{\partial}{\partial\phi}\left(\frac{\bar{u}}{\cos\phi}\right)\right.$$

$$+[V(n)V(-n)+V(-n)V(n)]\frac{1}{a}\frac{\partial\bar{v}}{\partial\phi}$$

$$+[U(n)\Omega(-n)+U(-n)\Omega(n)]\frac{\partial\bar{u}}{\partial p}$$

$$+[V(n)\Omega(-n)+V(-n)\Omega(n)]\frac{\partial\bar{v}}{\partial p}$$

$$\left.-[U(n)U(-n)+U(-n)U(n)]\bar{v}\frac{\tan\phi}{a}\right]dm$$

$$+\int_{M}\sum_{\substack{m=-\infty\\m\neq 0}}^{\infty}\left\langle U(m)\left\{\left[(U(n-m)\left(\frac{-inU(-n)}{a\cos\phi}\right)\right.\right.\right.$$

$$\left.+U(-n-m)\left(\frac{inU(n)}{a\cos\phi}\right)\right]$$

$$+\frac{1}{a}\left(V(n-m)\frac{\partial U(-n)}{\partial\phi}+V(-n-m)\frac{\partial U(n)}{\partial\phi}\right)$$

$$+\left(\Omega(n-m)\frac{\partial U(-n)}{\partial p}+\Omega(-n-m)\frac{\partial U(n)}{\partial p}\right)$$

$$
\begin{aligned}
&-\frac{\tan\phi}{a}\left[U(n-m)V(-n)+U(-n-m)V(n)\right]\Big\}\\
&+V(m)\left\{U(n-m)\left(\frac{-inV(-n)}{a\cos\phi}\right)+U(-n-m)\left(\frac{inV(n)}{a\cos\phi}\right)\right.\\
&+\frac{1}{a}\left(V(n-m)\frac{\partial V(-n)}{\partial\phi}+V(-n-m)\frac{\partial V(n)}{\partial\phi}\right)\\
&+\left(\Omega(n-m)\frac{\partial V(-n)}{\partial p}+\Omega(-n-m)\frac{\partial V(n)}{\partial p}\right)\\
&\left.+\frac{\tan\phi}{a}\left[U(n-m)U(-n)+U(-n-m)U(n)\right]\right\}\Big\rangle\, dm\\
&-g\int_M\left\langle\frac{1}{a\cos\phi}\left\{U(n)\left[-inZ(-n)\right]+U(-n)\left[inZ(n)\right]\right\}\right.\\
&\left.+\frac{1}{a}\left(V(n)\frac{\partial Z(-n)}{\partial\phi}+V(-n)\frac{\partial Z(n)}{\partial\phi}\right)\right\rangle dm\\
&-\int_M\left\{\left[U(n)P(-n)+U(-n)P(n)\right]\right.\\
&\left.+\left[V(n)Q(-n)+V(-n)Q(n)\right]\right\}dm\,. \qquad (10.104)
\end{aligned}
$$

Letting

$$
f_\lambda=\frac{\partial f}{\partial\lambda}=inF(n)\,,
$$

$$
f_\phi=\frac{\partial f}{\partial\phi}=\frac{\partial F(n)}{\partial\phi}\,,
$$

and

$$
f_p=\frac{\partial f}{\partial p}=\frac{\partial F(n)}{\partial p}\,,
$$

we can write (10.104) as

$$
\begin{aligned}
\frac{\partial}{\partial t}\int_M K(n)\,dm=&-\int_M\left[\Phi_{uv}(n)\frac{\cos\phi}{a}\frac{\partial}{\partial\phi}\left(\frac{\overline{u}}{\cos\phi}\right)\right.\\
&+\Phi_{vv}(n)\frac{1}{a}\frac{\partial\overline{v}}{\partial\phi}+\Phi_{u\omega}(n)\frac{\partial\overline{u}}{\partial p}\\
&\left.+\Phi_{v\omega}(n)\frac{\partial\overline{v}}{\partial p}-\Phi_{uu}(n)\overline{v}\frac{\tan\phi}{a}\right]dm\\
&+\int_M\sum_{\substack{m=-\infty\\ m\neq 0}}^{\infty}\left[U(m)\left(\frac{1}{a\cos\phi}\Psi_{uu_\lambda}(m,n)\right.\right.\\
&\left.+\frac{1}{a}\Psi_{vu_\phi}(m,n)+\Psi_{\omega u_p}(m,n)-\frac{\tan\phi}{a}\Psi_{uv}(m,n)\right)
\end{aligned}
$$

$$+V(m)\left(\frac{1}{a\cos\phi}\Psi_{uv_\lambda}(m,n)+\frac{1}{a}\Psi_{vv_\phi}(m,n)\right.$$

$$\left.\left.+\Psi_{\omega v_p}(m,n)+\frac{\tan\phi}{a}\Psi_{uu}(m,n)\right)\right]dm$$

$$-\int_M g\left(\frac{1}{a\cos\phi}\Phi_{uz_\lambda}(n)+\frac{1}{a}\Phi_{vz_\phi}(n)\right)dm$$

$$-\int_M\left[\Phi_{uA_1}(n)+\Phi_{vB_1}(n)\right]dm \tag{10.105}$$

where

$$\Phi_{fg}(n)=F(n)G(-n)+F(-n)G(n) \tag{10.106}$$

and

$$\Psi_{fg}(m,n)=F(n-m)G(-n)+F(-n-m)G(n). \tag{10.107}$$

As a particular case, one can obtain the time rate of change of kinetic energy of the mean flow by using the fact that

$$K(0)=\frac{\overline{\vec{V}}^2}{2}=\frac{|U(0)|^2+|V(0)|^2}{2}.$$

Thus setting $n=0$ in (10.105) and using the continuity equation (10.94), we obtain

$$\frac{\partial}{\partial t}\int_M\left(\frac{\overline{\vec{V}}^2}{2}\right)dm=-\int_M\sum_{m=1}^{\infty}\left[\Phi_{uv}(m)\frac{\cos\phi}{a}\frac{\partial}{\partial\phi}\left(\frac{\bar{u}}{\cos\phi}\right)\right. \tag{10.108}$$

$$+\Phi_{vv}(m)\frac{1}{a}\frac{\partial\bar{v}}{\partial\phi}+\Phi_{u\omega}(m)\frac{\partial\bar{u}}{\partial p}+\Phi_{v\omega}(m)\frac{\partial\bar{v}}{\partial p}$$

$$\left.-\Phi_{uu}(m)\bar{v}\frac{-\tan\phi}{a}\right]dm-\int_M\frac{g}{a}\bar{v}\frac{\partial\bar{z}}{\partial\phi}dm-\int_M\bar{c}\,dm.$$

Now let us try to attach physical meanings to the terms appearing in (10.105). The first integral, namely

$$-\int_M\left[\Phi_{uv}(n)\frac{\cos\phi}{a}\frac{\partial}{\partial\phi}\left(\frac{\bar{u}}{\cos\phi}\right)+\Phi_{vv}(n)\frac{1}{a}\frac{\partial\bar{v}}{\partial\phi}\right.$$

$$\left.+\Phi_{u\omega}(n)\frac{\partial\bar{u}}{\partial p}+\Phi_{v\omega}(n)\frac{\partial\bar{v}}{\partial p}-\Phi_{uu}(n)\bar{v}\frac{-\tan\phi}{a}\right]dm,$$

can be regarded as a *transformation function* which measures the transfer of energy between any individual scale of disturbance and the mean flow. The second integral, namely

$$\int_M\sum_{\substack{m=-\infty\\m\neq 0}}^{\infty}\left[U(m)\left(\frac{1}{a\cos\phi}\Psi_{uu_\lambda}(m,n)\right.\right.$$

$$\left.+\frac{1}{a}\Psi_{vu_\phi}(m,n)+\Psi_{\omega u_p}(m,n)-\frac{\tan\phi}{a}\Psi_{uv}(m,n)\right)$$

$$+V(m)\left(\frac{1}{a\cos\phi}\Psi_{uv_\lambda}(m,n)+\frac{1}{a}\Psi_{vv_\phi}(m,n)\right.$$

$$\left.\left.+\Psi_{\omega v_p}(m,n)+\frac{\tan\phi}{a}\Psi_{uu}(m,n)\right)\right]dm,$$

is a measure of the transfer of energy between a particular wavenumber n and all other wavenumbers due to nonlinear interactions. If we sum this integral over all wavenumbers (including wavenumber 0), the result must be zero. Fjørtoft (1953) showed that for two-dimensional divergent motions, if kinetic energy of one scale of motion is changed, then this will result in changes in the kinetic energy of both smaller- and larger-scale motions.

Using the continuity equation and hydrostatic equation, one can write the third integral on the right-hand side of (10.105) as

$$\int_M g\left(\frac{1}{a\cos\phi}\Phi_{uz_\lambda}(n)+\frac{1}{a}\Phi_{vz_\phi}(n)\right)dm$$

$$=-\int_M g\Phi_{\omega z_p}(n)dm=\int_M \frac{R}{p}\Phi_{\omega T}(n)\,dm. \tag{10.109}$$

This integral measures the conversion between eddy available potential energy and eddy kinetic energy of a particular wavenumber n. This integral shows that the baroclinic growth of a given wavenumber in a disturbance depends on the degree to which a particular wave in the vertical motion field is in phase with the same wave in the temperature field. The last integral in (10.105), namely

$$\int_M\left[\Phi_{uA_1}(n)+\Phi_{vB_1}(n)\right]dm,$$

represents frictional dissipation of various scales.

We shall next derive the equation for available potential energy in wavenumber domain. From (10.42), the available potential energy in wavenumber domain may be written as

$$A=\int_M\sum_{n=-\infty}^{\infty}\frac{|\Lambda(n)|^2}{2\overline{\sigma}}dm=\int_M\frac{|\Lambda(0)|^2}{2\overline{\sigma}}dm+\int_M\sum_{n=1}^{\infty}\frac{|\Lambda(n)|^2}{\overline{\sigma}}dm, \tag{10.110}$$

so that

$$\frac{\partial A}{\partial t}=\int_M\frac{1}{\overline{\sigma}}\left(\Lambda(0)\frac{\partial\Lambda(0)}{\partial t}\right)dm \tag{10.111}$$

$$+\int_M\sum_{n=1}^{\infty}\frac{1}{\overline{\sigma}}\left(\Lambda(-n)\frac{\partial\Lambda(n)}{\partial t}+\Lambda(n)\frac{\partial\Lambda(-n)}{\partial t}\right)dm,$$

where

$$\frac{\partial\overline{A}}{\partial t}=\int_M\frac{1}{\overline{\sigma}}\left(\Lambda(0)\frac{\partial\Lambda(0)}{\partial t}\right)dm$$

and

$$\frac{\partial A'}{\partial t}=\int_M\sum_{n=1}^{\infty}\frac{1}{\overline{\sigma}}\left(\Lambda(-n)\frac{\partial\Lambda(n)}{\partial t}+\Lambda(n)\frac{\partial\Lambda(-n)}{\partial t}\right)dm$$

are the time rate of change of zonal and eddy available potential energy in wavenumber domain.

$$\frac{\partial A(n)}{\partial t}=\int_M \frac{1}{\overline{\sigma}}\left(\Lambda(-n)\frac{\partial\Lambda(n)}{\partial t}+\Lambda(n)\frac{\partial\Lambda(-n)}{\partial t}\right)dm$$

is the time rate of change of eddy available potential energy for a particular wavenumber n.

To get the equation for available potential energy, we write (10.45) in wavenumber domain as

$$\frac{\partial\Lambda(n)}{\partial t}=-\sum_{m=-\infty}^{\infty}\left(U_\psi(n-m)\frac{im\Lambda(m)}{a\cos\phi}+V_\psi(n-m)\frac{\partial\Lambda(m)}{a\partial\phi}\right)$$
$$+\overline{\sigma}\Omega(n)+\frac{R}{c_p p}H(n)\,. \tag{10.112}$$

Similarly,

$$\frac{\partial\Lambda(-n)}{\partial t}=-\sum_{m=-\infty}^{\infty}\left(U_\psi(-n-m)\frac{im\Lambda(m)}{a\cos\phi}+V_\psi(-n-m)\frac{\partial\Lambda(m)}{a\partial\phi}\right)$$
$$+\overline{\sigma}\Omega(-n)+\frac{R}{c_p p}H(-n)\,. \tag{10.113}$$

Multiplying (10.112) by $\Lambda(-n)/\overline{\sigma}$, (10.113) by $\Lambda(n)/\overline{\sigma}$, adding, and integrating over the whole mass M, we get

$$\frac{\partial A(n)}{\partial t}=\int_M \frac{1}{\overline{\sigma}}\left(\Lambda(-n)\frac{\partial\Lambda(n)}{\partial t}+\Lambda(n)\frac{\partial\Lambda(-n)}{\partial t}\right)dm$$
$$=-\int_M\left(\frac{1}{\overline{\sigma}}\sum_{m=-\infty}^{\infty}\frac{im\Lambda(m)}{a\cos\phi}\left[\Lambda(-n)U_\psi(n-m)+\Lambda(n)U_\psi(-n-m)\right]\right.$$
$$\left.+\frac{1}{a}\frac{\partial\Lambda(m)}{\partial\phi}\left[\Lambda(-n)V_\psi(n-m)+\Lambda(n)V_\psi(-n-m)\right]\right)dm$$
$$+\int_M\left[\Lambda(-n)\Omega(n)+\Lambda(n)\Omega(-n)\right]dm$$
$$+\int_M\frac{R}{c_p\overline{\sigma}p}\left[\Lambda(-n)H(n)+\Lambda(n)H(-n)\right]dm\,. \tag{10.114}$$

Using the continuity equation for nondivergent flow,

$$\frac{i(n-m)}{a\cos\phi}U_\psi(n-m)+\frac{1}{a\cos\phi}|V_\psi(n-m)\cos\phi|_\phi=0$$

and

$$\frac{i(-n-m)}{a\cos\phi}U_\psi(-n-m)+\frac{1}{a\cos\phi}|V_\psi(-n-m)\cos\phi|_\phi=0\,,$$

and after some simplification, (10.114) may be written as

$$\frac{\partial A(n)}{\partial t}=-\int_M\frac{1}{a}\frac{\partial\Lambda(0)}{\overline{\sigma}\partial\phi}\left[\Lambda(-n)V_\psi(n)+\Lambda(n)V_\psi(-n)\right]dm$$
$$+\int_M\sum_{\substack{m=-\infty\\ m\neq 0}}^{\infty}\frac{1}{\overline{\sigma}}\left[\Lambda(m)\left(U_\psi(n-m)\frac{-in\Lambda(-n)}{a\cos\phi}\right.\right.$$

$$+U_\psi(-n-m)\frac{in\Lambda(n)}{a\cos\phi}\Bigg)$$

$$+\Lambda(m)\left(V_\psi(n-m)\frac{1}{a}\frac{\partial\Lambda(-n)}{\partial\phi}\right.$$

$$\left.+V_\psi(-n-m)\frac{1}{a}\frac{\partial\Lambda(n)}{\partial\phi}\right)\Bigg]\,dm$$

$$+\int_M\left[\Lambda(-n)\Omega(n)+\Lambda(n)\Omega(-n)\right]\,dm$$

$$+\int_M\frac{R}{c_p\bar{\sigma}p}\left[\Lambda(-n)H(n)+\Lambda(n)H(-n)\right]\,dm. \qquad (10.115)$$

Setting $n=0$ and noting that $V_\psi(0)=0$, we get the equation for zonal available potential energy as

$$\frac{\partial\bar{A}}{\partial t}=\int_M\frac{1}{\bar{\sigma}}\left(\Lambda(0)\frac{\partial\Lambda(0)}{\partial t}\right)dm$$

$$=\int_M\frac{1}{\bar{\sigma}}\sum_{\substack{m=-\infty\\ m\neq 0}}^{\infty}\Lambda(m)\left(V_\psi(-m)\frac{1}{a}\frac{\partial\Lambda(0)}{\partial\phi}\right)dm$$

$$+\int_M\Lambda(0)\Omega(0)dm+\frac{R}{c_p\bar{\sigma}p}\int_M\Lambda(0)H(0)\,dm$$

$$=\int_M\sum_{m=1}^{\infty}\frac{1}{a}\frac{\partial\Lambda(0)}{\bar{\sigma}\partial\phi}\left[\Lambda(-m)V_\psi(m)+\Lambda(m)V_\psi(-m)\right]dm$$

$$+\int_M\Lambda(0)\Omega(0)\,dm+\int_M\frac{R}{c_p\bar{\sigma}p}\Lambda(0)H(0)\,dm. \qquad (10.116)$$

Equation (10.115) is the eddy available potential energy equation for wavenumber *n* and (10.116) is the equation for zonal available potential energy.

We note that the first integral in (10.115) is equal and opposite in sign to the first integral for wavenumber *n* in (10.116). It therefore represents the exchange between zonal available potential energy and eddy available potential energy of wavenumber *n*. The second integral in (10.115) represents exchanges of eddy available potential energy among different waves to provide energy to wavenumber *n*. The two integrals

$$\int_M\Lambda(0)\Omega(0)\,dm$$

and

$$\int_M\left[\Lambda(-n)\Omega(n)+\Lambda(n)\Omega(-n)\right]dm$$

represent the conversion of zonal available potential energy to zonal kinetic energy and of eddy available potential energy to eddy kinetic energy (of wavenumber *n*), respectively. Finally,

$$\int_M\frac{R}{c_p\bar{\sigma}p}\Lambda(0)H(0)\,dm$$

and

$$\int_M \frac{R}{c_p \sigma p}[\Lambda(-n)H(n)+\Lambda(n)H(-n)]\, dm$$

represents generation of zonal available potential energy and eddy available potential energy (of wavenumber n), respectively.

10.4 Energy equations in Two-Dimensional Wavenumber Domain

In section 10.3 we derived the equations for atmospheric energetics in zonal wavenumber domain, where various atmospheric fields and their derivatives in the zonal direction were represented in terms of a truncated Fourier series. Finite differences were used for calculating derivatives in the meridional and vertical directions. Energy exchanges and energy conversions in wavenumber domain were determined using the orthogonality properties of the Fourier functions. This mixture of Fourier representation and finite differencing is well-suited for a limited domain in the meridional direction. For a complete global domain or a hemispheric domain the spectral energetics in spherical harmonics as basis functions are more appropriate.

Using the spectral properties of spherical harmonics discussed in Chapter 6, we derive here the equations for atmospheric energetics in two-dimensional wavenumber domain. The equation for available potential energy can be written in terms of specific volume α or potential temperature θ. In Section 10.3.2, this equation in one-dimensional wavenumber domain was obtained in terms of specific volume. In this section we derive the available potential energy equation in terms of potential temperature in the two-dimensional wavenumber domain.

As mentioned earlier, for a two-dimensional spectral representation it is convenient to represent the wind field in terms of vorticity and divergence (or streamfunction and velocity potential) as

$$\vec{V} = \vec{k} \times \nabla \psi + \nabla \chi \qquad (10.117)$$

$$\zeta = \vec{k} \cdot \nabla \times \vec{V} = \nabla^2 \psi \qquad (10.118)$$

$$D = \nabla \cdot \vec{V} = \nabla^2 \chi \, . \qquad (10.119)$$

The area-average kinetic energy is then given by

$$\overline{K} = \overline{K}_\psi + \overline{K}_\chi = \frac{1}{2}\overline{\nabla \psi \cdot \nabla \psi} + \frac{1}{2}\overline{\nabla \chi \cdot \nabla \chi} \, . \qquad (10.120)$$

As shown in Section 10.2.3, from (10.120) we may write

$$\overline{K} = \overline{K}_\psi + \overline{K}_\chi = -\frac{1}{2}\overline{\psi \nabla^2 \psi} - \frac{1}{2}\overline{\chi \nabla^2 \chi} \, . \qquad (10.121)$$

Using a triangular trunctation for the spectral representation as

$$\psi = \sum_{m=-N}^{N} \sum_{n=|m|}^{N} \psi_n^m Y_n^m(\lambda, \mu), \qquad (10.122)$$

$$\nabla^2 \psi = \sum_{m=-N}^{N} \sum_{n=|m|}^{N} \frac{-n(n+1)}{a^2} \psi_n^m Y_n^m(\lambda, \mu), \qquad (10.123)$$

we get

$$\overline{\overline{K}}_\psi = \frac{1}{4\pi g}\int_0^{p_0}\int_{-1}^{1}\int_0^{2\pi}\frac{1}{2}\left(\sum_{m=-N}^{N}\sum_{n=|m|}^{N}\psi_n^m(p)Y_n^m(\lambda,\mu)\right)$$
$$\times\left(\sum_{m_1=-N}^{N}\sum_{n_1=|m_1|}^{N}\frac{n_1(n_1+1)}{a^2}\psi_{n_1}^{m_1}(p)Y_{n_1}^{m_1}(\lambda,\mu)\right)d\lambda d\mu dp,$$

or for component Y_n^m

$$\overline{\overline{K}}_{\psi n}^{\,m} = \frac{1}{g}\int_0^{p_0}\frac{1}{2}\frac{n(n+1)}{a^2}\psi_n^m(p)\psi_n^{*m}(p)\,dp. \tag{10.124}$$

Similarly,

$$\overline{\overline{K}}_{\chi n}^{\,m} = \frac{1}{g}\int_0^{p_0}\frac{1}{2}\frac{n(n+1)}{a^2}\chi_n^m(p)\chi_n^{*m}(p)dp. \tag{10.125}$$

Likewise, the mean available potential energy is

$$\overline{\overline{A}} = \frac{1}{g}\int_0^{p_0}\int_{-1}^{1}\int_0^{2\pi}\frac{1}{2\overline{\sigma}}\alpha'^2 d\lambda\, d\mu\, dp$$
$$= \frac{1}{2}\frac{R}{g}\frac{1}{p_{00}^{\kappa}}\int_0^{p_0}\int_{-1}^{1}\int_0^{2\pi}\left(\frac{p^{\kappa-1}}{-\partial\overline{\theta}/\partial p}\right)\theta'^2 d\lambda\, d\mu\, dp.$$

Its spectral form may be written as

$$\overline{\overline{A}}_n^{\,m} = \frac{1}{g}\int_0^{p_0}\frac{1}{2\overline{\sigma}}\alpha'^m_n(p)\alpha'^{*m}_n(p)\,dp.$$
$$= \frac{1}{2}\frac{R}{g}\frac{1}{p_{00}^{\kappa}}\int_0^{p_0}\left(-\frac{p^{\kappa-1}}{\partial\overline{\theta}/\partial p}\right)\theta'^m_n(p)\theta'^{*m}_n(p)dp, \tag{10.126}$$

where $p_{00} = 1000$ mb and p_0 is the pressure at the bottom level.

We now derive the equations for the kinetic energy and available potential energy exchanges based on the quasi-nondivergent equations of Lorenz (1960a). These consist of only the vorticity, thermodynamic, and linear balance equations of the form

$$\frac{\partial\nabla^2\psi}{\partial t} + \vec{V}_\psi\cdot\nabla(\nabla^2\psi + f) + \nabla\cdot(f\nabla\chi) = \vec{k}\cdot\nabla\times\vec{F}, \tag{10.127}$$

$$\frac{\partial\theta'}{\partial t} + \vec{V}_\psi\cdot\nabla\theta' + \omega\frac{\partial\overline{\theta}}{\partial p} = \left(\frac{p_{00}}{p}\right)^{\kappa}\frac{1}{c_p}H', \tag{10.128}$$

and

$$\nabla^2\phi = \nabla\cdot(f\nabla\psi). \tag{10.129}$$

Equation (10.128) is an alternative form of

$$\frac{\partial\alpha'}{\partial t} + \vec{V}_\psi\cdot\nabla\alpha' - \overline{\sigma}\omega = \frac{R}{c_p p}H'. \tag{10.130}$$

Here $\overline{\sigma} = -\overline{\alpha}\partial\ln\overline{\theta}/\partial p$, $\omega = -\nabla^2\overline{\chi}$, and $\overline{\chi} = \int_0^p \chi\, dp$.

We can write (10.127) as

$$\frac{\partial}{\partial t}\nabla^2\psi = -J(\psi,\nabla^2\psi) - J(\psi,f) - f\nabla^2\chi - \nabla f\cdot\nabla\chi + \vec{k}\cdot\nabla\times\vec{F}. \tag{10.131}$$

The spectral representation of the various terms in (10.131) is

$$\frac{\partial}{\partial t}\nabla^2\psi = -\sum_{m_1=-N}^{N}\sum_{n_1=|m_1|}^{N}\frac{n_1(n_1+1)}{a^2}\frac{\partial\psi_{n_1}^{m_1}}{\partial t}Y_{n_1}^{m_1} \tag{10.132}$$

and

$$\begin{aligned}J(\psi,\nabla^2\psi) &= J\left(\sum_{m_2=-N}^{N}\sum_{n_2=|m_2|}^{N}\psi_{n_2}^{m_2}Y_{n_2}^{m_2}\right. \\ &\quad\left.\times\sum_{m_3=-N}^{N}\sum_{n_3=|m_3|}^{N}\frac{-n_3(n_3+1)}{a^2}\psi_{n_3}^{m_3}Y_{n_3}^{m_3}\right) \\ &= -i\sum_{m_2=-N}^{N}\sum_{n_2=|m_2|}^{N}\sum_{m_3=-N}^{N}\sum_{n_3=|m_3|}^{N}\frac{n_3(n_3+1)}{a^4} \\ &\quad\times\psi_{n_2}^{m_2}\psi_{n_3}^{m_3}\left(m_2Y_{n_2}^{m_2}\frac{\partial Y_{n_3}^{m_3}}{\partial\mu}-m_3Y_{n_3}^{m_3}\frac{\partial Y_{n_2}^{m_2}}{\partial\mu}\right).\end{aligned} \tag{10.133}$$

On interchanging indices in (10.133) we also get

$$\begin{aligned}J(\psi,\nabla^2\psi) &= i\sum_{m_2=-N}^{N}\sum_{n_2=|m_2|}^{N}\sum_{m_3=-N}^{N}\sum_{n_3=|m_3|}^{N}\frac{n_2(n_2+1)}{a^4} \\ &\quad\times\psi_{n_2}^{m_2}\psi_{n_3}^{m_3}\left(m_2Y_{n_2}^{m_2}\frac{\partial Y_{n_3}^{m_3}}{\partial\mu}-m_3Y_{n_3}^{m_3}\frac{\partial Y_{n_2}^{m_2}}{\partial\mu}\right).\end{aligned} \tag{10.134}$$

From (10.133) and (10.134) we get a symmetric form of this transform as

$$\begin{aligned}J(\psi,\nabla^2\psi) &= \frac{1}{2a^4}\sum_{m_2=-N}^{N}\sum_{n_2=|m_2|}^{N}\sum_{m_3=-N}^{N}\sum_{n_3=|m_3|}^{N} \\ &\quad i\left[n_2(n_2+1)-n_3(n_3+1)\right]\psi_{n_2}^{m_2}\psi_{n_3}^{m_3} \\ &\quad\times\left(m_2Y_{n_2}^{m_2}\frac{\partial Y_{n_3}^{m_3}}{\partial\mu}-m_3Y_{n_3}^{m_3}\frac{\partial Y_{n_2}^{m_2}}{\partial\mu}\right).\end{aligned} \tag{10.135}$$

Also,

$$\begin{aligned}J(\psi,f) &= J\left(\sum_{m_1=-N}^{N}\sum_{n_1=|m_1|}^{N}\psi_{n_1}^{m_1}Y_{n_1}^{m_1},2\Omega\mu\right) \\ &= \frac{2\Omega}{a^2}\sum_{m_1=-N}^{N}\sum_{n_1=|m_1|}^{N}im_1\psi_{n_1}^{m_1}Y_{n_1}^{m_1},\end{aligned} \tag{10.136}$$

$$\begin{aligned}f\nabla^2\chi &= 2\Omega\sum_{m_1=-N}^{N}\sum_{n_1=|m_1|}^{N}\frac{-n_1(n_1+1)}{a^2}\chi_{n_1}^{m_1}\mu Y_{n_1}^{m_1} \\ &= -\frac{2\Omega}{a^2}\sum_{m_1=-N}^{N}\sum_{n_1=|m_1|}^{N}\left[(n_1+1)(n_1+2)\in_{n_1+1}^{m_1}\chi_{n_1+1}^{m_1}\right. \\ &\quad\left.+n_1(n_1-1)\in_{n_1}^{m_1}\chi_{n_1-1}^{m_1}\right]Y_{n_1}^{m_1},\end{aligned} \tag{10.137}$$

$$\nabla f\cdot\nabla\chi = 2\Omega\nabla\mu\cdot\nabla\left(\sum_{m_1=-N}^{N}\sum_{n_1=|m_1|}^{N}\chi_{n_1}^{m_1}Y_{n_1}^{m_1}\right)$$

$$=\frac{2\Omega}{a^2}\sum_{m_1=-N}^{N}\sum_{n_1=|m_1|}^{N}\Big[(n_1+2)\in_{n_1+1}^{m_1}\chi_{n_1+1}^{m_1}$$

$$-(n_1-1)\in_{n_1}^{m_1}\chi_{n_1}^{m_1}\Big]Y_{n_1}^{m_1}, \tag{10.138}$$

and

$$F_r=\sum_{m_1=-N}^{N}\sum_{n_1=|m_1|}^{N}F_{n_1}^{m_1}Y_{n_1}^{m_1}. \tag{10.139}$$

While deriving spectral representations (10.137) and (10.138) we have made use of recurrence relations (1) and (2) in Section 6.5. Substituting (10.132) and (10.135) to (10.139) into (10.131) and multiplying both sides by $Y_{n_1}^{*m_1}$ and integrating over $\mu=-1,\ 1$ and $\lambda=0,\ 2\pi$ gives

$$-\frac{n_1(n_1+1)}{a^2}\frac{\partial\psi_{n_1}^{m_1}}{\partial t}=-\frac{i}{2a^4}\sum_{m_2=-N}^{N}\sum_{n_2=|m_2|}^{N}\sum_{m_3=-N}^{N}\sum_{n_3=|m_3|}^{N}$$

$$\big[n_2(n_2+1)-n_3(n_3+1)\big]\psi_{n_2}^{m_2}\psi_{n_3}^{m_3}J_{n_1n_2n_3}^{m_1m_2m_3}$$

$$-i\frac{2\Omega}{a^2}m_1\psi_{n_1}^{m_1}+\frac{2\Omega}{a^2}\big[n_1(n_1+2)\in_{n_1+1}^{m}\chi_{n_1+1}^{m_1}$$

$$+(n_1-1)(n_1+1)\in_{n_1}^{m_1}\chi_{n_1-1}^{m_1}\big]+F_{n_1}^{m_1}, \tag{10.140}$$

where

$$J_{n_1n_2n_3}^{m_1m_2m_3}=\int_{-1}^{1}P_{n_1}^{m_1}\left(m_2P_{n_2}^{m_2}\frac{\partial P_{n_3}^{m_3}}{\partial\mu}-m_3P_{n_3}^{m_3}\frac{\partial P_{n_2}^{m_2}}{\partial\mu}\right)d\mu,$$

with $m_1=m_2+m_3$.

It can also be shown that

$$J_{n_1n_2n_3}^{m_1^*m_2^*m_3^*}=-J_{n_1n_2n_3}^{m_1m_2m_3}. \tag{10.141}$$

From (10.124) we have

$$\frac{\partial\overline{\overline{K}}_{\psi n}^{m}}{\partial t}=\frac{1}{g}\int_0^{p_0}\frac{1}{2}\frac{n_1(n_1+1)}{a^2}\left(\psi_{n_1}^{*m_1}\frac{\partial\psi_{n_1}^{m_1}}{\partial t}+\psi_{n_1}^{m_1}\frac{\partial\psi_{n_1}^{*m_1}}{\partial t}\right)dp. \tag{10.142}$$

From (10.140), (10.141), and (10.142) we get

$$\frac{\partial\overline{\overline{K}}_{\psi n}^{m}}{\partial t}=\frac{1}{g}\int_0^{p_0}\frac{i}{4a^4}\sum_{m_2=-N}^{N}\sum_{n_2=|m_2|}^{N}\sum_{m_3=-N}^{N}\sum_{n_3=|m_3|}^{N}$$

$$\big\{[n_2(n_2+1)-n_3(n_3+1)]$$

$$\times\psi_{n_1}^{*m_1}\psi_{n_2}^{m_2}\psi_{n_3}^{m_3}-\psi_{n_1}^{m_1}\psi_{n_2}^{*m_2}\psi_{n_3}^{*m_3}\big\}J_{n_1n_2n_3}^{m_1m_2m_3}\,dp$$

$$-\frac{1}{g}\int_0^{p_0}\frac{\Omega}{a^2}\Big[n_1(n_1+2)\in_{n_1+1}^{m_1}\left(\chi_{n_1+1}^{m_1}\psi_{n_1}^{*m_1}+\chi_{n_1+1}^{*m_1}\psi_{n_1}^{m_1}\right)$$

$$+(n_1-1)(n_1+1)\in_{n_1}^{m_1}\left(\chi_{n_1-1}^{m_1}\psi_{n_1}^{*m_1}+\chi_{n_1-1}^{*m_1}\psi_{n_1}^{m_1}\right)\Big]dp$$

$$-\frac{1}{g}\int_0^{p_0}\left(F_{n_1}^{m_1}\psi_{n_1}^{*m_1}+F_{n_1}^{*m_1}\psi_{n_1}^{m_1}\right)dp. \tag{10.143}$$

This is the spectral form of the kinetic energy equation for nondivergent flow in two-dimensional wavenumber domain. To obtain the equation for available potential energy we write the thermodynamic equation (10.128) in its spectral form as

$$\frac{\partial \theta_{n_1}^{m_1}}{\partial t} = -\frac{i}{a^2} \sum_{m_2=-N}^{N} \sum_{n_2=|m_2|}^{N} \sum_{m_3=-N}^{N} \sum_{n_3=|m_3|}^{N} \psi_{n_2}^{m_2} \theta_{n_3}^{m_3} J_{n_1 n_2 n_3}^{m_1 m_2 m_3}$$

$$-W_{n_1}^{m_1} \frac{\partial \bar{\theta}}{\partial p} + \frac{1}{c_P}\left(\frac{p_{00}}{p}\right)^{\kappa} H_{n_1}^{m_1}. \qquad (10.144)$$

Differentiating (10.126) with respect to time gives

$$\frac{\partial \overline{\overline{A}}_n^m}{\partial t} = -\frac{1}{2}\frac{R}{g}\frac{1}{p_{00}^{\kappa}} \int_0^{p_0} \frac{p^{\kappa-1}}{\partial \bar{\theta}/\partial p}\left(\theta^{*m_1}_{n_1} \frac{\partial \theta_{n_1}^{m_1}}{\partial t} + \theta_{n_1}^{m_1} \frac{\partial \theta^{*m_1}_{n_1}}{\partial t}\right) dp. \qquad (10.145)$$

Substituting for $\partial \theta_n^m / \partial t$ and $\partial \theta_n^{*m} / \partial t$ from (10.144) we get

$$\frac{\partial \overline{\overline{A}}_n^m}{\partial t} = \frac{R}{2g}\frac{1}{p_{00}{}^{\kappa}} \int_0^{p_0} \frac{p^{\kappa-1}}{\partial \bar{\theta}/\partial p}\frac{1}{a^2} \sum_{m_2=-N}^{N} \sum_{n_2=|m_2|}^{N} \sum_{m_3=-N}^{N} \sum_{n_3=|m_3|}^{N}$$

$$\left(\theta^{*m_1}_{n_1} \psi_{n_2}^{m_2} \theta_{n_3}^{m_3} - \theta_{n_1}^{m_1} \psi^{*m_2}_{n_2} \theta^{*m_3}_{n_3}\right) J_{n_1 n_2 n_3}^{m_1 m_2 m_3}\, dp$$

$$+\frac{R}{2g}\frac{1}{p_{00}{}^{\kappa}} \int_0^{p_0} p^{\kappa-1}\left(W_{n_1}^{m_1} \theta^{*m_1}_{n_1} + W^{*m_1}_{n_1} \theta_{n_1}^{m_1}\right) dp$$

$$\times \frac{R}{2gc_p} \int_0^{p_0} \frac{1}{p(\partial \bar{\theta}/\partial p)}\left(\theta^{*m_1}_{n_1} H_{n_1}^{m_1} + \theta_{n_1}^{m_1} H^{*m_1}_{n_1}\right) dp. \qquad (10.146)$$

The first integral on the right-hand side of (10.143) represents the nonlinear exchanges of kinetic energy between different waves as well as between waves and the zonal flow. Likewise, the first integral on the right-hand side of (10.146) describes such nonlinear exchanges of available potential energy. Following Fjørtoft (1953), it can be shown that the sum of these kinetic energy exchanges and potential energy exchanges over the complete truncated spectrum vanishes.

We shall show that the second integral on the right-hand side of (10.143) and (10.146) represents conversion between available potential energy and kinetic energy. Differentiating the linear balance equation (10.129) with respect to p and making use of the hydrostatic relation, we get

$$\frac{R}{p_{00}^{\kappa}} p^{\kappa-1} \nabla^2 \theta = \nabla \cdot \left(f \nabla \frac{\partial \psi}{\partial p}\right). \qquad (10.147)$$

Following (10.137) and (10.138), the spectral transform of (10.147) can be shown to be

$$\frac{R}{p_{00}^{\kappa}} p^{\kappa-1} n_1(n_1+1)\theta_{n_1}^{m_1} = -\frac{2\Omega}{a^2}\left(n_1(n_1+2) \in_{n_1+1}^{m_1} \frac{\partial \psi_{n_1+1}^{m_1}}{\partial p}\right.$$

$$\left. +(n_1-1)(n_1+1) \in_{n_1}^{m_1} \frac{\partial \psi_{n_1-1}^{m_1}}{\partial p}\right),$$

so that

$$\frac{R}{p_{00}^{\kappa}}\int_{o}^{p_0} p^{\kappa-1}\frac{n_1(n_1+1)}{a^2}\left(W_{n_1}^{m_1}\theta^{*m_1}_{n_1}+W^{*m_1}_{n_1}\theta_{n_1}^{m_1}\right)dp$$

$$=-\frac{2\Omega}{a^2}\int_{o}^{p_0}\left[n_1(n_1+2)\in_{n_1+1}^{m_1}\left(W_{n_1}^{m_1}\frac{\partial\psi^{*m_1}_{n_1+1}}{\partial p}+W^{*m_1}_{n_1}\frac{\partial\psi^{m_1}_{n_1+1}}{\partial p}\right)\right.$$

$$\left.+(n_1-1)(n_1+1)\in_{n_1}^{m_1}\left(W_{n_1}^{m_1}\frac{\partial\psi^{*m_1}_{n_1-1}}{\partial p}+W^{*m_1}_{n_1}\frac{\partial\psi^{m_1}_{n_1-1}}{\partial p}\right)\right]dp$$

$$=-\frac{2\Omega}{a^2}\int_{o}^{p_0}\left[n_1(n_1+2)\in_{n_1+1}^{m_1}\frac{\partial}{\partial p}\left(W_{n_1}^{m_1}\psi^{*m_1}_{n_1+1}+W^{*m_1}_{n_1}\psi^{m_1}_{n_1+1}\right)\right.$$

$$+(n_1-1)(n_1+1)\in_{n_1}^{m_1}\frac{\partial}{\partial p}\left(W_{n_1}^{m_1}\psi^{*m_1}_{n_1-1}+W_{n_1}^{m_1}\psi^{m_1}_{n_1-1}\right)$$

$$-n_1(n_1+2)\in_{n_1+1}^{m_1}\left(\frac{\partial W_{n_1}^{m_1}}{\partial p}\psi^{*m_1}_{n_1+1}+\frac{\partial W^{*m_1}_{n_1}}{\partial p}\psi^{m_1}_{n_1+1}\right)$$

$$\left.-(n_1-1)(n_1+1)\in_{n_1}^{m_1}\left(\frac{\partial W_{n_1}^{m_1}}{\partial p}\psi^{*m_1}_{n_1-1}+\frac{\partial W^{*m_1}_{n_1}}{\partial p}\psi^{m_1}_{n_1-1}\right)\right]dp. \quad (10.148)$$

Assuming that the vertical vorticity for all waves vanishes at the top and bottom of the atmosphere, (10.148) reduces to

$$\frac{1}{2g}\frac{R}{p_{00}^{\kappa}}\int_{o}^{p_0} p^{\kappa-1}\left(W_{n_1}^{m_1}\theta^{*m_1}_{n_1}+W^{*m_1}_{n_1}\theta_{n_1}^{m_1}\right)dp \quad (10.149)$$

$$=\frac{1}{g}\frac{\Omega}{a^2}\int_{o}^{p_0}\left[n_1(n_1+2)\in_{n_1+1}^{m_1}\left(\chi_{n_1}^{m_1}\psi^{*m_1}_{n_1+1}+\chi^{*m_1}_{n_1}\psi^{m_1}_{n_1+1}\right)\right.$$

$$\left.+(n_1-1)(n_1+1)\in_{n_1}^{m_1}\left(\chi_{n_1}^{m_1}\psi^{*m_1}_{n_1-1}+\chi^{*m_1}_{n_1}\psi^{m_1}_{n_1-1}\right)\right]dp.$$

Thus we see that the second integral on the right-hand side of the kinetic energy equation (10.143) is equal and has opposite sign to the second integral in the available potential energy equation (10.146). These terms therefore represent conversion of available potential energy to kinetic energy. The last integral in (10.143) is the dissipation rate of kinetic energy due to frictional forces, while in (10.146) it represents generation of available potential energy resulting from the covariance of temperature and heat sources and sinks.

In actual calculations of atmospheric energetics in two-dimensional wavenumber domain using the primitive equations of the atmosphere, it is convenient to use the transform method in calculating the nonlinear terms of the energetics equations. These terms represent the energy transformation between various two-dimensional wavenumbers. Also it is convenient to replace the momentum equation with vorticity and divergence equations. In the vorticity and divergence equations one can then separate the terms involving the rotational and divergent part of the flow and those involving products of the rotational and divergent parts as discussed in Section 10.2.3. Using such a framework, we show the energetics in two-dimensional wavenumber domain for the troposphere between 200 and 1000 mb during January 1989 for *T*–15 spectral resolution.

Table 10.2. Spectral distribution of the mean kinetic energy of the rotational part of the wind in units of 10^2 J m^{-2}. $\overline{\overline{K}}_\psi = 0.1059 \times 10^7$ J m^{-2}.

n \ m	0	1	2	3	4	5	6	7	8	9	10	11	12	13	14	15
15	15	11	13	11	15	12	8	7	10	5	7	6	4	3	4	1
14	7	21	20	9	9	13	10	10	7	6	7	5	6	2	1	
13	22	16	16	15	26	23	15	9	9	11	9	9	4	1		
12	11	35	30	31	23	20	21	19	14	13	9	5	2			
11	16	45	52	37	25	29	52	29	24	8	12	3				
10	155	187	15	68	38	58	64	29	21	17	4					
9	18	113	77	97	66	83	81	46	25	6						
8	75	40	53	59	160	96	57	30	7							
7	92	308	99	107	78	70	48	17								
6	91	371	51	160	81	57	14									
5	822	87	39	72	25	19										
4	157	73	43	18	17											
3	1541	52	15	3												
2	80	37	2													
1	2926	1														
0	0															

Table 10.3. Spectral distribution of the mean kinetic energy of the divergent part of the wind in units of 10^2 J m^{-2}. $\overline{\overline{K}}_\chi = 0.8422 \times 10^4$ J m^{-2}.

n \ m	0	1	2	3	4	5	6	7	8	9	10	11	12	13	14	15
15	5	14	9	16	12	14	17	14	13	17	14	15	27	23	19	22
14	9	18	16	16	19	18	22	18	20	18	28	29	21	21	21	
13	10	17	18	20	18	22	25	23	28	28	31	30	21	20		
12	11	16	22	27	29	32	24	30	26	21	52	33	25			
11	17	38	21	29	33	29	38	40	43	52	35	31				
10	15	30	25	37	35	45	83	65	66	46	45					
9	24	30	29	56	62	80	80	81	62	47						
8	20	45	60	53	69	81	171	134	61							
7	74	29	42	63	72	86	154	63								
6	269	93	101	82	117	151	118									
5	93	86	93	110	122	111										
4	164	45	72	58	260											
3	173	100	70	188												
2	15	95	259													
1	211	1169														
0	0															

Table 10.4. Spectral distribution of the mean potential energy of the divergent part of the wind in units of 10^2 J m^{-2}. $\overline{\overline{\text{APE}}} = 0.5031 \times 10^7$ J m^{-2}.

	15	3	6	6	6	5	4	5	2	2	3	3	1	2	1	0	0
	14	8	8	6	5	6	5	6	3	4	2	3	2	1	1	0	
	13	6	10	8	11	7	7	7	6	5	4	3	2	1	0		
	12	10	14	18	12	11	8	11	9	6	3	3	2	1			
	11	13	22	12	17	15	16	21	13	6	6	2	1				
	10	9	27	20	32	27	32	34	15	2	4	1					
	9	29	97	34	43	45	51	36	16	5	3						
	8	117	54	75	40	46	53	35	16	6							
n	7	70	156	61	99	103	63	19	6								
	6	98	137	165	128	37	35	11									
	5	49	114	194	95	86	22										
	4	131	326	140	70	32											
	3	122	206	101	51												
	2	41587	189	34													
	1	3860	187														
	0	0															
		0	1	2	3	4	5	6	7	8	9	10	11	12	13	14	15
									m								

A lower resolution has been purposely chosen so that the results can be easily shown in tabular form.

Tables 10.2 and 10.3 show the spectral distribution of the mean kinetic energy of the rotational and divergent components of the flow, respectively. Table 10.4 shows the spectral distribution of mean available potential energy. The energy distribution in these tables is in units of 10^2 J m^{-2}. The total energy contents ($\overline{\overline{\text{K}}}_\psi$, $\overline{\overline{\text{K}}}_\chi$, and $\overline{\overline{\text{APE}}}$) for the whole spectrum are also shown in these tables.

The available potential energy is about five times the kinetic energy. The kinetic energy of the divergent flow is two orders of magnitude smaller than the kinetic energy of the divergent part of the flow. Nearly 80% of the total available potential energy is found to reside in component Y_2^0. The kinetic energy of the rotational flow is maximum (about 30% of the total) in component Y_1^0, which is a measure of mean angular momentum. The component Y_1^1 contains about 15% of the total kinetic energy of the divergent flow, while the rest is distributed over different components, mostly up to wavenumbers 8 to 10.

Tables 10.5 and 10.6 show the spectral distribution of nonlinear exchanges of the kinetic energy of the rotational flow and available potential energy, respectively, between different two-dimensional scales. The sum of the nonlinear exchanges of kinetic energy as well as of available potential energy over the whole spectrum is found to vanish. From Table 10.5, we notice that in accordance with the Fjørtoft theorem [Fjørtoft (1953)], the small-and large-scale waves gain kinetic energy at the expense of medium-scale waves though wave-wave energy exchange. In the case of available potential energy exchanges, nearly all wave components gain available potential energy at the expense of zonal

Table 10.5. Spectral distribution of nonlinear exchanges of the kinetic energy of the rotational flow between different two-dimensional wavenumbers in units of 10^{-4} W m^{-2}.

	15	-11	15	3	28	31	65	-14	11	32	14	44	50	50	11	5	-2
	14	-23	61	29	23	-13	84	13	40	36	33	29	25	19	13	7	
	13	45	-66	-5	49	61	100	96	15	10	-16	-26	-24	-25	-2		
	12	-4	19	52	1	90	76	-30	-10	13	18	-6	-6	0			
	11	18	-10	26	44	-68	12	-44	-15	-29	7	-17	-12				
	10	-92	-270	-50	166	-44	-106	-106	-133	-121	57	14					
	9	50	-50	16	346	27	172	-230	-191	-28	5						
	8	120	-150	-54	-38	-13	-167	-378	-123	-19							
n	7	-269	51	152	-520	123	-296	-107	-69								
	6	-242	396	-211	-138	-98	-52	-83									
	5	223	-272	-71	-96	-155	-32										
	4	52	85	-189	-60	-32											
	3	2519	-149	35	16												
	2	129	124	11													
	1	0	0														
	0	0															
		0	1	2	3	4	5	6	7	8	9	10	11	12	13	14	15
									m								

Table 10.6. Spectral distribution of nonlinear exchanges of available potential energy between different two-dimensional wavenumbers in units of 10^{-4} W m^{-2}.

	15	55	64	45	50	54	52	73	19	46	40	25	12	12	-2	2	0
	14	23	28	56	44	59	72	80	31	36	43	55	20	1	0	-1	
	13	78	57	39	137	77	130	164	58	77	33	39	16	-2	2		
	12	60	137	33	146	56	118	106	96	58	21	4	-10	-4			
	11	68	114	105	86	84	124	287	194	91	12	-23	0				
	10	108	198	159	172	96	234	584	255	122	-3	-8					
	9	60	646	283	118	354	266	470	215	28	-12						
	8	505	-8	10	167	360	419	609	179	-5							
n	7	325	69	283	53	282	162	250	34								
	6	-387	694	260	364	393	257	0									
	5	2	545	556	436	-60	-178										
	4	686	-223	454	249	13											
	3	220	334	134	-54												
	2	-16271	171	130													
	1	-762	-12														
	0	0															
		0	1	2	3	4	5	6	7	8	9	10	11	12	13	14	15
									m								

Table 10.7. The conversion of available potential energy to kinetic energy of the divergent flow in units of 10^{-4} W m^{-2}. Total $< \mathrm{APE\ to\ K}_\chi > = -0.832$ W m^{-2}.

n \ *m*	0	1	2	3	4	5	6	7	8	9	10	11	12	13	14	15
15	-19	-18	-19	-17	-47	-35	-75	-23	-29	-50	-27	-14	-28	-2	-7	-2
14	-12	-45	-44	-55	44	-34	-39	-24	-37	-18	-61	-14	-4	-2	-2	
13	-27	15	-4	-44	-53	-67	-127	-49	-67	-45	-27	-16	0	4		
12	-5	-35	28	-89	-23	-71	-84	-65	-56	-2	35	5	-8			
11	110	14	-72	-57	-89	-136	-170	-152	-73	-4	13	-1				
10	-90	-73	-51	-63	21	-161	-329	-176	-91	11	0					
9	-52	-117	-156	90	-167	-164	-271	-190	-19	13						
8	-119	51	138	-297	-116	-240	-379	-170	-2							
7	121	254	-76	-122	102	1	-214	-9								
6	931	-187	-371	-95	-332	-199	6									
5	149	-6	-571	-324	147	103										
4	-407	-127	-93	-46	-190											
3	-164	-352	59	-84												
2	1634	-70	-289													
1	-1063	-883														
0	0															

Table 10.8. A summary of mean energetics. In units of W m^{-2}.

$\overline{\dfrac{\partial K_\psi}{\partial t}}$	=			$\overline{< K_\chi \Rightarrow K_\psi >}$	$+\mathrm{dis.}K_\psi$
0.096	=			0.577	-0.481
				$\Uparrow$	
$\overline{\dfrac{\partial K_\chi}{\partial t}}$	=	$\overline{< A \Rightarrow K_\chi >}$	+	$\overline{< K_\chi \Rightarrow K_\psi >}$	$+\mathrm{dis.}K_\chi$
0.003	=	0.832		-0.577	-0.252
		$\Uparrow$			
$\overline{\dfrac{\partial A}{\partial t}}$	=	$\overline{< A \Rightarrow K_\chi >}$			$+\mathrm{gen.}\bar{A}$
-0.067	=	-0.832			+0.765

components Y_1^0 and Y_2^0. Medium-scale waves gain maximum available potential energy through nonlinear exchanges.

Table 10.7 shows the conversion of available potential energy to kinetic energy of the divergent flow. Except for some zonal components, the available potential energy is converted to kinetic energy of the divergent flow over all scales. Maximum conversion takes place at medium-and large-scale waves.

A summary of mean energetics based on these calculations is given in Table 10.8. The available potential energy is generated at the rate of 0.765 W m^{-2}. In this particular case, 0.832 W m^{-2} are converted to kinetic energy of the divergent flow, resulting in a net decrease in available potential energy of the atmosphere at the rate of 0.067 W $m^{-2.}$ About 70% of the divergent kinetic energy converted from available potential energy gets transformed to kinetic energy of the rotational flow while the remaining 30% is dissipated by frictional forces.

Most of the kinetic energy of the rotational part of the flow received from its divergent part gets dissipated by frictional forces, while a small part is used to strengthen the rotational flow. It is interesting to note that even though the kinetic energy of the rotational part of the flow is about two orders of magnitude greater than that of the divergent flow, the dissipation of the rotational kinetic energy is only twice that of the divergent kinetic energy. Thus dissipation of the divergent part of the kinetic energy by frictional forces is very high as compared to its rotational counterpart.

Chapter 11

Limited Area Spectral Model

11.1 Introduction

As we have seen, global spectral models have some definite advantages over grid point models. On account of this, global spectral models are used by most of the numerical weather prediction centers for short and medium-range weather forecasts. The accuracy of the numerical forecasts by these models increases as the model resolution is increased, since at a higher resolution they are able to capture finer scales that are necessary to define and forecast smaller regional-scale weather systems properly. However, at very high resolution the global spectral models become computationally very expensive. The current computational resources at many centers tend to limit the horizontal resolution of global spectral models to about T-255, which is equivalent to a 50 km grid resolution near the equator. Often our interest in a very high-resolution forecast is over some specified limited area over the globe. This can be achieved effectively by running a very high-resolution limited area model in conjunction with a relatively low-resolution global model, rather than increasing the resolution of the global model. Until recently, most of the limited area models have used grid point or finite difference methods. In spite of careful formulation of finite differencing schemes, the grid point models continue to have some problems, such as phase and aliasing errors and nonlinear instability. Computing space derivatives with higher-order finite differencing schemes also does not resolve these problems. Besides, economical time integration methods, such as semi-implicit time integration schemes, are not very convenient to implement in a grid point model.

To overcome these problems and exploit the advantages of spectral methods, a number of limited area spectral models have been developed recently (Tatsumi 1986, Hoyer 1987, Juang and Kanamitsu 1994, Cocke 1998). These limited area models are currently one-way nested with the global model. In one-way nesting the output from the global model provides boundary values or basic large-scale fields to the limited area model. The output from the limited area model is not used in global model integration.

The spectral methods for a limited area model have some difficulties in spectral representation of fields with time-dependent boundary conditions. Tatsumi (1986) overcame this problem by representing time-dependent boundary values by non-orthogonal functions, while using orthogonal Fourier series to represent interior fields. Haugen and Machenhaur (1993) solved the lateral boundary problem by extending the domain beyond lateral boundaries to handle the periodicity of sine and cosine basis functions.

Another approach is the perturbation technique for a limited area forecast. Such a technique has been used by Hoyer (1987) for the ECMWF Spectral Limited Area Model, by Juang and Kanamitsu (1994) for the NCEP Regional Spectral Model, and by Cocke (1998) for the Florida State University Nested Regional Spectral Model. In these models, the finer regional scales are represented by perturbations or deviations with respect to a coarser resolution global model. We shall discuss in detail the perturbation technique for limited area modeling as applied by Cocke (1998) to the FSU Nested Regional Spectral Model.

11.2 Map Projection

For limited area modeling over lower and middle latitudes, the Mercator projection is very suitable. The transformation of spherical coordinates to the Mercator projection coordinates, with the equator as the standard latitude, is given by

$$x = \lambda ,$$

and

$$y = \int \cos^{-1}\theta d\theta = \frac{1}{2}\ln\left(\frac{1+\sin\theta}{1-\sin\theta}\right). \tag{11.1}$$

Here λ = longitude and θ = latitude are the spherical coordinates, and x, y are the coordinates on the Mercator projection. The zonal and meridional distance increments are related as

$$\delta x = \delta\lambda \quad \text{and} \quad \delta y = \cos^{-1}\theta\delta\theta . \tag{11.2}$$

An inverse coordinate transformation (from Mercator to spherical coordinates) is

$$\lambda = x \quad \text{and} \quad \theta = \sin^{-1}\left(\frac{e^{2y}-1}{e^{2y}+1}\right), \tag{11.3}$$

and the distance increments are

$$\delta\lambda = \delta x \quad \text{and} \quad \delta\theta = \cos\theta\delta y . \tag{11.4}$$

Following Cocke (1998), we define $m_F = \cos^{-2}\theta$ as the map factor, which is the ratio of the area $\delta x \cdot \delta y$ on the Mercator projection to the area on earth's surface.

The Mercator projection is a conformal map projection wherein, like on earth's surface, the latitudes and longitudes run orthogonal to each other. The longitudinal lines appear equispaced while the separation of latitudinal lines increases with latitude. The regional model has equal grid spacing on the Mercator map projection. In the physical space, corresponding spacing will not be equal; it is related to map projection spacing through the map factor. This results in distortion of weather patterns at higher latitudes. From computational considerations, this distortion from varying grid spacing has consequences with respect to the CFL condition at higher latitudes due to the decrease in physical grid spacing with latitude. It is therefore desirable to restrict the meridional extent of the regional model to within about 45°S and 45°N. In case the model domain is at a higher latitude, a rotation of coordinates is recommended so that the central parts of the regional model are placed over the equator of the rotated coordinates.

11.3 Model Equations

The primitive equations for the regional spectral model are the same as used in the global spectral model in the σ coordinate system and, as detailed in Chapter 7, are written as:

- The divergence equation

$$\frac{\partial D}{\partial t}+\nabla^2 P=\alpha\left(B,-A\right)-a^2\nabla^2 E, \tag{11.5}$$

- The vorticity equation

$$\frac{\partial \zeta}{\partial t}=-\alpha\left(A,B\right), \tag{11.6}$$

- The thermodynamics equation

$$\frac{\partial}{\partial t}\left(\frac{\partial}{\partial \sigma}\left(\frac{\sigma}{R\Gamma^*}\frac{\partial}{\partial \sigma}P\right)\right)-D=\frac{\partial}{\partial \sigma}\left[\frac{1}{\Gamma^*}\left(\alpha\left(UT',VT'\right)-B_T\right)\right]-G, \tag{11.7}$$

- The continuity equation

$$\frac{\partial q}{\partial t}+\hat{D}=-\hat{G}, \tag{11.8}$$

- The moisture equation

$$\frac{\partial S}{\partial t}=\alpha\left(US,VS\right)+B_s, \tag{11.9}$$

where

$$A\equiv(\zeta+f)U+\dot{\sigma}\frac{\partial V}{\partial \sigma}+\frac{RT'}{a^2}\cos\theta\frac{\partial q}{\partial \theta}-\frac{\cos\theta F_\theta}{a}, \tag{11.10}$$

$$B\equiv(\zeta+f)V-\dot{\sigma}\frac{\partial U}{\partial \sigma}-\frac{RT'}{a^2}\frac{\partial q}{\partial \lambda}+\frac{\cos\theta F_\lambda}{a}, \tag{11.11}$$

$$\alpha\left(A,B\right)\equiv\frac{1}{\cos^2\theta}\left[\frac{\partial A}{\partial \lambda}+\cos\frac{\partial B}{\partial \theta}\right], \tag{11.12}$$

$$E\equiv\frac{U^2+V^2}{2\cos^2\theta}, \tag{11.13}$$

$$G\equiv\frac{1}{\cos^2\theta}\left[U\frac{\partial q}{\partial \lambda}+V\cos\theta\frac{\partial q}{\partial \theta}\right], \tag{11.14}$$

$$T'\equiv T-T^*, \tag{11.15}$$

$$P\equiv\Phi+RT^*q, \tag{11.16}$$

$$B_T\equiv DT'+\Gamma'\dot{\sigma}-\frac{RT'}{C_p}\left(\hat{G}+\hat{D}\right)+\frac{RT}{C_p}G+r^*(G^{N\sigma}-G^\wedge)+H_T, \tag{11.17}$$

$$B_S \equiv SD - \dot{\sigma}\frac{\partial S}{\partial \sigma} + t\left(\frac{RT}{c_p} - \frac{RTd^2}{\in L(Td)}\right)\left(\frac{\dot{\sigma}}{\sigma} + G - \hat{G} - \hat{D}\right) + H_T - H_M, \quad (11.18)$$

$$\Gamma \equiv \frac{RT}{C_p} - \frac{\partial T}{\partial \sigma}, \quad (11.19)$$

$$\Gamma^* \equiv \Gamma\left(T^*\right), \quad (11.20)$$

$$\Gamma' \equiv \Gamma - \Gamma^*, \quad (11.21)$$

$$\dot{\sigma} = (\sigma - 1)\left(\hat{D} + \hat{G}\right) + \left(\hat{D}^\sigma + \hat{G}^\sigma\right), \quad (11.22)$$

$$\hat{D} \equiv \int_0^1 D \, d\sigma, \quad (11.23)$$

$$\hat{D}^\sigma \equiv \int_0^1 D \, d\sigma, \quad (11.24)$$

and

$$U = \frac{u\cos\theta}{a} \text{ and } V = \frac{v\cos\theta}{a}. \quad (11.25)$$

For a regional model on a Mercator projection, the terms involving horizontal space derivatives get modified. The Laplacian operator (∇^2) in spherical coordinates is,

$$\nabla^2 = \frac{1}{\cos^2\theta}\left[\frac{\partial^2}{\partial \lambda^2} + \frac{\cos\theta\partial}{\partial \theta}\left(\frac{\cos\theta\partial}{\partial \theta}\right)\right] = \cos^{-2}\left(\frac{\partial^2}{\partial x^2} + \frac{\partial^2}{\partial y^2}\right) \quad (11.26)$$

or

$$\tilde{\nabla}^2 = m_F^{-1}\nabla^2.$$

Here

$$\tilde{\nabla}^2 = \frac{\partial^2}{\partial x^2} + \frac{\partial^2}{\partial y^2}$$

is the Laplacian operator in Mercator projection coordinates, ∇^2 is the Laplacian operator in spherical coordinates, and $m_F = \cos^{-2}\theta$ is the map factor. Similarly the divergence and the vorticity become

$$\tilde{D} = m_F^{-1}D \quad \text{and} \quad \tilde{\zeta} = m_F^{-1}\zeta. \quad (11.27)$$

Also, α and G in equations (11.12) and (11.14) become

$$\tilde{\alpha} = m_F^{-1}\alpha \quad \text{and} \quad \tilde{G} = m_F^{-1}G. \quad (11.28)$$

The derivatives $\partial/\partial\lambda$ and $\cos\theta\,\partial/\partial\theta$, where they appear, change to $\partial/\partial x$ and $\partial/\partial y$. The modified divergence and vorticity, $\tilde{D}$ and $\tilde{\zeta}$, are related to wind components as

$$\tilde{D} = \frac{\partial U}{\partial \lambda} + \cos\theta\frac{\partial V}{\partial \theta} = \frac{\partial U}{\partial x} + \frac{\partial V}{\partial y} \quad (11.29)$$

and

$$\tilde{\zeta} = \frac{\partial V}{\partial \lambda} - \cos\theta \frac{\partial U}{\partial \theta} = \frac{\partial V}{\partial x} - \frac{\partial U}{\partial y}. \tag{11.30}$$

With these, the above primitive equations for a Mercator projection can be written as

$$\frac{\partial D}{\partial t} + \tilde{\nabla}^2 P = m_F^{-1}\mathrm{RHS}D, \tag{11.31}$$

$$\frac{\partial \zeta}{\partial t} = \tilde{\alpha}(A,B) = m_F^{-1}\mathrm{RHS}\zeta, \tag{11.32}$$

$$\frac{\partial}{\partial t}\left(\frac{\partial}{\partial \sigma}\left(\frac{\sigma}{R\Gamma^*}\frac{\partial P}{\partial \sigma}\right)\right) - m_F\tilde{D} = \mathrm{RHS}P, \tag{11.33}$$

$$\frac{\partial q}{\partial t} + m_F\hat{\tilde{D}} = \mathrm{RHS}q, \tag{11.34}$$

and

$$\frac{\partial S}{\partial t} = \mathrm{RHS}S, \tag{11.35}$$

where RHSD, RHSζ, RHSP, RHSq, and RHSS are the nonlinear terms on the right-hand side of the divergence, vorticity, thermodynamic, continuity, and moisture equations, respectively. The left-hand side of these equations contains the linear terms. Such separation of linear and nonlinear terms is necessary for integrating the model using the semi-implicit time integration scheme. However, the terms $m_F\tilde{D}$ and $m_F\hat{\tilde{D}}$ on the left-hand side of equations (11.33) and (11.34), respectively, are nonlinear. To linearize these equations fully we take $m_F = m_o + m'$, where m_o is the domain average value of m_F and m' is the deviation from it. With this, the above set of equations take the form

$$\delta_t\tilde{D} + \tilde{\nabla}^2\overline{P} = m_F^{-1}\mathrm{RHS}D, \tag{11.36}$$

$$\delta_t\tilde{\zeta} = m_F^{-1}\mathrm{RHS}\zeta, \tag{11.37}$$

$$\delta_t\mathbf{A}P - m_o\overline{\tilde{D}} = \mathrm{RHS}P + m'\tilde{D}, \tag{11.38}$$

$$\delta_t q + m_o\mathbf{I}\overline{\tilde{D}} = \mathrm{RHS}q - m'\mathbf{I}\tilde{D}, \tag{11.39}$$

and

$$\delta_t S = \mathrm{RHS}S. \tag{11.40}$$

From semi-implicit time integration considerations, we have replaced P by $\overline{P}$ in equation (11.36) and $\tilde{D}$ by $\overline{\tilde{D}}$ on the left-hand side of equations (11.38) and (11.39), where

$$\overline{D} = \left[D(t+\Delta t) + D(t-\Delta t)\right]/2 \tag{11.41}$$

and

$$\delta_t D = \left[D(t+\Delta t) - D(t-\Delta t)\right]/2\Delta t \tag{11.42}$$

is the centered time-differencing approximation for $\partial D/\partial t$.

The operators **A** and **I** are the second-order vertical finite differencing operator and the vertical finite integration operator, respectively.

The model variables $\tilde{D}$, $\tilde{\zeta}$, P, q, and S in equations (11.36)-(11.40) are full variables, i.e. the sum of the regional perturbation and the large-scale base field from the global model. The purpose of the regional model is to predict the perturbation fields only. For this, we split various variables on the left-hand side of equations (11.36)-(11.40) into perturbation and base (global) fields. The equations then take the form

$$\delta_t \tilde{D}' + \tilde{\nabla}^2 \overline{P}' = -\delta_t \tilde{D}_g - \tilde{\nabla}^2 \overline{P}_g + m_F^{-1} \cdot \mathrm{RHS}D = \mathrm{RHS}D', \tag{11.43}$$

$$\delta_t \tilde{\zeta}' = -\delta_t \tilde{\zeta}_g + m_F^{-1} \cdot \mathrm{RHS}\zeta = \mathrm{RHS}\zeta', \tag{11.44}$$

$$\delta_t \mathbf{A}P' - m_o \overline{\tilde{D}}' = -\delta_t \mathbf{A}P_g + m_o \overline{\tilde{D}}_g + \mathrm{RHS}P + m'\tilde{D} = \mathrm{RHS}P', \tag{11.45}$$

$$\delta_t q' + m_o \mathbf{I}\overline{\tilde{D}}' = -\delta_t q_g - m_o \mathbf{I}\overline{\tilde{D}}_g + \mathrm{RHS}q + m'\mathbf{I}\tilde{D} = \mathrm{RHS}q', \tag{11.46}$$

and

$$\delta_t S' = -\delta_t S_g + \mathrm{RHS}S = \mathrm{RHS}S'. \tag{11.47}$$

The terms on the right-hand side of the above equations are calculated on the transform grid. The terms based on global base variables (indicated by subscript g) are based on global model output. The rest of the terms are calculated via nonlinear products. The right-hand side terms are then Fourier analyzed using two-dimensional trigonometric (sine and cosine) functions and the transform method. With this, the above set of equations can be written in spectral form as

$$\delta_t \tilde{D}'_{mn} - a^{-2}\left[(mf_x)^2 + (nf_y)^2\right]\overline{P}'_{mn} = [\mathrm{RHS}D']_{mn}, \tag{11.48}$$

$$\delta_t \tilde{\zeta}'_{mn} = [\mathrm{RHS}\zeta']_{mn}, \tag{11.49}$$

$$\delta_t \mathbf{A}P'_{mn} - m_o \overline{\tilde{D}}'_{mn} = [\mathrm{RHS}P']_{mn}, \tag{11.50}$$

$$\delta_t q'_{mn} + m_o \mathbf{I}\overline{\tilde{D}}'_{mn} = [\mathrm{RHS}q']_{mn}, \tag{11.51}$$

and

$$\delta_t S'_{mn} = [\mathrm{RHS}S']_{mn}. \tag{11.52}$$

In the above equations, subscripts m and n indicate Fourier wavenumbers in the zonal and meridional direction, respectively. The details of Fourier functions and wavenumber truncation used for spectral representation are given in sections 11.5 and 11.6.

The vorticity equation and the moisture equation in the above system of equations are integrated explicitly using centered time-differencing schemes. The divergence equation, the thermodynamic equation, and the continuity equation are integrated using an implicit time integration scheme. Using the time-differencing and time-averaging equations (11.41) and (11.42), we can eliminate $\tilde{D}'_{mn}$ from equations (11.48) and (11.50). This results in a second-order differential equation in the vertical for $\overline{P}_{mn} = \left(P_{mn}(t+\Delta t) + P_{mn}(t-\Delta t)\right)/2$. This equation is written in a linear tridiagonal system of equations in the vertical and solved for $P'_{mn}(t+\Delta t)$ for each (m, n). The forecast value $P'_{mn}(t+\Delta t)$ is then substituted into equation (11.48) to provide $\tilde{D}'_{mn}(t+\Delta t)$ which, in turn, is used in equation (11.51) to obtain $q'_{mn}(t+\Delta t)$. This procedure of semi-implicit time integration is similar to that used in the global model and has been discussed in detail in Chapter 7.

11.4 Orography and Lateral Boundary Relaxation

A finer orography is included in the regional model by means of a perturbation orography, ϕ'_s, which is obtained as the deviation of orography of a higher resolution global model data set from that of the lower resolution global model. The initial perturbation fields for the atmospheric variables are also likewise obtained from these models. Thus, a consistency between the atmospheric variable perturbations and orography perturbations is maintained. In practice, these perturbation fields are usually obtained this way, but they can also be obtained from a high-resolution regional analysis.

The effect of perturbation orography in the model dynamics is incorporated via a pseudo-pressure perturbation,

$$P' = \phi' + RT^* q', \tag{11.53}$$

where

$$\phi' = \phi'_s + \int_1^\sigma RT' d\sigma. \tag{11.54}$$

The perturbation orography is blended at the lateral boundaries to the global model orography to provide continuity between the regional and the global orography. A similar continuity between the regional model and the global model variables is ensured at the boundaries. For this, a simple method is used to relax the perturbation variables to zero value at the lateral boundaries. An extra term is added to each tendency equation such that $\frac{\partial A'}{\partial t} \Rightarrow \frac{\partial A'}{\partial t} - \alpha\left(A' - A_{ref}\right)$, where $\alpha \to 0$ in the interior of the regional domain and $\alpha \to \tau^{-1}$ near the boundaries, τ being the relaxation time typically of the order of one hour. Generally, α is very small or zero over most of the interior so the relaxation is effective essentially only near the boundaries.

The blending of regional orography with global orography at the boundaries and relaxation of perturbations of model variables to zero value at the boundaries is done in such a way that the consistency between orography perturbations and model variable perturbations is maintained.

11.5 Spectral Representation and Lateral Boundary Conditions

In Chapter 6, we saw that the spherical harmonics used as the basis functions for the global spectral model are the eigenfunctions of the Laplace equation in spherical coordinates. The Laplacian on the Mercator projection is $\tilde{\nabla}^2 = \frac{\partial^2}{\partial x^2} + \frac{\partial^2}{\partial y^2}$. This has two-dimensional Fourier (sine and cosine) functions as its eigenfunctions, which form the basis functions for the regional model.

For this model, the regional domain is taken as π-periodic. If the number of east-west and north-south grid points in the regional model are J and I respectively, then the spectral representation of any perturbation field $A(x, y)$ is given by

$$A(x,y)=\sum_n\sum_m a_{mn} f\left(\frac{i\pi m}{I}\right) g\left(\frac{j\pi n}{J}\right), \tag{11.55}$$

where f and g are either sine or cosine functions, m and n are the zonal and meridional wavenumbers, respectively, and (i, j) are grid point coordinates. It is necessary that the spectral representation of various model variables is consistent with lateral boundary conditions of the model.

The regional model has slip-wall lateral boundary conditions with respect to perturbations, i.e., there can be perturbation wind flow along the boundary but not across the boundary. Also, no advection of perturbations of scalar fields like temperature, pressure, and moisture is allowed across the boundaries. This is possible if the spectral representation of these fields ensures vanishing of their gradients normal to boundaries. For a π-periodic domain, these conditions can be fulfilled by selecting the basis functions as follows:

$$U' \to \sin\frac{i\pi m}{I}\cdot\cos\frac{j\pi n}{J}, \tag{11.56}$$

$$V' \to \cos\frac{i\pi m}{I}\cdot\sin\frac{j\pi n}{J}, \tag{11.57}$$

$$\zeta' \to \sin\frac{i\pi m}{I}\cdot\sin\frac{j\pi n}{J}, \tag{11.58}$$

$$D' \to \cos\frac{i\pi m}{I}\cdot\cos\frac{j\pi n}{J}, \tag{11.59}$$

and

$$T', P', S' \text{ and } q' \to \cos\frac{i\pi m}{I}\cdot\cos\frac{j\pi n}{J}. \tag{11.60}$$

11.6 Spectral Truncation

As in the global model, the spectral resolution of the regional model is also truncated at an appropriate wavenumber. The number of grid points on the transform grid for such truncation ensure that the calculation of nonlinear quadratic terms by the transform method are free from any aliasing errors. In Chapter 7, we saw that for the aliasing-free Fourier transform of quadratic terms on a 2π-periodic domain the minimum number of grid points is $3M+1$, where M is the wavenumber at which the Fourier series is truncated. If I is the number of east-west grid points on a π-periodic domain, then the number of points to represent a 2π-periodic domain will be $2(I-1)$. Therefore, if the east-west Fourier functions are truncated at wavenumber M, then for aliasing-free transform we have

$$3M+1=2(I-1)$$

or

$$M = (2I - 3)/3. \tag{11.61}$$

Similarly,

$$N = (2J - 3)/3. \tag{11.62}$$

Thus, for aliasing-free calculations of nonlinear terms using the transform method with spectral truncation of the zonal and meridional wavenumbers at wavenumbers *M* and *N*, respectively, the minimum number of grid points *I* and *J* in the zonal and meridional directions on a π-periodic domain are given by

$$I = (3M + 3)/2 \tag{11.63}$$

and

$$J = (3N + 3)/2. \tag{11.64}$$

As with the global model, the regional model also can have two types of truncations – the rectangular truncation and the elliptic truncation. In the rectangular truncation,

$$m \le M \quad \text{and} \quad n \le N, \tag{11.65}$$

while in the elliptic truncation

$$\left(\frac{m}{M}\right)^2 + \left(\frac{n}{N}\right)^2 \le 1. \tag{11.66}$$

The elliptic truncation, being isotropic, provides a better description of a field than the rectangular one. Both types of truncations are available in the FSU regional spectral model, however, the elliptic truncation is more commonly used.

11.7 Model Physics and Vertical Structure

The physics parameterization and the vertical structure of the regional spectral model is similar to those in the global spectral model. As with the global model, the regional model physics include short and longwave radiation incorporating the effect of clouds and surface heat balance, the convective and large-scale precipitation, the shallow convection, boundary layer fluxes of sensible heat, moisture, and momentum based on similarity theory, and the horizontal and vertical diffusion processes.

In the vertical, the $\sigma = p/p_s$ coordinate is used with 14 discrete σ levels: 0.05, 0.07, 0.10, 0.20, 0.30, 0.40, 0.50, 0.60, 0.70, 0.80, 0.85, 0.90, 0.95, and 0.99, which are the same levels used in the global model. The base field and initial perturbations are spectrally transformed from the global model to the regional model on σ surfaces. This eliminates the need for any interpolation from grid points and thus reduces interpolation errors. As currently implemented, the regional and global models share the same physics, but it is not a requirement of the spectral perturbation method. Even now there are some differences in the physics of the two models. The regional model calls radiation every 1 hour while the global model calls radiation every 3 hours. There are also some modifications in cumulus parameterization due to very high resolution of the regional model.

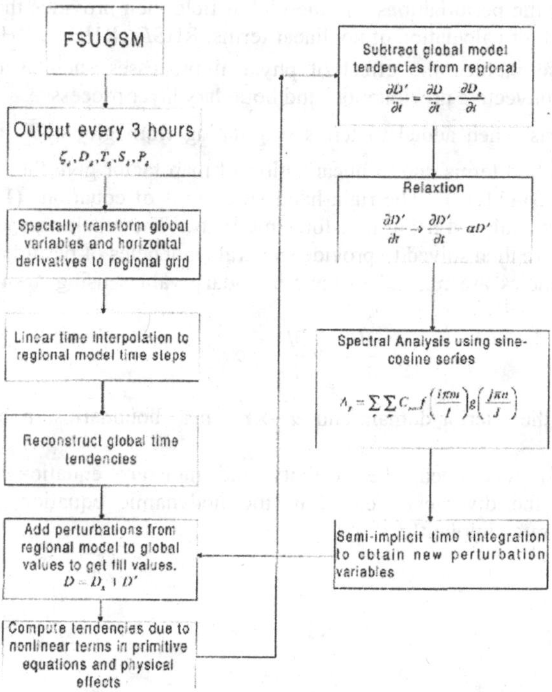

Figure 11.1. Regional Spectral Model integration – a schematic representation.

11.8 Regional Model Forecast Procedure

The regional spectral model predicts only the perturbations fields. Fig. 11.1 outlines the procedure for integration of the regional spectral model for their prediction.

The coarse resolution (*T*-106 or *T*-126) global model is run for a specified forecast period and its output (relevant to the regional model) is saved at the interval of every 3 hours to provide the base field for the regional model. The difference between the initial fields from a high-resolution global (or regional) analysis and the coarse resolution analysis provides the initial perturbation fields for the regional model. The perturbation orography is also obtained in a similar way. At the lateral boundaries, perturbations are set to zero. These perturbation fields are blended in such a way that at the lateral boundaries, perturbations smoothly go to their zero value. These perturbation fields are then spectrally analyzed consistent with slip-wall boundary conditions and recast on grid points.

The global fields and their horizontal derivatives are spectrally transformed to the regional grid and then are linearly interpolated to the regional model time step. The global time tendencies, δ_{tg}, are also reconstructed from the global field at the regional grid. The sum of the perturbations and the global field then provides the full values of various fields used for calculation of nonlinear terms: RHSD, RHSζ, RHSP, RHSq, and RHSS, which also include the effect of physical processes such as radiation, deep convection, non-convective precipitation, and boundary layer processes.

These terms when added to terms originating from global $\overline{P}_g$ and $\bar{\bar{D}}_g$ fields, global tendencies, and terms due to linearization of map-factor give the right-hand sides of equations (11.43)-(11.47). The right-hand side terms of equations (11.43) – (11.47) are then spectrally analyzed using transform methods. The resulting spectral equations (11.48) – (11.52) are then solved to provide spectral tendencies of perturbation quantities. These time tendencies are relaxed to their boundary values using a simple relaxation scheme

$$\frac{\partial D'}{\partial t} \to \frac{\partial D'}{\partial t} - \alpha D',$$

where $\alpha \to 0$ in the interior domain and $\alpha \to \tau^{-1}$ near boundaries, τ being relaxation time.

As already mentioned, the vorticity and moisture equations are integrated explicitly, while the divergence equation, thermodynamic equation, and continuity equation are integrated implicitly.

Chapter 12

Ensemble Forecasting

12.1 Introduction

Since the atmosphere is a chaotic dynamic system, any small errors in the initial condition can lead to growing errors in the forecast. In a numerical weather prediction system, these errors may be due to observational errors, errors in data transmission, or the errors resulting from the analysis scheme. Any systematic errors can normally be corrected if the nature and the source of such errors are known. But, the random errors are hard to correct as we have little knowledge about their source. These errors, however, are small within the range of observational or analysis error. Since numerical weather prediction is an initial value problem, even if the prediction models are perfect the forecasts are sensitive to any errors in the initial input. Lorenz (1969) has shown that the limited errors in small scales grow rapidly inducing errors in larger scales. These in turn grow into still larger scales and in about two days the errors have invaded the synoptic scales. Likewise, the initial errors in larger scales invade into smaller scales and synoptic scales. Thus all scales of motion are affected by the initial errors, eventually leading to total loss of predictive information. The rate of the error growth and hence the lead time at which predictability is lost depends on factors such as the circulation regime, season, and geographical location. It is possible to know the effect of initial errors and hence have information on the inherent predictability by running the model with a number of initial conditions, which differ from the control analysis within the uncertainty limits of the analysis. In their pioneering works, Epstein (1969) and Leith (1974) showed that a mean based on an ensemble of such forecasts provides a better forecast than that from the control analysis, as long as the initial states of the ensemble represent the uncertainty present in our control analysis. This important finding is the basis of ensemble forecasting. The various ensemble forecast techniques differ primarily from the way the initial perturbed state of the ensemble members is defined. In view of its classical significance, we first describe below the Monte Carlo method.

12.2 Monte Carlo Method

The processes which involve an element of random chance are referred to as Monte Carlo processes. The random errors in atmospheric observations and analyses fall in this category. In general, one does not know beforehand the location of errors in the initial data which are of importance for the eventual accuracy of the forecast. The idea of the

Monte Carlo method is to perturb all data simultaneously with random numbers of a realistic magnitude. The resulting forecast will differ almost throughout the forecast domain. Repeating the experiment many times with different sets of numbers, one can get the idea of the forecast errors that are due to the uncertainty of the observations and analyses.

The Monte Carlo method for ensemble forecasting was first applied by Leith (1974) in a perfect model environment. He generated a set of perturbations which were normally distributed with a zero mean and the perturbation sets were orthogonal to each other. If $X(0)$ is the initial analysis and E_i is the *i*th set of random errors (random numbers), then the *i*th perturbed initial state $X(0)$ is given by

$$X_i^E(0) = X(0) + E_i, \tag{12.1}$$

where the multivariable vector E_i satisfies the conditions

$$\langle E_i \rangle = 0 \tag{12.2}$$

and

$$\langle E_i E_j^* \rangle = \sigma \delta_{ij} \tag{12.3}$$

where σ is the variance of random errors.

If $X_i^E(t)$ is the forecast of the *i*th ensemble member after time *t*, then

$$\overline{X}(t) = \frac{1}{N} \sum_{i=1}^{N} X_i^E(t) \tag{12.4}$$

is the ensemble mean forecast and

$$\sigma_x = \sqrt{\frac{1}{N} \sum_{i=1}^{N} \left(X_i(t) - \overline{X}(t) \right)^2} \tag{12.5}$$

is the forecast variance, a measure of the spread of the ensemble forecast.

The Monte Carlo technique has been applied for ensemble forecasts by Tribbia and Baumhefner (1988). Kuo and Low-Nam (1990) and Mullen and Baumhefner (1994) used this technique for storm forecasting.

Although the ensemble mean forecast is found to become closer to the truth as the ensemble size increases, the random perturbations generated in the Monte Carlo method are not an efficient way of creating initial perturbed states, as the errors of operational analysis are rarely random. In the operational analysis cycle, the previous 6 hour forecast is generally used as the first guess for the analysis. The final analysis, therefore, has errors of forecast as well as of observations. The errors of forecast contain fast growing baroclinic unstable modes as well as slow growing or non-growing modes, depending upon synoptic conditions and geographical locations. The observational errors are generally of random nature. The operational analysis, thus, contains a combination of all these types of errors. The growing type of errors are much more important than the non-growing errors that affect the forecast skill, hence, the ensemble forecasts, based on initial perturbed states are worthy of examination. In the recent years, techniques have been developed to identify the growing modes of perturbations in order to define the initial perturbed states for the ensemble forecast. We describe below the techniques developed at the National Center for Environmental Prediction (NCEP), Florida State University (FSU), the European Center for Medium Range Weather Forecasts

(ECMWF), and finally the Superensemble Method developed at the Florida State University.

12.3 National Center for Environmental Prediction (NCEP) Method

12.3.1 *Generation of Perturbations*

The ensemble forecast method used at NCEP is a combination of the Lagged-Average-Forecasting (LAF) (Hoffmann and Kalnay 1983) and the Breeding of Growing Modes (Toth and Kalnay 1991) methods.

The forecast carries regions where error growth is large and also regions where error growth is small. The differences between the predicted field and the corresponding verification analysis will be large over regions of large error growth. The LAF method thus allows the selection of preferred growing modes. Because of the error growth over some regions, forecasts started at an earlier time may grow into larger amplitudes compared to forecasts that were started at a later time. This is a feature of the lagged average forecasting. This deficiency was corrected by Hoffman and Kalnay (1983) by using different weights for different members of the ensemble. Later, in the Scaled Lagged-Forecasting (SLAF) method, the correction was done by rescaling of perturbations by their 'age' factor, where smaller weights are assigned to perturbations from forecasts started at earlier times and larger weights for those started more recently, resulting in similar sized perturbations from all forecasts (Ebisuzaki and Kalnay 1991). To further increase the growing component in the perturbations, Kalnay and Toth (1991) used the difference between short-range forecasts (SRFD) started at earlier times, but verified at the initial time of ensemble. Experiments by Toth and Kalnay (1991) on different methods of generating the perturbation showed a clear increase in the growth rate of perturbations from random perturbation to SLAF and from SLAF to SRFD. These increases were also accompanied by an increase in the quality of the ensemble mean forecasts. Figure 12.1, after Toth and Kalnay (1993), shows schematically the generation of perturbations by the various methods described above.

12.3.2 *The Breeding of Fast-growing Modes*

The Breeding of Growing Modes (BGM) method, developed by Toth and Kalnay (1991), identifies the fast growing errors during the analysis/forecast cycle. During the analysis/forecast cycle of the data assimilation system, the perturbations periodically get rescaled at each analysis due to blending of observations with the first guess. Since observations are sparse they cannot eliminate all errors from the short-range forecast that becomes the first guess for the next analysis. Obviously, any errors that grow in the previous short-range forecast will have a larger chance of remaining in the latest analysis. These growing errors will then start growing quickly in the next short-range forecast. Thus the analysis contains fast growing errors that are dynamically created by repetitive

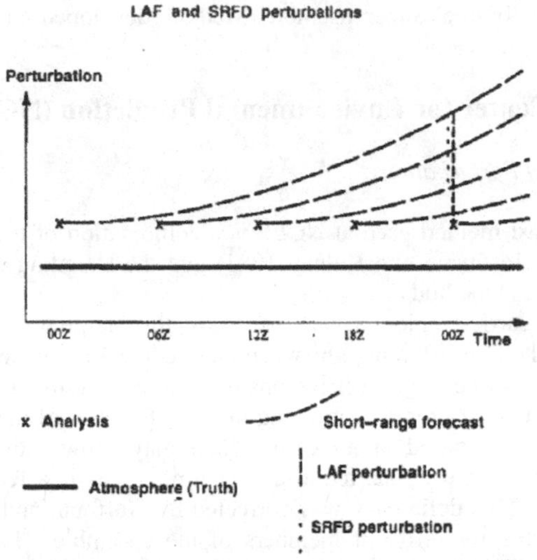

Figure 12.1. Schematic of the creation of LAF SRFD perturbations. Note that the LAF perturbation includes not only the short-range forecast errors but also the random errors of the latest analysis, whereas the SRFD perturbation is not affected by the random errors of the latest analysis. This results in a significant reduction of the random errors and therefore in a higher growth rate for the SRFD perturbations. Figure taken from Toth and Kalnay (1993).

use of the model to create the first guess field. This is what is referred to as Breeding Growing Modes (BGM). Toth and Kalnay (1991) developed a simple method for identifying Breeding Growing Modes consisting of the following steps:

a) add a small arbitrary perturbation to the atmospheric analysis,
b) integrate the model for 6 hours with both the unperturbed (control) and perturbed initial condition,
c) subtract the 6 hour control (analysis cycle) forecast from the perturbed forecast, and
d) scale down the difference field so that in the root mean square (RMS) sense it has the same size as the initial perturbation.

This perturbation is now added to the following 6-hour analysis as in (a) and the process is repeated. NCEP uses seven independent breeding cycles to generate the 14 initial ensemble perturbations. Each breeding cycle begins with an analysis/forecast cycle, which differs from the others only in initially prescribed random errors ('seed'). These initial random errors are added and subtracted from the control analysis so that each breeding cycle generates a pair of perturbed analyses. From this point on the perturbation patterns evolve freely in each breeding cycle. These perturbations are just the difference between short-range forecast started from last perturbed analysis and the 'control' analysis, rescaled to the magnitude of the seed perturbation which is a small percentage

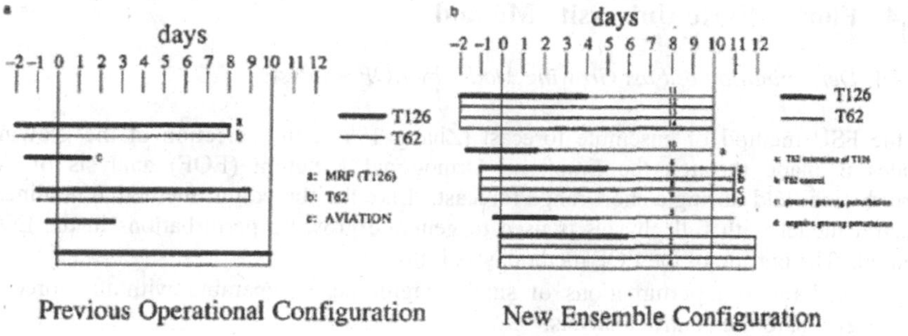

Figure 12.2. (a) Operational configuration of global predictions before 7 December 1992 and (b) the new ensemble configuration. In (b), individual ensemble member are identified by numbers 1-14. Figure obtained from NOAA.

of (10-12%) the climatological RMS variance of the concerned field. Since the short-range forecasts are just the early part of the extended range prediction, the generation of perturbations is basically cost free with respect to the forecast/analysis system.

12.3.3 *Operational Ensemble Scheme*

Due to the non-availability of additional computational resources, at the end of 1992 NCEP designed an operational ensemble scheme that made maximum use of the normal operational forecast products for the ensemble forecast (shown schematically in Fig. 12.2). The NCEP operational product configuration, as it was before 7 December 1992 and which was utilized for the new operational ensemble forecast scheme, is shown in Fig 12.2(a). It consisted of 10-day operational medium range forecasts at *T*-126 and *T*-62 resolutions starting at 00 UTC of days -2, -1, and 0 and the *T*-126 aviation forecast up to 3 days starting at the 12 UTC analysis of day -2. These operational products were staggered to provide a 14 member family out of the 10-day ensemble forecast scheme within the existing computational sources as illustrated in Fig 12.2(b). This scheme combines breeding growing mode (BGM) forecasts with overlapping predictions from time lagging (LAF) forecasts where initial differences are model short-range forecast errors. As seen from Fig 12.2(b), all predictions are for a 12-day duration so that with a time lag of up to two days we get 14 forecasts available every day for ten days. The *T*-126 forecast started at day 0 is truncated to *T*-62 resolution on day 6 and integrated for another six days up to day 12. Other members that begin on day 0 and that are integrated for twelve days are the *T*-62 resolution, 12-day forecast and two BGM forecasts, one with positive growing mode and the other with negative growing mode. Two similar sets of four 12-day forecasts are prepared starting at day -1 and day -2. In addition, the *T*-126 aviation forecasts starting at 12 UTC of day -2 and day -1 are integrated with *T*-62 resolution beyond day 3. Thus, all together these forecasts provide 14 members of the ensemble forecast valid at day 10.

12.4 Florida State University Method

12.4.1 *Determination of Fast Growing Modes by EOF Analysis*

In the FSU method of ensemble forecast (Zhang 1997), the selection of the growing modes is made through the Empirical Orthogonal Function (EOF) analysis of the perturbation field during a short-range forecast. Like the 'breeding' method, a nonlinear spectral model with full physics is used to generate growing perturbations in the EOF method. The outline of the EOF method is as follows:

- Add random perturbations of small magnitude, comparable with the forecast errors, to the control analysis.
- Integrate the model with full physics for 36 hours starting with both the control and the perturbed initial states. Output the forecast results every three hours from these runs.
- Subtract the control forecast from the perturbed forecast at the corresponding times.
- Perform an EOF analysis of the time series of difference fields to determine the modes (eigenvectors) whose EOF coefficients increase rapidly with time. These fast growing modes constitute the optimal perturbations.
- Choose the first few modes to construct perturbation fields.
- Add/subtract the perturbation to/from the control analysis to form initial states for ensemble forecasts.

The EOF analysis is flexible enough to be applied to any domain (global or tropical, regular grid or Gaussian grid). At FSU the method has been applied to the tropics to get better estimates of the perturbation fields over the tropical region. As the EOF perturbations are calculated from the model, which includes all physical processes, the optimal perturbations from the EOF method may therefore include the effects of various instability mechanisms caused by the model physics, such as the interactions between cumulus convection and the large-scale flow.

The EOF analysis is performed only for the temperature and wind fields. In the case of wind, a complex vector wind perturbation

$$\delta\omega_{s,t} = \delta v_{s,t} + i\delta u_{s,t}, \tag{12.6}$$

where $u_{s,t}$ and $v_{s,t}$ are the perturbations of the *u* and *v* components of the wind at location *s* and time *t*, is defined as the difference of *U* and *V* fields from the perturbed and control runs.

The entire data set then can be expressed as an *S* x *T* rectangular matrix (*S* being the total grid points and *T* being the total time intervals at which perturbations were output):

$$W = \begin{bmatrix} \delta\omega_{11} & \delta\omega_{12} & \dots & \dots & \delta\omega_{1T} \\ \delta\omega_{21} & \delta\omega_{22} & \dots & \dots & \delta\omega_{2T} \\ \dots & \dots & \dots & \dots & \dots \\ \delta\omega_{S1} & \delta\omega_{S2} & \dots & \dots & \delta\omega_{ST} \end{bmatrix}. \tag{12.7}$$

Eigenvectors and eigenvalues are calculated from the covariance matrix

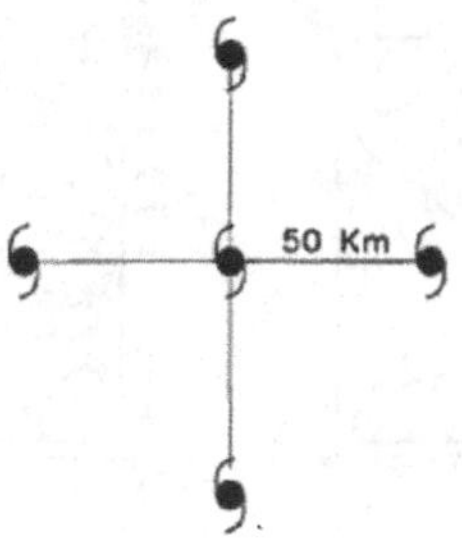

Figure 12.3. Schematic diagram of hurricane initial position perturbations. Figure taken from Zhang (1997).

$$H = \frac{1}{T} W^* W, \tag{12.8}$$

where W^* is the complex conjugate transpose of W and H is a symmetric matrix composed of complex elements, with the exception of the diagonal elements which are real and proportional to perturbation kinetic energy for each grid point.

If $\vec{e}_i$ and $\vec{\lambda}_i$ are eigenvectors and eigenvalues, respectively, of matrix H and the eigenvectors $\vec{e}_i$ are in descending order according to the magnitude of $\vec{\lambda}_i$, then the matrix W can be expanded with respect to base eigenvectors $\vec{e}_i$ as

$$W = YE, \tag{12.9}$$

where matrix E consists of row vectors of $\vec{e}_i$, which are function of space only. These are called EOFs. The matrix Y contains coefficients of different eigenvectors at different times and these are called principal components (PC). The fast growing modes are selected by time evolution of the EOF coefficients. Those EOF modes whose coefficients increase rapidly constitute the optimal growing mode. In actual practice, the EOF eigenmodes of the first order were found sufficient to construct the perturbed initial states for the ensemble forecast.

12.4.2 *Application of the EOF Method to Hurricane Forecasting*

The EOF method has been applied at FSU for ensemble forecasting of hurricane intensity and track. The initial field for the control run is from the physically initialized ECMWF data. A synthetic hurricane, based on observed estimates of maximum wind and central surface pressure, is inserted at the analysis location of the hurricane (Trinh and Krishnamurti 1992). For generalizing the perturbed states, this initial hurricane analyzed position is perturbed by displacing its location by 50 km (assumed as typical error in hurricane position) to the north, south, east and west (Fig. 12.3). This gives us five members in terms of position perturbations in the initial data. For the initial analysis corresponding to each of these positions, optimal perturbations are calculated using the EOF method. The original initial state and two perturbed initial states obtained by adding/subtracting the EOF perturbations to/from the original initial state give initial

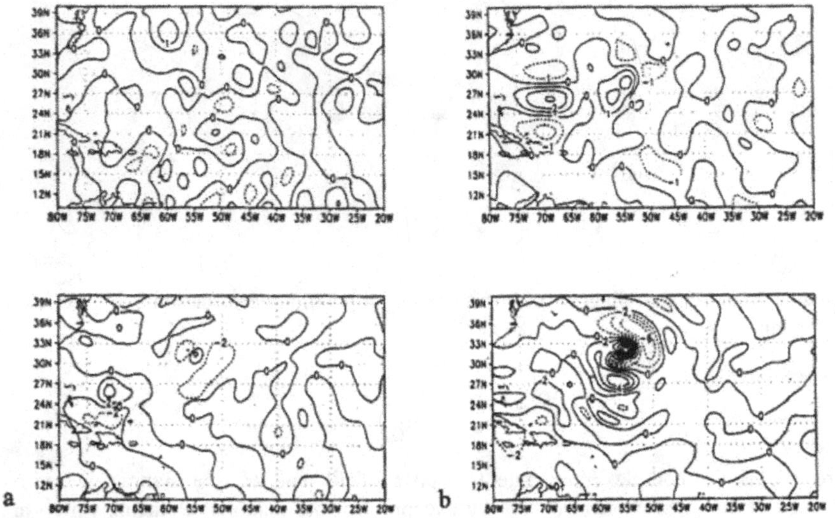

Figure 12.4. Hurricane Florence. (a) Random wind perturbations at 1000 mb at hour 0 (top) and at hour 36 (bottom). (b) EOF-based wind perturbations at 1000 mb at hour 0 (top) and at hour 36 (bottom). The contours are isotachs of wind speed. Figure taken from Zhang (1997).

states for three ensemble members corresponding to each position. Among these, the unperturbed initial state at the non-displaced (central) location of the hurricane is the initial state for the control run. The other 14 ensemble members start with a perturbed state (with respect to location or initial perturbation or both). The EOF perturbations of the u, v, and T fields are calculated from 36-hour forecasts carried out from initial random perturbations applied at all the five hurricane locations. A clear advantage of the EOF method in selecting the growing mode can be seen from Fig. 12.4. Figure 12.4(a) shows the random wind perturbations for hurricane Florence (4 November 1994) at 1000 mb at hour 0 and at hour 36, while Fig. 12.4(b) shows the EOF based wind perturbations for this hurricane at the same level at these two hours. As seen from Fig 12.4(a), the random perturbations are very weak and without any preferred pattern. On the other hand, the EOF perturbations are strong with well-defined patterns in the proximity of the hurricane location. Such preferred growing modes define the initial perturbed states for ensemble members in the EOF method.

As an example of the EOF method of ensemble forecasting we show here some aspects of the forecast of hurricane Gilbert, which occurred in September 1998. Figure 12.5 and Fig. 12.6 show the 850 mb streamlines of the ensemble members at hours 0 (12 UTC 11 September 1988) and hour 48 (12 UTC 13 September 1988), respectively, for hurricane Gilbert. The pattern at the top left corner is for the control experiment. The initial (hour 0) difference in 850 mb streamline patterns of various ensemble members is

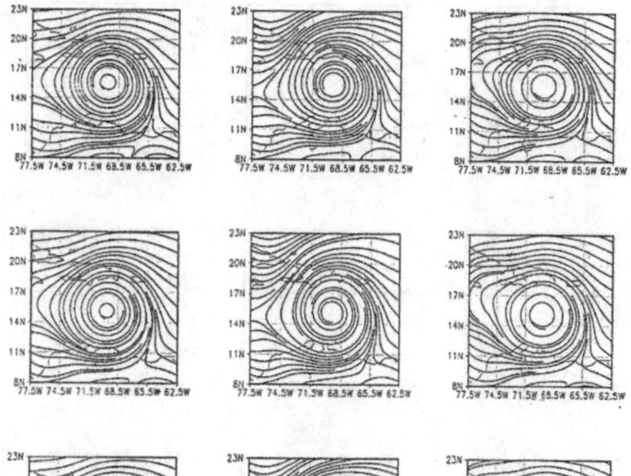

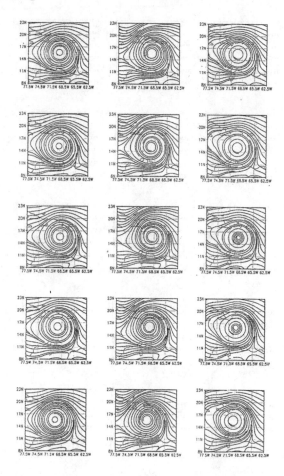

Figure 12.5. Hurricane Gilbert, 850 mb streamlines and isotachs at day 0 from individual ensemble members. Figure taken from Zhang (1997).

very small. Looking closely, we notice the hurricane position of various ensemble members displaced slightly from the control with very small differences in the flow fields. By hour 48, the 850 mb flow field, which was initially nearly circular in all members, has become very distorted and differs significantly in intensity and orientation. The difference in the flow fields of ensemble members was found to increase further with the time of integration at all levels. A similar behavior in the surface pressure and upper-level temperature fields was noticed in the ensemble forecasts. The hurricane's control pressure, which in various members was about 980 mb at the start, varied between 974

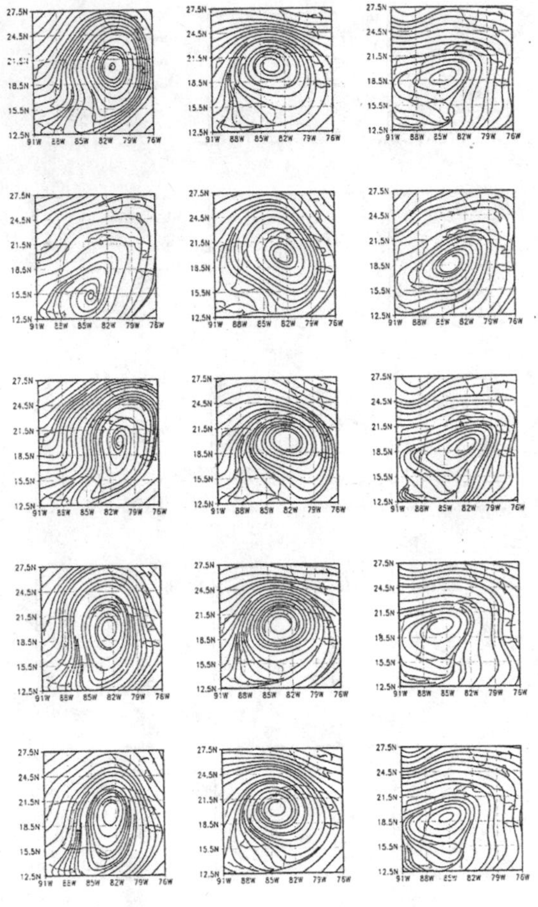

Figure 12.6. Hurricane Gilbert, 850 mb wind fields at day 2 from individual ensemble members. Figure taken from Zhang (1997).

mb and 998 mb at hour 48 of the forecast with pressure gradients differing among ensemble members.

Figure 12.7 shows the predicted tracks of hurricane Gilbert starting at 12 UTC 11 September 1988. The top panel shows the track predicted by various ensemble members along with the observed track and that predicted by the control run. The hurricane forecast positions by all ensemble members including the control are very close to each other for the first 24 hours of the forecast. After that they start diverging. On day 4, the predicted hurricane tracks are scattered over a very large area. Among these, the control track has taken a more northerly direction and is a poor forecast when compared to the

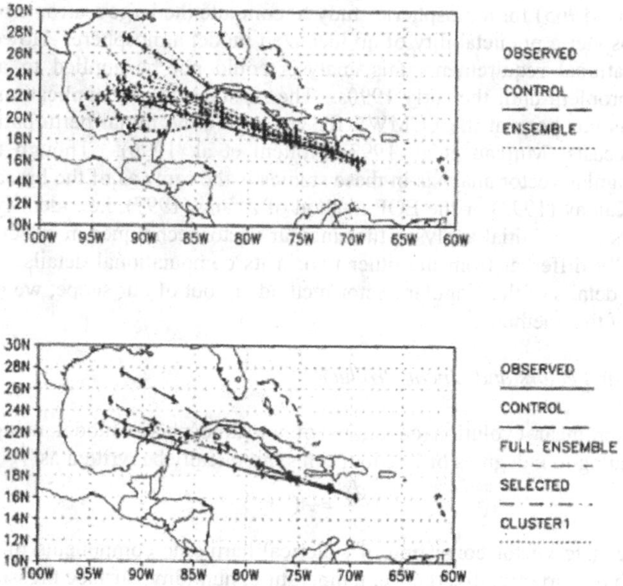

Figure 12.7. Hurricane Gilbert track prediction, starting from 12 UTC 11 September 1998. Top panel shows the best track (solid line), the track from the control experiment (long dash), and the track forecasts from different ensemble members (short dash line). Bottom panel contains best track (solid line) compared with full ensemble mean (dash dot line), cluster 1 mean (short dash), track prediction from control experiment (long dash line), and selected track mean (dot dash line). Figure taken from Zhang (1997).

observed track, which has a less northerly component. The ensemble mean track is much closer to the observed track than the track predicted by the control run. This clearly demonstrates the advantage of ensemble forecasting. The lower panel shows a 'selected' mean and cluster mean, along with observed, control, and ensemble mean tracks. The 'selected' mean is the mean track based on ensemble members whose 12-hour forecast position is sufficiently close to the observed position at that time. For such hurricane track forecasts, we need to have knowledge of the observed hurricane position at hour 12. The cluster mean is calculated from ensemble members by cluster analysis and based on high correlation among them, and in this case, contain almost 90% of the total number of ensemble members.

12.5 European Center for Medium Range Forecasts (ECMWF) Method

Ensemble forecasting at the European Center for Medium Range Weather Forecasts (ECMWF) is based on singular vector analysis. The singular vector analysis was first used by Lorenz (1965) for atmospheric study to compute the largest error growth rates to estimate atmospheric predictability of an idealized model atmosphere. However, due to large computational requirements this analysis could not be applied to any realistic atmospheric problem until the early 1990s. The most successful application of singular vector analysis has been at the ECMWF for generating the initial perturbation state for ensemble forecasts (Mureau et al. 1993; Molteni et al. 1996). Though the inherent purpose of singular vector analysis in these studies is the same as of the breeding method of Toth and Kalnay (1993) or the EOF method of Zhang (1997), i.e., identifying the fast growing errors in the initial analysis, the singular vector techniques in selecting growing modes is totally different from the other two in its computational details. As rigorous mathematical details of the singular vector method are out of our scope, we give below a brief outline of this method.

12.5.1 *Singular Vectors and Linear Product*

A set of n nonlinear evolution equations of an atmospheric model using a spectral expansion leading to n degrees of freedom can symbolically be written as

$$\frac{dx}{dt} = A(x)\,. \tag{12.10}$$

Here, x is the state vector consisting of spherical harmonic components of atmospheric variables such as vorticity, divergence, temperature, humidity, surface pressure, etc. The operator A represents the effect of nonlinear dynamical, as well as physical (parameterized), processes.

Over a sufficiently small time interval the evolution of a small perturbation x' of the state vector x can be described by a linearized approximation

$$\frac{dx'}{dt} = \left.\frac{dA}{dt}\right|_{x(t)} x'\,. \tag{12.11}$$

This system of linear ordinary differential equations is the tangent linear model in differential form. Its solution between time t_0 and t_1 can be obtained by time integration of (12.11) as

$$x'(t_1) = L(t_1,t_0)x'(t_0)\,. \tag{12.12}$$

L is an $(n \times n)$ real matrix referred to as the resolvent or forward propagator of the tangent linear model. Because of linearization, L depends on the basic nonlinear trajectory $x(t)$, the solution of the basic nonlinear model, but it does not depend on the perturbation x'. Thus $L(t_1,t_0)$ maps small perturbation x' along the basic nonlinear trajectory from an initial time t_0 to some future time t.

Singular vector decomposition is a linear algebra problem where any $n \times n$ real matrix L can be written as the product of an $n \times n$ orthogonal matrix U, an $n \times n$ diagonal matrix S, and the transpose of an $n \times n$ orthogonal matrix V as

$$L = USV^T\,. \tag{12.13}$$

As U and V are orthogonal matrices

$$UU^T = I \text{ and } VV^T = I\,. \tag{12.14}$$

Alternatively, pre-multiplying (12.13) by U^T and post-multiplying it by V gives us

$$U^T LV = S = \begin{bmatrix} \sigma_1 & 0 & 0 & . & . & 0 \\ 0 & \sigma_2 & 0 & . & . & 0 \\ . & . & . & . & . & . \\ . & . & . & . & . & \sigma_n \end{bmatrix} \tag{12.15}$$

and U, V may be written as

$$U = [u_1, u_2, ..., u_n], \qquad V = [v_1, v_2, ..., v_n]. \tag{12.16}$$

Elements of (12.15) satisfy the relation

$$\sigma_1 \geq \sigma_2 ... \geq \sigma_n \geq 0. \tag{12.17}$$

Multiplying the left-hand side of (12.15) by U gives

$$LV = US, \text{ i.e., } L[v_1, v_2, ..., v_n] = [\sigma_1 u_1, \sigma_2 u_2, ..., \sigma_n u_n], \tag{12.18}$$

or

$$Lv_i = \sigma_i u_i, \tag{12.19}$$

where v_i are the right singular vectors or initial singular vectors of L. Multiplying the right-hand side of (12.15) by V^T gives

$$U^T L = SV^T, \tag{12.20}$$

which on transposing gives

$$L^T U = VS, \text{ i.e., } L^T [u_1, u_2, ..., u_n] = [\sigma_1 v_1, \sigma_2 v_2, ..., \sigma_n v_n] \tag{12.21}$$

or

$$L^T u_i = \sigma_i v_i, \tag{12.22}$$

where u_i are the left singular vectors or final singular vectors of L. From (12.19) and (12.22) we get

$$L^T L v_i = \sigma L^T u_i = \sigma_i^2 v_i. \tag{12.23}$$

Therefore, the initial singular vectors, v_i, can be obtained as eigenvectors of $L^T L$, a normal matrix whose eigenvalues are squares of the singular values. For the final perturbation $x'(t_1)$ we get

$$x'(t_1) = L(t_0, t) x(t_0) = \sum_{i=1}^{n} \langle x_0, v_i \rangle \sigma_i u_i, \tag{12.24}$$

where $\langle x, v \rangle$ is the inner product of x and v. Taking the inner product of (12.24) with u_i we get

$$\langle x'(t_1), u_i \rangle = \sigma_i \langle x(t_0), v_i \rangle. \tag{12.25}$$

Thus, initial vector v_i will be stretched by an amount equal to singular value σ_i (or contracted if $\sigma_i < 1$) and the direction will be rotated to that of the evolved vector u_i.

The selection of the dominant singular vector is based on the maximum energy growth over the time $t_1 - t_0$ as determined by the norm or inner product of the perturbation. The inner product of the perturbation at time t is defined as

$$\|x'(t)\|^2 = (x'(t); x'(t)) = (x(t_0); L^T L x(t_0)). \tag{12.26}$$

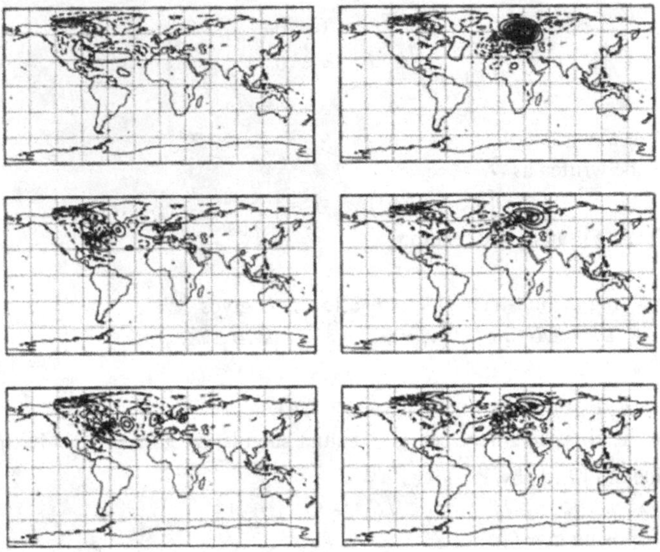

Figure 12.8. Streamfunction associated with the dominant singular vector (with 3-day global energy optimization) for 9 January 1993 at (left) initial time and (right) optimization time. Top row: model level 7 (about 200 mb); middle row: model level 13 (about 700 mb); and bottom row: model level 15 (about 850 mb). Contour interval at optimization time is 20 times larger than at initial time. Figure taken from Buizza and Palmer (1995).

The components of perturbation state vector x' represent the perturbations in the vorticity, divergence, temperature, etc., fields. At the initial time, the eigenvalues $v_i(t_0)$ of $L^T L$ can be chosen to form a complete orthogonal basis of n-dimensional tangent space of linear perturbations with real eigenvalues $\sigma_i^2 \geq 0$.

Thus from (12.22)

$$L^T L v_i(t_0) = \sigma_i^2 v_i(t_0) . \tag{12.27}$$

At a future time, these eigenvectors evolve to $v_i(t_1) = L v_i(t_0)$, which gives us

$$L L^T v_i(t_1) = \sigma_i^2 v_i(t_1) . \tag{12.28}$$

From (12.26) and (12.28) we get

$$\|v_i(t_1)\|^2 = \left(v_i(t_0); L^T L v_i(t_0)\right) = \sigma_i^2 . \tag{12.29}$$

Since any $\dfrac{x(t_1)}{\|x(t_0)\|}$ can be written as linear combination of $v_i(t)$,

$$\max x'(t_0) \# o \left(\frac{\|x(t_1)\|}{\|x(t_0)\|} \right) = \sigma_1 . \tag{12.30}$$

The maximum energy growth over the time $t_1 - t_0$ is therefore associated with dominant singular vector $v_1(t_0)$ at the initial time, and $v_1(t_1)$ at the end optimization time. As an example, Fig. 12.8 from Buizza and Palmer (1995) shows the streamfunction associated with the dominant singular vector at the initial time on 9 January 1993, as well as at the optimizing time (with 3-day global energy optimization).

After the dominant perturbation modes have been selected by the singular vector method, the initial perturbation fields for the ensemble forecasts are constructed by adding or subtracting them from the base analysis.

12.6 Superensemble Methodology and Results

12.6.1 *Introduction*

The superensemble approach is a recent contribution to the general area of weather and climate forecasting. This approach was developed at FSU and has been discussed in a series of publications, Krishnamurti et al. (1999), (2000a), (2000b), and (2001). The novelty of this approach lies in the methodology, which differs from ensemble analysis techniques used elsewhere. This approach yields forecasts with considerable reduction in forecast error compared to the errors of the member models, the 'bias-removed' ensemble averaged forecasts, and the ensemble mean. This technique entails the partition of a time line into two parts. One part is a 'training' phase where forecasts by a set of member models are compared to the observed or analysis fields with the objective of developing a statistic on the least squares fit of the forecasts to the observations. Specifically, observed anomalies are fit to the member model forecasts according to the classical prescription

$$O' = \sum_{i=1}^{N} a_i (F_i - \overline{F}_i) + \varepsilon_i \,, \tag{12.31}$$

where F is the ith model forecast (out of N total models), $\overline{F}_i$ is the mean of the ith forecast over the training period, O' is the observed anomaly relative to the observed mean over the training period, the a_i values represent the regression coefficients, and ε_i is an error term. The a_i terms are determined by requiring the summed squared error integrated over the training period $E = \sum_{i=1}^{N} \varepsilon_i^2$ to be as small as possible. A fit of this sort is performed for all model variables and at all model grid points for which reanalysis observations are available and typically yields close to 7 million regression parameters. This large number arises from the number of transform grid points, number of vertical levels, number of basic variables, and the number of models. Over all such locations we have noted diverse performance characteristics of the member models. That arises from differences in horizontal and vertical discretization, treatment of physics, handling of inhomogenity of the land surface, orography, lakes, water bodies, surface physics, and boundary conditions. All such peculiarities tend to leave their signature in the error distributions and hence on these weights. These may be thought of as bias correction weights. The second part of the time line is composed of genuine model predictions, i.e., the forecasts of the member models. The superensemble approach combines each of these forecasts according to the weights determined during the training period through the formulation

$$S = \overline{O} + \sum_{i=1}^{N} a_i (F_i - \overline{F}_i), \qquad (12.32)$$

where the notation is defined above. The determination of a_i follows the well-known Gauss-Jordan elimination method. The prediction S is referred to as the superensemble forecast. This forecast should be contrasted with the more standard anomaly forecasts known as the biased-removed ensemble mean (E) or ensemble mean ($\hat{E}$) forecasts,

$$E = \frac{1}{N}\sum_{i=1}^{N}(F_i - \overline{F}_i) \quad \text{or} \quad \hat{E} = \frac{1}{N}\sum_{i=1}^{N}(F_i - \overline{O}). \qquad (12.33)$$

The distinction between them comes in the weighting. Assigning all models a weight of $1/N$ (where N is the number of models) in equation (12.32) illustrates the connection between the forecasts and also illustrates how the training period attempts to identify and highlight good model performance.

The skill of the multi-model superensemble method significantly depends on the error covariance matrix since the weights of each model are computed from a designed covariance matrix. The current method for the construction of the superensemble utilizes a least square minimization principle within a multiple regression of model output against observed 'analysis' estimates. This entails a matrix inversion that is currently solved by the Gauss-Jordan elimination technique. That matrix can be ill-conditioned and singular depending on the interrelationships of the member models of the superensemble. We have recently designed a singular value decomposition (SVD) method that overcomes this problem and removes the ill-conditioning of the covariance matrix entirely (Yun and Krishnamurti 2002). Early tests of this method have shown great skill in weather and seasonal climate forecasts compared to the Gauss-Jordan elimination method.

In medium range, real-time global weather forecasts, the largest skill improvements are seen for precipitation forecasts both regionally and globally. The overall skill of the superensemble is 40-120% higher than the precipitation forecast skills of the best global models. For forecasts of variables other than precipitation, the superensemble exhibited major improvements in skill for the divergent part of the wind and the temperature distributions. Tropical latitudes show major improvements in daily weather forecasts. For most variables, we have used the operational ECMWF analysis at 0.5° latitude/longitude as the observed benchmark fields for the training phase. The observed measures of precipitation are derived from the 2A12 algorithm of NASA Goddard that is described in some detail in Krishnamurti et al. (2001) and within the references stated therein.

The area of seasonal climate simulations has only been addressed recently in the context of atmospheric climate models where the sea surface temperatures and sea ice are prescribed, such as the AMIP data sets. In this context, given a training period of some 8 years and a training data base from the ECMWF, the results exhibit improved skill compared to the member models and the ensemble mean. Preliminary work in this area (Krishnamurti et al. 2002) examines the difficulties involved with prediction of seasonal precipitation anomalies. Most individual member models perform poorly compared to climatology, whereas the superensemble appears to demonstrate precipitation skills slightly above those of climatology. The effectiveness of weather and seasonal climate forecasts from superensemble methodology has also been assessed from measures of standard skill scores such as correlation against observed fields, root mean square (RMS) errors, anomaly correlations, and the so-called Brier skill scores (Stefanova and Krishnamurti 2001) for climate forecasts (assessing skills above those of a climatology).

Training is a major component of this forecast initiative. We have compared training with the best quality 'observed' past data sets versus training deliberately with poorer data sets. This has shown that forecasts are improved when higher quality training data sets are employed for the evaluation of the multi-model bias statistics. It was felt that the skill during the forecast phase could be degraded if the training was executed with either a poorer analysis or poorer forecasts. This was noted in our recent work on precipitation forecasts where we showed that the use of poorer rainfall estimates during the training period affected the superensemble forecasts during the forecast phase (Krishnamurti et al. 2001). In addition, issues on optimizing the number of training days has been addressed from an examination of training with days of high forecast skill versus training with low forecast skill, and training with the best available rain rate datasets versus those from poor representations of rain. We have learned to improve the forecast skill by selectively improving the distribution of weights for the training phase.

Why does the superensemble generally have higher skill compared to all participating multi-models and the ensemble mean? At each location and for all variables the ensemble mean assigns a weight of $1/N$ to all N member models, which includes several poorer performing models. As a result, assigning the same weight of $1/N$ to some poorer models was noted to degrade the skill of the ensemble mean. It is possible to remove the bias of models individually (at all locations and for all variables) and to perform an ensemble mean of the bias removed models. That, too, has somewhat lower skill compared to the superensemble, which carries selective weights distributed in space among all multi-models and for all variables. A poorer model simply does not reach the levels of the best models after its bias removal.

12.6.2 *Experimental Real-time Global Weather Forecasts Based on Superensemble*

We have developed an experimental, real-time NWP capability for the forecast of all basic variables such as winds, temperature, surface pressure, geopotential heights, and precipitation. These are multianalysis-multimodel superensemble forecasts where eleven models are currently being used on a daily basis. These include the daily operational forecasts from the NCEP, Canadian Weather Service RPN, Australian model from the BMRC, U.S. Navy's NOGAPS, the Japanese model from JMA, and different versions of our in-house FSU global spectral model that are physically initialized using different rain rate algorithms. In some sense, the construction of the superensemble is a post-processing of multi-model forecasts. This is still a viable forecast product that is being prepared experimentally in real-time at FSU.

The forecast product has been started with an aim to provide near real-time, multimodel superensemble-based weather forecasts over the entire globe up to six days in advance. Forecasts for the mean sea-level pressure, 500 mb heights, surface temperature, and winds (isotachs) at the surface, 850 mb, and 200 mb are displayed for the whole globe as well as nine different regions of the globe. These

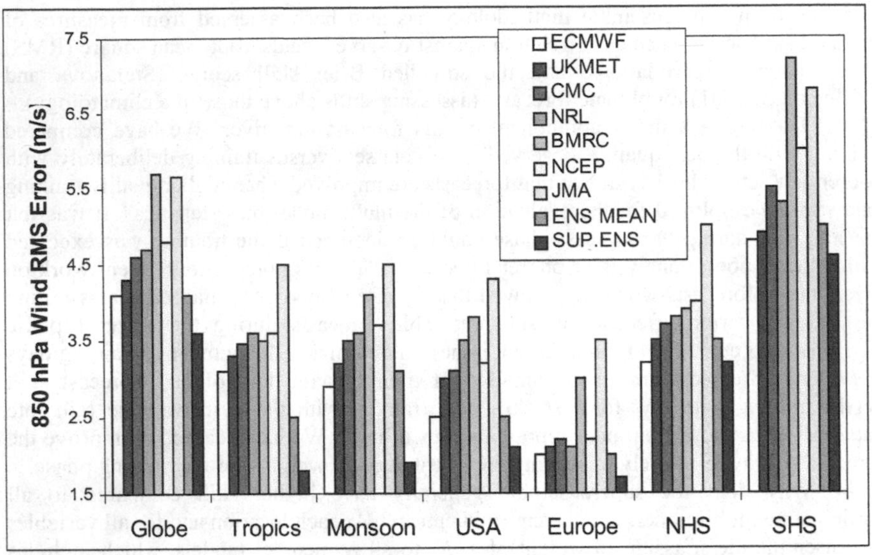

Figure 12.9. RMS error (m s^{-1}) of 850 mb winds over different parts of the globe, day 3 forecast, August 1998.

forecasts can be viewed by clicking on a specific region over the world map provided on the forecast page. Apart from these dynamical variables, 5-day forecasts of the 24-hour total precipitation and a new 5-day accumulated flood potential forecast is also shown for the entire globe, which again can be viewed over a specific region of interest. Different skill scores computed from these data sets are also shown on the forecast page, including RMS and systematic errors for forecasts of winds, mean sea level pressure and winds at 850 mb and 200 mb, and equitable threat scores, RMS errors, and correlation coefficients for the precipitation forecasts. The website also features the archives for up to ten previous days and provides links to recent publications based on the superensemble technique.

12.6.3 *The Multimodel Superensemble for Numerical Weather Prediction*

As many as seven global models are being used (Krishnamurti et al. 2000a) for the prediction of weather on a real-time basis. Figure 12.9 illustrates typical superensemble-based weather prediction skills derived from this product. Here the mean RMS forecast errors of 850 mb winds on day 3 of the forecast for various regions of the globe for the month of August 1998 is shown. The results for member models, an ensemble mean of these member models, and that for the superensemble are presented in this figure. Large improvements in reduction of wind forecast errors can be seen over the tropical belt from the superensemble. These results convey what has been stated above on the performance

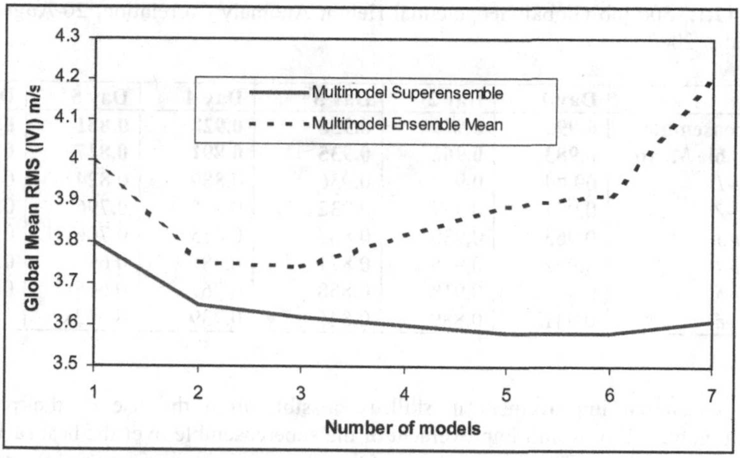

Figure 12.10. Mean RMS error (m s^{-1}) of total wind at 850 mb during August 1998 for day 3 forecast over monsoon domain.

of the superensemble. These results have been confirmed for each month since 1998 to the present.

To assess how many models are minimally needed to improve the skill of the multimodel superensemble, we examined the issue sequentially using one to seven models. Results for the mean global wind RMS errors at 850 mb during August 1998 for day 3 forecasts are shown in Fig. 12.10. The models with lower and lower skill are added sequentially while proceeding from one model to seven models. The dashed line shows the error for the ensemble mean and the solid line indicates that of the superensemble. The superensemble skill is higher than that of the ensemble mean for any selection of the number of ensemble members. The skill of the superensemble between four and seven models is small, i.e., around 3.6 m s^{-1}. The ensemble mean error increases as we add more ensemble members beyond three. This is due to the gradual addition of models with lower skill. That rapid increase is not seen for the superensemble since it automatically assigns low weights to the models with low skill. It is also worth noting that half the skill improvement comes from a single model for this procedure.

Anomaly correlation of 500 mb geopotential heights is another stringent measure for assessing the performance of the superensemble in the medium-range weather forecasts. Table 12.1 provides some recent results of anomaly correlations at 500 mb for the global belt obtained from real-time superensemble. Here the entries for the anomaly correlation skills covering a forecast period from 20 August to 17 September 2000 are presented. Results for the member models, the ensemble mean, and the superensemble are included here. Results for forecast days 1 through 6 are provided in this table. A consistent high skill around 0.75 to 0.8 for the superensemble for day 6 is noted in these experimental runs. Also shown in this table are the entries for the ensemble mean, which lie roughly halfway in between the best model and the superensemble. Thus it appears

Table 12.1. 500 mb Global Geopotential Height Anomaly Correlation: 20 August – 17 September 2000.

	Day 1	Day 2	Day 3	Day 4	Day 5	Day 6
Superensemble	**0.992**	**0.979**	**0.958**	**0.928**	**0.881**	**0.799**
Ensemble Mean	**0.983**	**0.962**	**0.935**	**0.891**	**0.827**	**0.756**
Model-1	09.84	0.967	0.936	0.889	0.824	0.713
Model-2	0.981	0.957	0.932	0.880	0.796	0.623
Model-3	0.963	0.930	0.885	0.815	0.706	0.579
Model-4	0.962	0.925	0.871	0.786	0.697	0.578
Model-5	0.956	0.918	0.858	0.767	0.665	0.549
Model-6	0.941	0.889	0.846	0.739	0.632	

that a substantial improvement in skill is possible from the use of the proposed superensemble. The overall improvement of the superensemble over the best (available) model is around 10%. This improvement of the superensemble is a result of the selective weighting of the available models during the training phase. We have also noted that the skills over the Southern Hemisphere reach those of the Northern Hemisphere from this procedure.

12.6.4 *Precipitation Forecasts from TRMM–SSM/I Based Multianalysis Superensemble*

A major improvement in tropical precipitation forecasts has emerged from the use of a TRMM – SSM/I based multi-analysis superensemble (Krishnamurti et al. 2000b). "Multi-analysis" refers to different initial analyses contributing to forecasts from the same model. In this study, the multi-analysis component is based on the FSU global spectral model (FSUGSM) initialized with TRMM and SSM/I data sets via a number of rain rate algorithms. Five different initial analyses for each day are deployed that define the multianalysis component. Those are based on different versions of rain rate estimates derived from TRMM and the DMSP-SSM/I satellites. These rain rate initializations of the different rain rate estimates follow the physical initialization procedures outlined in Krishnamurti et al. (1991). The differences in the analyses arise from the use of these rain rates within a physical initialization. The resulting initialized fields have distinct differences among their initial divergence, heating, moisture, and rain rate descriptions.

The impact of physical initialization on the improvement of precipitation forecast skills was examined in detail by Treadon (1996) where he used the GPI based rain rates for physical initialization. Figure 12.11 illustrates the correlation of rainfall (observed versus modeled) plotted against the forecast days. Here a very high nowcasting skill of the order of 0.9 is seen. This was a feature of the physical initialization also noted by Krishnamurti et al. (1994b). However, the forecast skill degrades to 0.6 by day 1 of the forecast and it degrades even more by days 2 and 3 to values such as 0.5 and 0.45, respectively. Using the proposed superensemble approach, it is possible to improve the forecast skills when the TRMM-SSM/I based rain rates are used as a benchmark for the definition of the superensemble statistics and forecast verification.

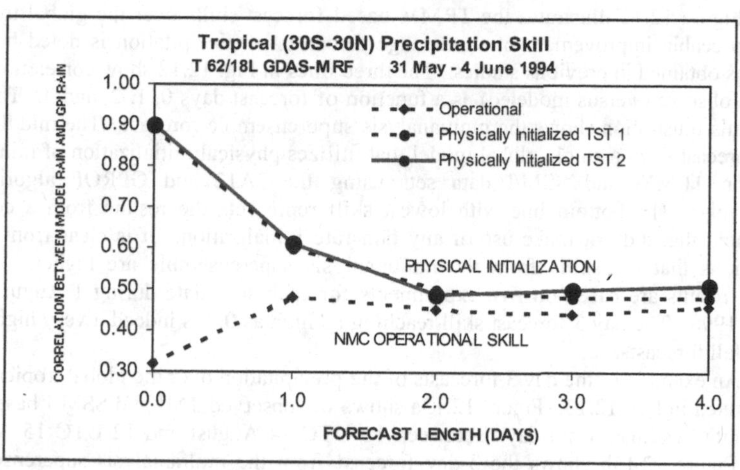

Figure 12.11. Skill of precipitation forecasts over global tropics based on point correlation (Treadon 1996).

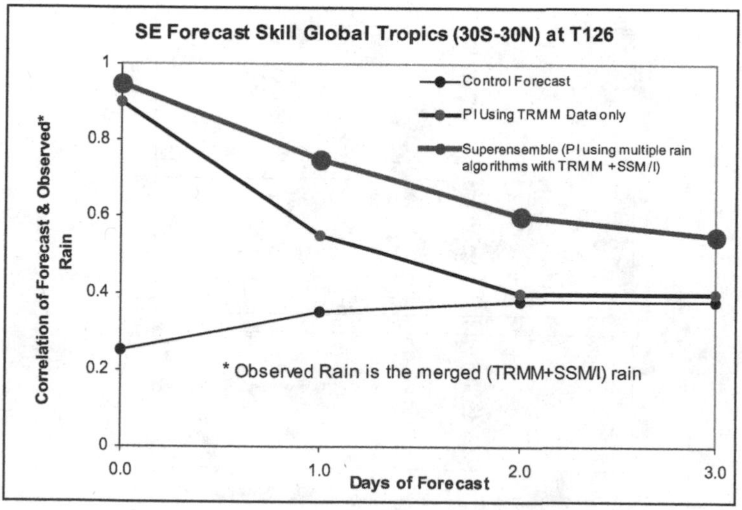

Figure 12.12. Skill of precipitation forecasts over global tropics (30°S – 30°N) for the control forecast, physical initialization using TRMM data only, and superensemble forecasts where the TRMM plus SSM/I rain rates are used as a benchmark.

Figure 12.12 illustrates the TRMM based forecast skills over the global tropics. Here noticeable improvement of short-range forecasts of precipitation is noted beyond what was obtained in previous studies. The three lines in Fig. 12.12 show correlations of rainfall (observed versus modeled) as a function of forecast days 0, 1, 2, and 3. The top line in this illustration shows the multianalysis superensemble forecast. The middle line is the forecast from a single global model that utilizes physical initialization of rain rates based on TRMM and SSM/I data sets using the 2A12 and GPROF algorithms, respectively. The bottom line with lowest skill represents the results from a control experiment that did not make use of any rain rate initialization. It is clear from these illustrations that the skills from the multianalysis superensemble are higher. These forecast results are based on five experiments for each start date during 1 August to 5 August 1998. The day 3 forecast skill reaching as high as 0.7 is indeed a very high skill for rainfall forecasts.

An example of the day 3 forecasts of the precipitation over the global tropical belt is illustrated in Fig. 12.13. Figure 12.13a shows the observed TMI and SSM/I based 24-hour rainfall estimate (mm day^{-1}) between 12 UTC 14 August and 12 UTC 15 August 1999. Figure 12.13b shows the 3-day forecast from the multianalysis superensemble valid for the same period, while Fig. 12.13c shows the corresponding results from a single best model. The global tropical correlation between the observed and the multianalysis superensemble is 0.55 where the correlation of the best model with respect to the observed estimate is 0.30 for these day 3 forecasts. This reflects a major improvement in rainfall forecasts over the global tropics.

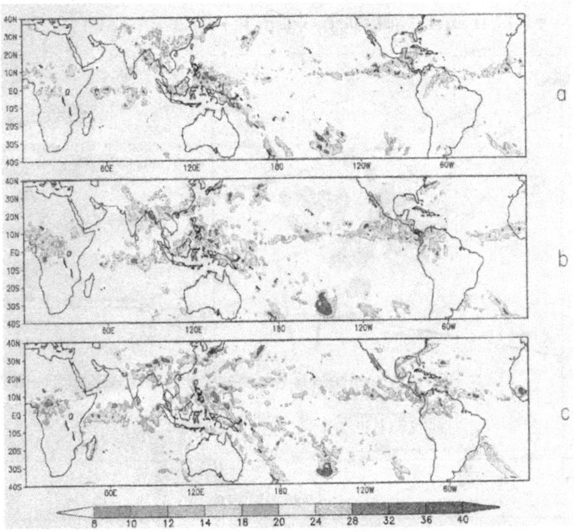

Figure 12.13. Day 3 rainfall forecast over the global tropical belt, 12 UTC 15 August 1999: the accumulated rainfall (mm day^{-1}) by (a) observation based on TRMM and SSM/I, (b) multianalysis superensemble, and (c) a best single model.

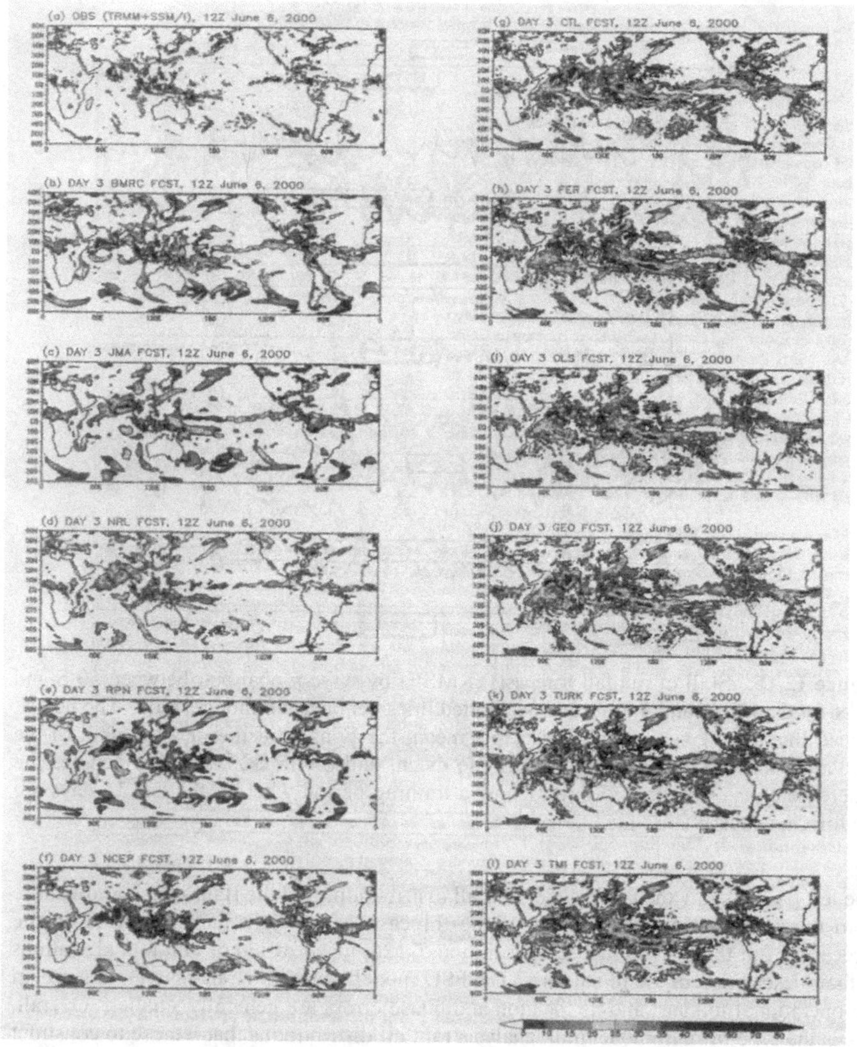

Figure 12.14. The observed rainfall estimate from the TMI-2A12 and SSM/I-GPROF algorithm for 5 June 2000 is compared with the day 3 forecasts from the 11 member models of the multimodel-multianalysis system.

The next area of our research was multianalysis/multimodel superensemble (Krishnamurti et al. 2001). The 12 panels of Fig. 12.14 illustrate the day 3 rainfall forecasts valid on 6 June 2000. Here the observed rain is shown in the top left panel.

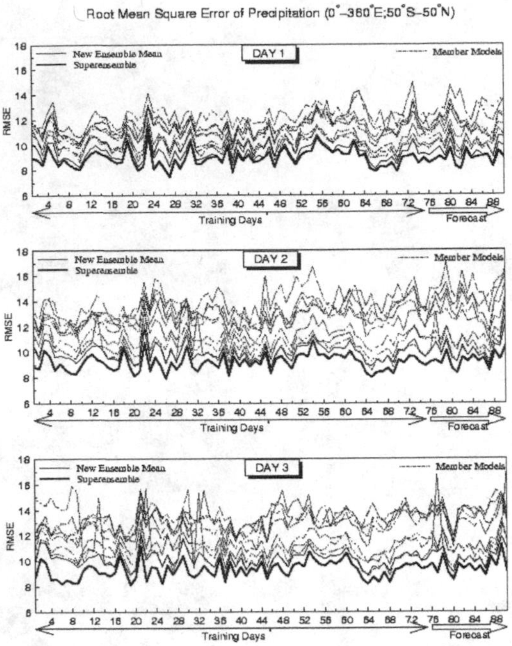

Figure 12.15. Skill of rainfall forecasts (RMSE) over the global belt between 50°S and 50°N for days 1, 2, and 3 of forecasts. Dotted lines denote multimodel skills. The heavy, dashed line denotes skill of the ensemble mean, the thin, solid line denotes skill of the individual model's bias removed ensemble mean, and the thick, black line denotes the superensemble. The first 75 days denote a training period whereas the last 15 days are the forecast days.

The left panels show the multimodel rainfall distributions and the right panels show those from the multianalysis components of the forecasts. The right panel is based on the forecasts from the FSU model at a resolution *T*-126, using different rain rate algorithms in their description of the initial rain. The FSU model's rainfall is, in general, larger than the operational models, and its location and phase errors are generally smaller. Overall, this is the type of multimodel/multianalysis rainfall distributions that we use to construct the superensemble forecasts in our experimental real-time forecasts.

Some important results from the 11-model superensemble are presented here. We calculate three measures of skill on a regular basis: i) correlation of model predicted daily rainfall totals and observed estimates; ii) RMS errors of model predicted daily rainfall totals; iii) equitable threat scores for different thresholds of observed and predicted rain. The root mean square errors (RMSE) in precipitation forecasts over the global belt, 50° S to 50° N, covering a forecast period from 1 April to 15 April 2000 are shown in Fig. 12.15. The training period for these forecasts included the preceding 75 days. The thick black lines denote the RMSE for the multianalysis/multimodel

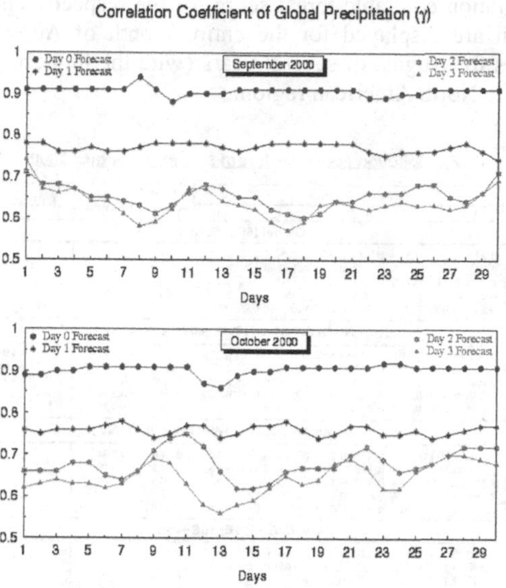

Figure 12.16. Forecast skill based on correlation of observed rainfall estimates from TRMM-2A12 and the SSM/I-GPROF and the superensemble for day 1, day 2, and day 3 forecasts during September and October 2000.

superensemble. The dotted lines show the skills for the selected individual member models, whose skills were high. The thin, solid line shows results for the ensemble mean, with bias removal for individual models. Overall, these results over the global belt show great promise for the 3-day forecasts of precipitation. It should be pointed out that these results are fairly robust and we see the same skills in the day-to-day real-time runs. There is some noticeable improvement in the skill for the superensemble over the ensemble mean. That arises from the fact that the poorer models are assigned weights of 1.0 over the entire globe, whereas the superensemble is more selective regionally (and vertically) for each variable and for each model. Its weights are fractional positive or negative based on the member models' past performance.

We can also look at the correlation of the observed rain (24-hourly totals ending on days 0, 1, 2, and 3 of forecasts) derived from the TRMM-2A12 plus the SSM/I-GPROF based rainfall against the global gridded forecasts of the superensemble-based rains. Those are shown in Fig. 12.16 for the months September and October 2000. The global forecast correlation skills for days 0, 1, 2, and 3 lie roughly around 0.9, 0.8, 0.62, and 0.55 for these months. These are higher skills compared to what were seen for a single model shown in Fig. 12.10. Similar results are noted for all the recent months.

As was summarized in Krishnamurti et al. (2001) the one to three-day forecast skills of the daily precipitation totals for the three metrics used here are indeed the highest for the superensemble. Table 12.2 illustrates the results of the threat scores for eight participating members of the real time multimodel/multianalysis system. The threat

Table 12.2. Precipitation equitable threat score for the respective member models over the identical domain are displayed for the entire month of August 2000. The ETA model's threat scores for August of several years (with the highest scores) are shown in the last column for the North American region.

Precipitation Equitable Threat Scores for August 2000

Pr mm	Member Models 1	2	3	4	5	6	7	8	Ens Mean	Super Ensemble	ETA Model
	Global (50S–50N)										
0.2	0.313	0.295	0.343	0.302	0.296	0.268	0.276	0.273	0.386	0.568	
10	0.237	0.157	0.195	0.132	0.190	0.152	0.174	0.157	0.219	0.312	
25	0.215	0.117	0.153	0.089	0.165	0.114	0.136	0.119	0.148	0.257	
50	0.171	0.088	0.112	0.064	0.145	0.081	0.092	0.080	0.112	0.198	
75	0.073	0.057	0.012	0.000	0.037	0.044	0.055	0.044	0.011	0.272	
	North America (120W–65W, 20N–50N)										
0.2	0.202	0.256	0.200	0.171	0.180	0.222	0.232	0.215	0.305	0.641	0.308/1999
10	0.088	0.062	0.020	0.021	0.014	0.072	0.092	0.076	0.066	0.458	0.288/1995
25	0.054	0.045	0.000	0.012	0.000	0.038	0.066	0.049	0.006	0.425	0.221/1995
50	0.033	0.005	0.000	0.012	0.000	0.021	0.036	0.028	0.008	0.142	0.199/1995
75	0.013	0.000	0.000	0.012	0.000	0.000	0.020	0.014	0.000	0.039	0.131/1991
	South America (110W–10W, 50S–15N)										
0.2	0.340	0.261	0.309	0.325	0.266	0.240	0.248	0.247	0.369	0.594	
10	0.298	0.160	0.222	0.130	0.189	0.161	0.171	0.153	0.243	0.333	
25	0.251	0.118	0.153	0.083	0.148	0.119	0.135	0.114	0.133	0.276	
50	0.166	0.079	0.071	0.040	0.102	0.080	0.087	0.071	0.053	0.216	
75	0.115	0.052	0.026	0.012	0.057	0.048	0.053	0.041	0.018	0.151	
	Asia (50E–120E, 15S–45N)										
0.2	0.390	0.474	0.543	0.426	0.458	0.428	0.459	0.440	0.589	0.636	
10	0.306	0.172	0.270	0.197	0.236	0.165	0.200	0.177	0.246	0.352	
25	0.267	0.131	0.211	0.132	0.170	0.122	0.155	0.133	0.175	0.279	
50	0.198	0.092	0.160	0.045	0.122	0.088	0.112	0.072	0.132	0.198	
75	0.153	0.077	0.117	0.020	0.060	0.075	0.090	0.055	0.041	0.172	
	Africa (20W–55E, 35S–40N)										
0.2	0.411	0.462	0.457	0.416	0.396	0.431	0.447	0.439	0.569	0.692	
10	0.249	0.246	0.189	0.143	0.167	0.261	0.295	0.274	0.248	0.357	
25	0.217	0.167	0.137	0.105	0.131	0.190	0.216	0.204	0.151	0.286	
50	0.141	0.096	0.064	0.052	0.055	0.111	0.101	0.109	0.087	0.185	
75	0.097	0.065	0.052	0.017	0.036	0.085	0.075	0.078	0.025	0.145	
	Australia (110E–160E,40S–0)										
0.2	0.363	0.341	0.368	0.380	0.340	0.292	0.324	0.322	0.364	0.425	
10	0.271	0.191	0.264	0.192	0.242	0.197	0.199	0.185	0.273	0.332	
25	0.218	0.146	0.219	0.114	0.194	0.156	0.165	0.141	0.193	0.285	
50	0.145	0.119	0.130	0.049	0.140	0.111	0.121	0.102	0.130	0.185	
75	0.094	0.116	0.085	0.029	0.111	0.088	0.089	0.076	0.073	0.155	

* Pr mm denotes precipitation class intervals for rainfall rates greater than the indicated amount in column 1. The threat score for the respective member models over the indicated domain are displayed for the entire month of August 2000. The ETA models threat scores for August of several years (with the highest scores) are shown in the last column for the North American region. The 98 days training period ends with 1 August 2000.

scores are evaluated covering the precipitation rate intervals greater than 0.2, 10.0, 25.0, 50.0, and 75.0 mm day^{-1}. The size of the individual domains is identified within the table. The BIAS for the member models, ensemble mean, and superensemble were found to be comparable (not shown). The threat scores for the superensemble for all rainfall intervals are the highest compared to the member models and the ensemble mean rainfall. We have also shown the threat scores for the ETA model over North America in the last column. The forecasts for the member models and superensemble are for August 2000. This covers a 31-day period. The ETA model's equitable threat scores for August over different years (shown by the ETA entry) are shown with their highest scores included. Here, again, the superensemble threat scores are higher than those for the ETA. The superensemble was cast at the resolution T-126 (i.e., roughly 90 km horizontal resolution), whereas the operational ETA model had a resolution of 32 km. Considering those differences in resolution, the performance of the superensemble (for these experiments) appears impressive. Although the improvement in the equitable threat

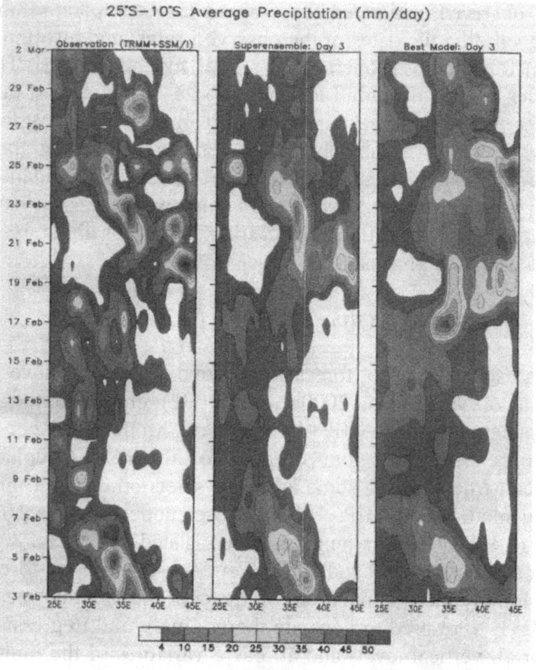

Figure 12.17. A Hovmoller diagram of daily precipitation (mm day^{-1}) on day 3 of the forecast during the Mozambique floods. Ordinate shows days (bottom to top); abscissa denotes longitude. The three panels denote (left) observed rain (from TRMM-2A12 plus SSM/I GPROF); (middle) superensemble forecasts; (right) best operational model.

scores appear quite large, they should still be regarded as modest. Heavy rain events in excess of 75 mm day^{-1} are not handled very well by any of the models. The superensemble also underestimates the rain by roughly a factor of 2. We have examined such cases in some detail and it is clear that much further improvement is needed from the member models in order to improve the superensemble based rain. This may require higher resolution modeling for the member models with improved physics and initialization of rain.

It is of considerable interest to ask whether the superensemble forecasts of rainfall can provide any useful guidance for floods. Most of the heavy rains that resulted in the Mozambique floods during February and March 2000 resulted from heavy rains over Mozambique and Zimbabwe. The headwaters of the Limpopo River over Zimbabwe experienced the heaviest rainfall that resulted in the cresting of the river over southern Mozambique where the flooding was most severe. Forecasts of rain from this study were projected on Hovmoller diagrams (longitude versus time) and are shown in Fig. 12.17. Here we show the daily rainfall for the belt 10°S to 15°S covering the longitudes 24°E to 45°E for the entire month of February (dates are plotted from the bottom up). The three panels denote the 'observed' estimates, those based on the superensemble forecasts (for day 3 of forecasts) valid on the dates of the observed rains, and also those predicted (for day 3 of forecasts) from the best operational model for this region. The best model is determined from the RMSE of rainfall for each model. The units of rainfall are in mm day^{-1}. It is clear that the multimodel superensemble carries the 3-day forecasts of heavy rains associated with the Mozambique floods very well. Given the higher rainfall forecast skills from the superensemble, it appears that useful guidance of heavy rain events resulting in floods may be possible from this approach. We have examined the performance of superensemble in flood forecasting issues for about 10 case studies and the results have been equally promising.

12.6.5 *Seasonal Climate Forecasts from Multimodel Superensemble*

In the area of seasonal climate forecasting, several papers have been published (Krishnamurti et al. 1999, 2000a, 2000b, and 2002) on the initial development of strategies and application with AMIP (Atmospheric Model Inter-comparison Project) data sets and a first effort with four versions of the FSU Coupled Models. The superensemble is constructed using some arbitrary selection of eight models from about 31 different global models of AMIP. All of these models have a 10-year integration period from 1979 to 1988. The training period consisted of the last 8 years of the data sets while the first two years (1979 and 1980) were subjected to the forecast phase of the superensemble. Monthly mean simulations along with the monthly mean analysis fields provided by ECMWF were used to generate the anomaly multiregression coefficients at each grid point for all vertical levels and all basic variables of the multimodels. Figure 12.18 shows the time sequence of the RMS error for the meridional wind over the global tropics. One can observe a marked improvement in the skill scores achieved using the superensemble approach.

Further to this study, several types of model-generated data sets are examined to address the question of seasonal prediction of precipitation over the Asian and the North American monsoon systems (Krishnamurti et al. 2002). The main question asked is if there is any useful skill in predicting seasonal anomalies (beyond those of climatology). The superensemble methodology is applied here to the anomalies of the predicted multimodel data sets and the observed (analysis) fields. We noted that the superensemble based anomaly forecasts have somewhat higher skill compared to the ensemble mean of member models, individually bias removed ensemble mean of the member models, the climatology, and the member models that are being used in this exercise. The illustration for the seasonal correlations (of model rainfall anomalies) with respect to the observed anomalies is presented in Fig. 12.19a and Fig. 12.19b for the Asian Monsoon domain and North American Monsoon domain, respectively. The highest anomaly correlations for the seasonal precipitation forecast are generally seen for the multimodel superensemble (heavy line). The other heavy line shows the results for the ensemble mean, while the remaining thinner lines show the skill of the member models of AMIP1 and AMIP2. The calculations carried out here used the cross-validation technique, i.e., all years (except the one being forecasted) were used to develop the training data statistics. The skill of

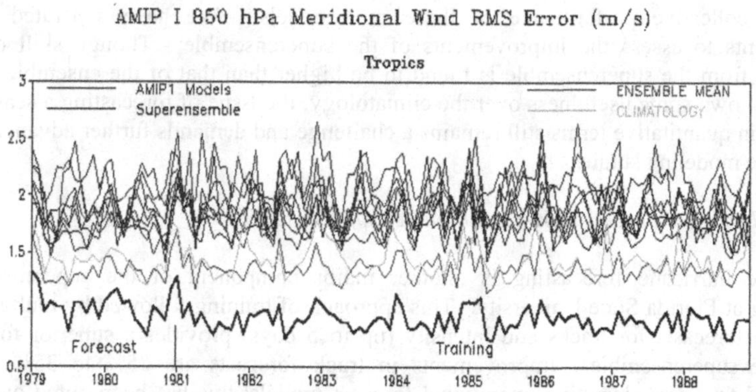

Figure 12.18. 850 mb meridional wind RMS error (m s^{-1}) from AMIP1 data sets. Training phase is from 1981 to 1988 and forecast phase is from 1979 to 1980. Results from AMIP1 model forecasts, ensemble mean, superensemble, and climatology are shown (after Krishnamurti et al. 2000a).

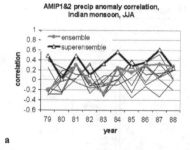

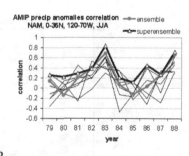

Figure 12.19. (a) Correlation of seasonal forecasts of precipitation anomalies with respect to observed anomaly estimates based on Xie and Arkin (1996) from the mix of AMIP1 and AMIP2 data sets. Heavy line at top: superensemble. Other heavy line: bias removed ensemble mean. Thin lines: member multimodels. Ten years of summer monsoon results are shown here. (b) Same as 12.20 (a) but for North American Monsoon Domain.

forecasts from the superensemble come partly from the forecast performance of multimodels and partly from the training component built into this system that is based on past collective performance of these multimodels. We have separated these components to assess the improvements of the superensemble. Though skill of the forecasts from the superensemble is found to be higher than that of the ensemble mean and has shown some usefulness over the climatology, the issue of forecasting a season in advance in quantitative terms still remains a challenge and demands further advancement in climate modeling studies.

12.6.6 *Hurricane Forecasts from Multimodel Superensemble*

Real-time hurricane forecasting is another major component of the superensemble modeling at Florida State University. This approach of training followed by multimodel real-time forecasts for tracks and intensity (up to 5 days) provides a superior forecast from the superensemble. Improvements in track forecasts are 25% to 35% better compared to the participating operational forecast models; this has been noted over the Atlantic Ocean basin. The intensity forecasts for hurricanes are only marginally better than those of the best models. Recent real-time tests, during 1999 to 2001, showed marked skills in the forecasts of difficult storms such as Floyd and Lenny where the performance of the superensemble was considerably better than those of the member models (Williford et al. 2002). An example of the superensemble track forecasts for hurricane Lenny is shown in Fig. 12.20. Here the observed best track for hurricane Lenny is compared to those of a few member models and the superensemble. In these track forecasts, we note improvements for days 1, 2, and 3 of forecasts for the storm

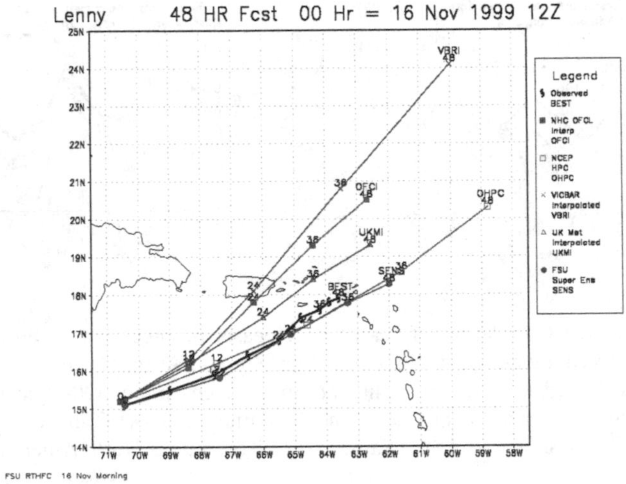

Figure 12.20. Superensemble track forecast of Lenny. Here the predicted tracks of some member models, ensemble mean, and superensemble are shown.

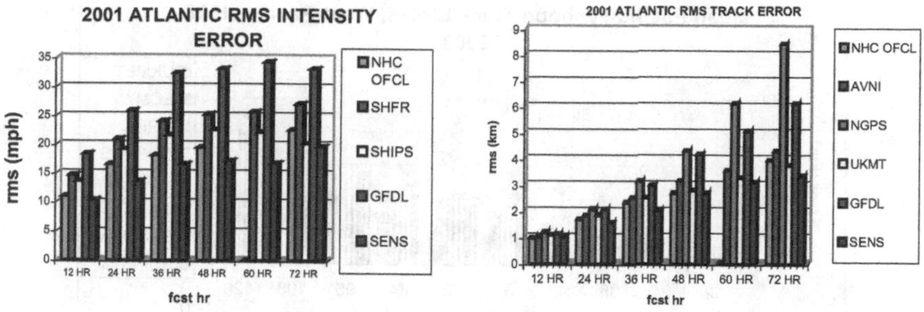

Figure 12.21. A seasonal summary on the performance of hurricane forecast skill during the year 2001 from various models including the FSU superensemble (SENS). Here the errors for the intensity (mph) and track (km) are displayed for 3-day forecasts.

positions, which were of the order of 125, 200, and 350 km. This is an example showing a marked improvement for position forecasts. The illustrations in Fig. 12.21 show the forecast performance during the year 2001 for the Atlantic hurricane track and intensity. The least error for the superensemble in both categories is a consistent feature compared to all the participating models. This is an area where the use of multimodels (from diverse global modeling units and from FSU) has shown the most promise for forecasts on imminent landfall, tracks, and intensity. A superensemble forecast constructed with a suite of some of the finest global models that are currently available holds great promise for the improvements in short-range predictions for the landfall and tracks of hurricanes.

Similar studies on superensemble based track and intensity forecasts for the Pacific region (Vijaya Kumar et al. 2002) have also revealed the usefulness of this methodology displaying considerable improvement of the forecast skills. A summary of the Pacific typhoon track and intensity errors for the years 1998-2000 is provided in Fig. 12.22. Here the position and intensity errors at 12-hour intervals are shown where the skill from the superensemble is consistently high compared to the member models and the ensemble mean. In all of the aforementioned work, preservation of the member model features is an essential requirement during the training and the forecast phases. If drastic changes occur in the member models then the proposed statistical component of the superensemble is invalidated. It is apparent that if no major model changes are invoked during the training and the forecast phases of these forecasts, then we can obtain skill improvements of the order of 61, 138, 159, and 198 km for the typhoon position errors over the best models for forecasts at the end of days 1, 2, 3, and 4, respectively. The corresponding intensity forecast skills (RMS errors) at days 1, 2, 3, and 4 of forecasts from the superensemble exceed those of the best models by 5, 10, 13, and 20 knots.

12.6.7 *Prospects for Future Research and Applications*

Currently a number of regional mesoscale models (such as regional spectral models of FSU, NCEP, ETA, various versions of MM5, ARPS, WRF, RAMS, and

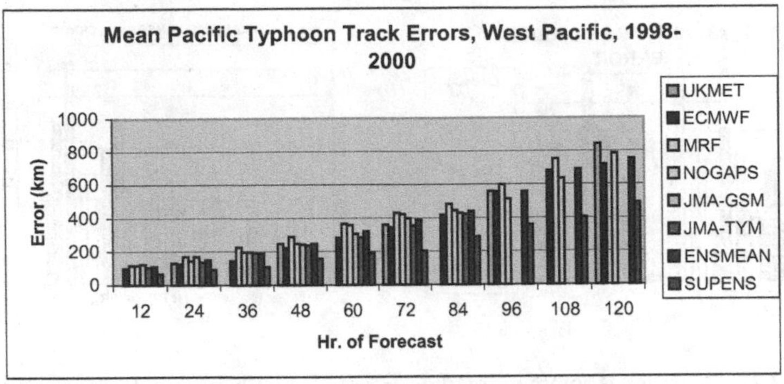

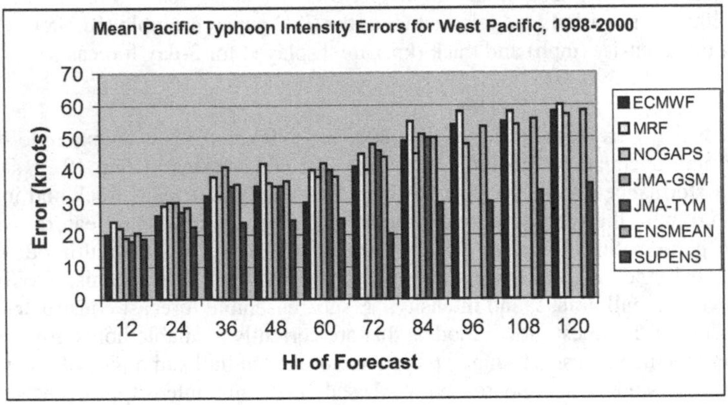

Figure 12.22. (a) Mean typhoon track errors for the West Pacific (km), 1998-2000. (b) Mean typhoon intensity errors for the West Pacific (km), 1998-2000.

others) are available that carry out real-time forecasts over regional domains. In principle, the superensemble methodology can be extended to this class of models and we expect much progress towards more accurate forecasts of severe weather events.

Chapter 13

Adaptive Observational Strategies

13.1 Introduction

While carrying out a forecast it is possible to suggest observations over certain targeted regions that would improve the forecast. There are some well-known strategies for deploying special observations over such targeted regions. This is an important area in numerical weather prediction and various research groups have proposed several strategies in recent years. A simple adaptive observational strategy was developed by Zhang and Krishnamurti (2000) that is designed for applications to hurricane forecasts. Zhang and Krishnamurti (1997) developed an EOF-based perturbation technique for hurricane ensemble predictions that utilized ensemble forecasts. This technique worked towards improving the forecast skill and thus reducing the uncertainties of the initial states. Ensemble studies have shown that given a basic initial analysis, the pattern of the ensuing NWP forecast error variance is not homogeneous. In other words, the prediction in some areas tends to be more sensitive to the initial state compared to others or the inaccuracy of analysis has a greater affect on the growth of forecast errors over some locations as compared to others. A high-resolution observational network would be too expensive and not feasible. Zhang and Krishnamurti (2000) proposed a method that provided some guidance for aircraft reconnaissance missions in a hurricane environment. This was based on the mapping of the spread of forecast errors from the construction of the variance of some 50-member ensemble forecasts with respect to a single reference, i.e. a control run. The variables at locations of maximum variance were back correlated to an initial field to locate regions from where the error emanates thus identifying regions of initial data uncertainties. If data is deployed over such regions then a marked improvement for the hurricane forecasts could be demonstrated. The method is simple and takes only a limited amount of resources compared to several other methods. This method tags regions where additional data sets are needed in order to obtain a better initial analysis.

Several forecasters had noted that observations over particular locations near the center of a storm were very helpful in predicting the future course of that storm. Bowie (1922) showed that observations on the northwest side of the cyclone were important for track prediction. Aberson (2002) noted that merely obtaining data around a tropical storm would not improve the forecasts, rather more observations taken at particular regions are needed. Gregg (1920), Riehl and Shafer (1944), and others felt that observations at certain vertical levels would improve the forecasts. Aberson and Franklin

(1999) noted that intensity forecasts of hurricanes in which all dropwindsonde data was included rather than data below 400 mb were much better than those that did not utilize dropwindsonde data provided by special research aircraft.

Time and space are both important for targeted observations. Bristor (1958) showed that initial analysis errors do not grow to an important degree as long as the spacing between synoptic reports confines error fields to a scale smaller than those of the synoptic disturbances. Sampling the environment only partially around the vortex may introduce an asymmetry in the flow, forcing the storm to move with an incorrect velocity (Derber and Bouttier 1999).

13.2 Techniques for Targeted Observations

Different research groups have developed several techniques for targeted observations. Bishop and Toth (1999) used an ensemble transformation technique to assess targeting areas where observations were needed. An ensemble transform technique uses nonlinear ensemble forecasts to construct ensemble-based approximations for the prediction error covariance matrices associated with a wide range of different possible deployments of observational resources. Expected forecast errors are obtained from distinct deployment of observational resources. Optimum deployment is where the forecast error is minimized. Bishop et al. (2001) introduced the concept of Ensemble Transform Kalman Filter (ETKF). This utilizes an ensemble transformation and a normalization to rapidly obtain the predicted error covariance matrix associated with a particular deployment of observations. This enables a large number of future feasible sequences of observational networks to reduce forecast error variances. This method appears to have an edge over the ensemble transform technique. The dominant singular vectors of the integral linear propagator for a nonlinear dynamical system provide information about maximum perturbation growth (measured by a given norm) during finite time intervals, and can be used to estimate the evolution of initial errors during the course of a forecast (Lacarra and Talagrand 1988; Farrell 1990; Borges and Hartmann 1992). Singular vectors are currently used at the European Centre for Medium-Range Weather Forecasts (ECMWF) in the construction of the initial perturbations of the Ensemble Prediction system (Buizza and Palmer 1995; Molteni et al. 1996). Breeding of growing modes consists of one additional, perturbed short-range forecast, introduced on top of the regular analysis in an analysis cycle. The difference between the control and (perturbed six-hour) first guess forecast is scaled back to the size of the initial perturbation and then reintroduced onto the new atmospheric analysis. Thus, the perturbation evolves along with the time dependent analysis fields, ensuring that after few days of cycling the perturbation field consists of a superposition of fast-growing modes corresponding to the contemporaneous atmosphere, akin to local Lyapunov vectors (Toth and Kalnay 1993).

Reynolds et al. (1994) and Zhu et al. (1996) indicated that, in the midlatitudes, most synoptic-scale errors in global numerical weather prediction models are not due primarily to model deficiencies and that the largest forecast improvements are likely to be achieved by decreasing the analysis error.

In 1982, the NOAA Hurricane Research Division (HRD) began to conduct synoptic flow experiments using Omega dropwindsondes deployed from the NOAA WP-3D (P-3) aircraft. The aim was to improve forecasts of hurricane tracks by obtaining vertical profiles of wind, temperature, and humidity below approximately 400 mb within 1000 km of the cyclone center. Burpee et al. (1996) showed that significant improvements occurred (16-30% error reduction for a 12-60 hour forecast) that provided primary numerical guidance for the (National Hurricane Center) NHC official track forecast. Since 1997, the Gulfstream IV-SP jet aircraft (G-IV), procured by NOAA in the previous year, has been deployed in "synoptic surveillance" missions for improvements of forecasts of tropical cyclones affecting the coastal United States, Puerto Rico, U.S. Virgin Islands, and Hawaii. A new type of dropwindsonde developed at the National Center for Atmospheric Research (NCAR), which uses a Global Positioning System (GPS), was utilized from 1997. They transmit data in real-time to the aircraft and the processed observations are then beamed to operational forecast centers for assimilation into numerical model analyses. The flight tracks for synoptic surveillance missions were generally still drawn up subjectively, although new objective strategies are being developed in a research mode. Aberson and Franklin (1999) showed that dropwindsondes observations reduce mean track forecast errors by 32% and the intensity forecasts by 20%.

13.2.1 *Random perturbation based method*

Most of the above methods are somewhat more complex than the EOF based method. Here we shall describe the workings of the simple method based on perturbations and include some illustrations.

a) A control experiment plus ensemble forecast: This is a first step in this exercise. One can take a single run from a global model, i.e. a forecast out to roughly 24 hours and call it a control experiment. This experiment utilizes all the conventional weather observations. An ensemble of some fifty experiments are next carried out where the initial state of the control run are simply perturbed by introducing random numbers whose amplitudes are scaled to typical observational errors for variables such as winds, temperature, pressure, and humidity. The 50-member ensemble uses different distributions of these random members (added to the initial state of the control run).
b) A next step in this analysis is to obtain a variance field from the 24-hour forecast field of the 50-member ensemble with respect to the control run for a selected variable. Local maxima over certain regions of interest can be expected because of the spread of the forecasts within the 50-member ensemble. If deploying this over a hurricane domain, a maxima in this variance field would indicate a spread of forecasts and would identify a region where forecasts errors are amplifying.
c) A back correlation of data from the region of maximum variance to the initial state field distribution for the same variable would reveal the possible source regions of such error growth. As an example, sea level pressures of the 50 ensemble members (at the location of the maximum variance at hour 24) can be back correlated to the 50 values at hour zero at each and every transform grid point. Such a field of back correlations at time 0 identifies regions where such error growth would have occurred.

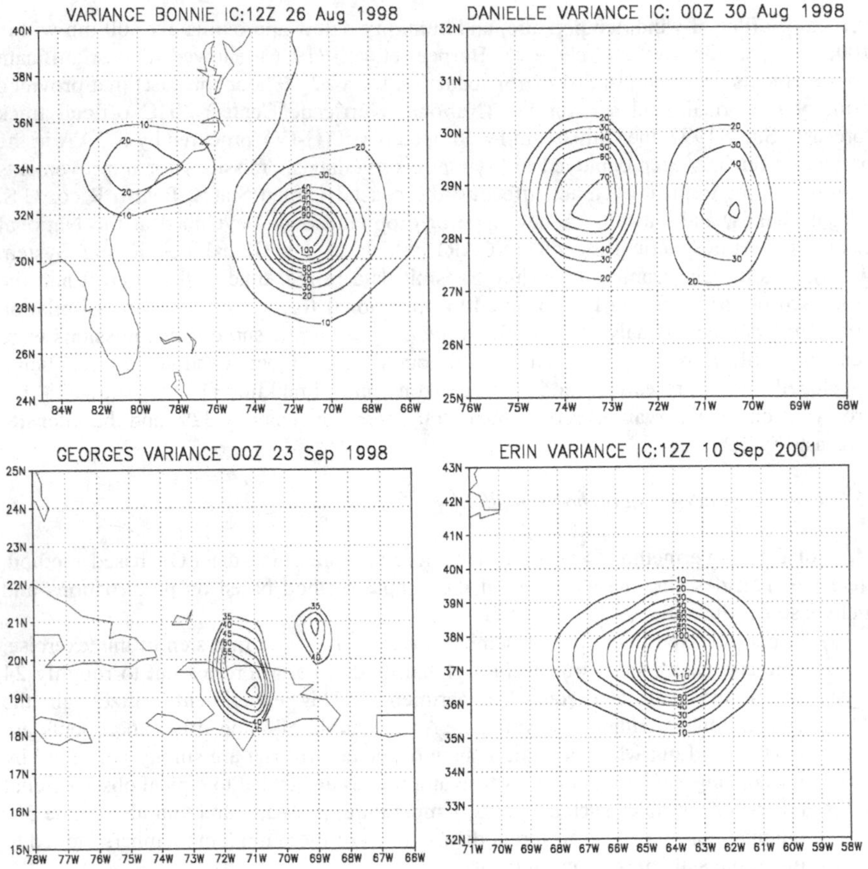

Figure 13.1. Distribution of variances for mean sea-level pressure for hurricanes Bonnie, Georges, Danielle, and Erin at forecast time $t = 24$ hours.

d) A next step in the analysis is to introduce new data sets within that targeted area of high correlations. Those targeted data sets are assimilated with those of the control run to obtain a new data analysis.

This appears to be a very powerful strategy for the deployment of adaptive observations.

13.2.2 *Examples*

We shall next illustrate the workings of the above steps from illustrations of hurricane forecasts using a high-resolution global spectral model. We illustrate the results from four experiments for hurricane Bonnie, Georges, Danielle, and Erin. The start dates for

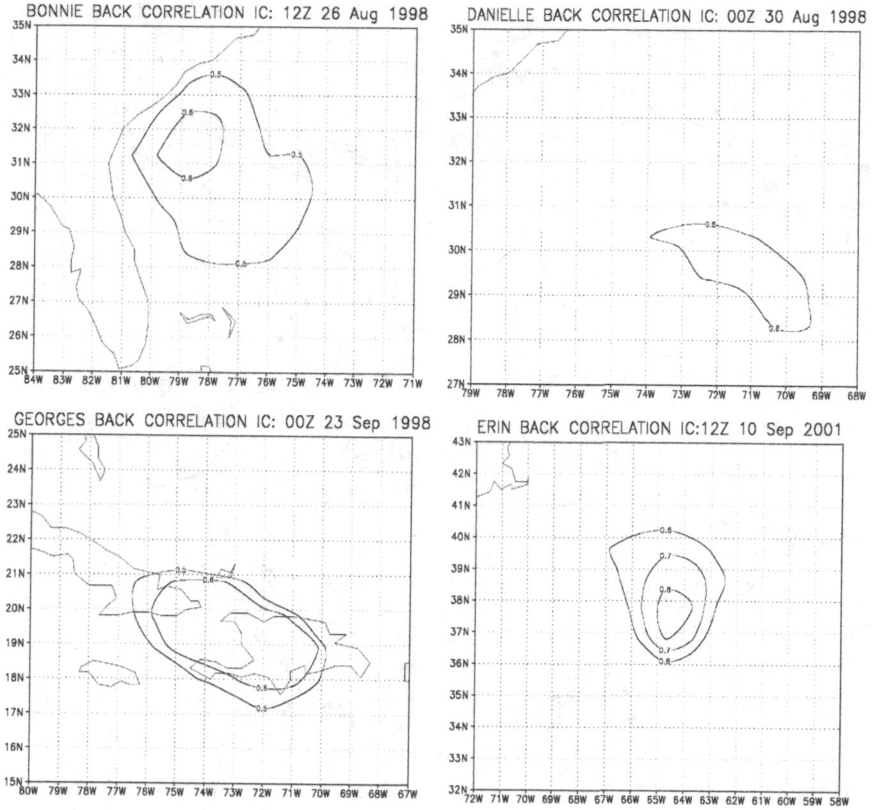

Figure 13.2. Fields of back correlations for hurricanes Bonnie, Georges, Danielle, and Erin at time $t = 0$.

these experiments are 12 UTC 26 August 1998, 00 UTC 23 September 1998, 00 UTC 30 August 1998, and 12 UTC 10 September 2001, respectively. The control initial states for these experiments were obtained from the operational analysis of ECMWF. The FSU model used in our studies is described in chapter 12 and has a horizontal resolution of *T*-126 (with a transform grid resolution of ~80 km) and 14 vertical levels.

Figure 13.1 illustrates the distribution of the variances for mean sea-level pressure calculated from the ensemble spread of 50 forecast experiments (24-hour forecasts) in each case. These variances cover the region where the storm was expected to move in 24 hours. A large spread of the variances was noted and this suggests that model forecasts have a considerable sensitivity to the initial states.

The fields of back correlations for these four storms are illustrated in Fig. 13.2. These are analyzed to locate regions of large correlations that signify possible regions from where the error spread emanates. This identifies a region for targeted observations.

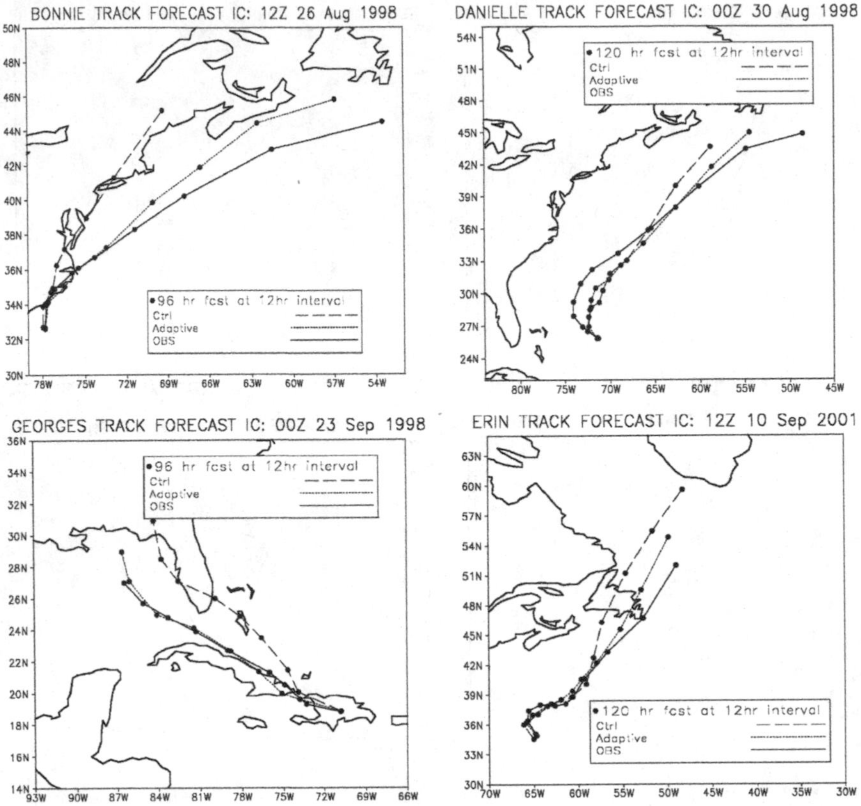

Figure 13.3. Predicted and the observed tracks for hurricanes Bonnie, Georges, Danielle, and Erin. Dashed line shows control forecast, dotted line shows forecasts with targeted observations, and solid line show the observed track.

Special observations of humidity over these regions of high back correlations are assimilated along with those of the control run to prepare for the adaptive observation based forecast experiments.

In Fig.13.3 the predicted and the observed tracks for the four hurricanes where adaptive observations were deployed are compared. Here in each panel we compare the observed tracks with those from the control experiment and the experiment where research aircraft based adaptive observations were included. Based on all four sets of experiments it appears that the inclusion of the adaptive observations leads to some major improvements for the three-day hurricane track forecasts. This demonstrates the promise of adaptive observational strategies.

In Fig. 13.4 the position errors for the hurricane forecast tracks for these four hurricanes are illustrated. Errors (in kilometers) at different hours of the forecast are shown. These simply provide quantitative estimates for what is clearly apparent in Fig. 13.3. These examples of hurricane track forecasts show that the impact of targeted observations is quite clear. Such studies have been carried out by various weather services using such data impact. This is a promising area for future research.

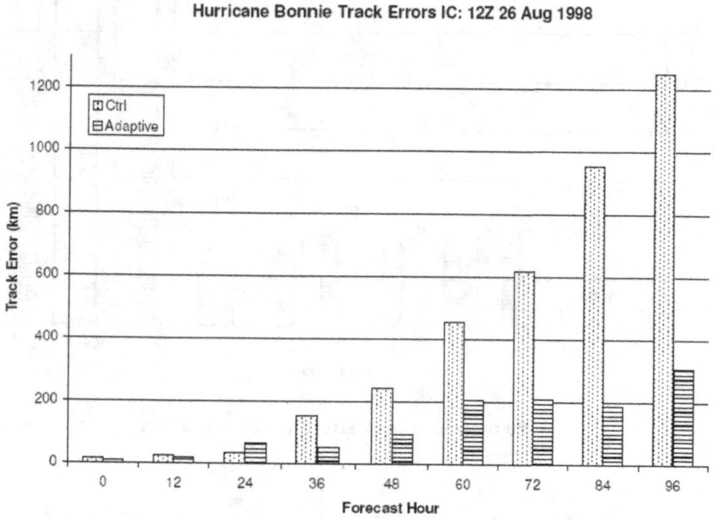

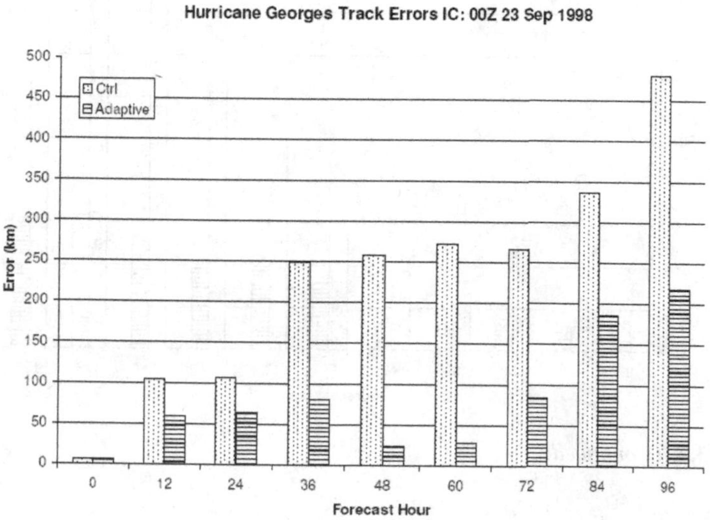

Figure 13.4. Position errors for the hurricane forecast tracks for hurricanes Bonnie, Georges, Danielle, and Erin. (See next page.)

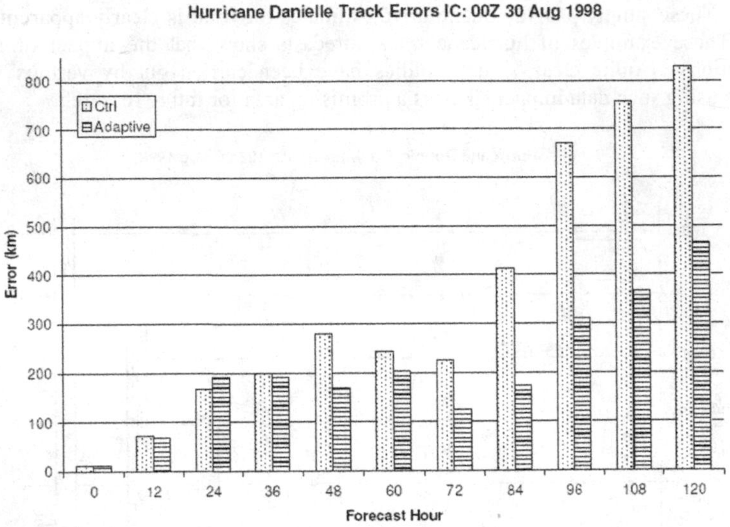

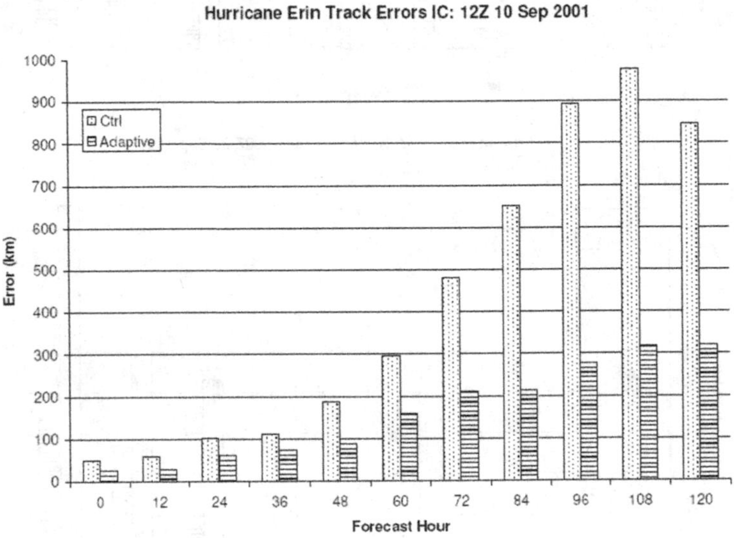

Figure 13.4. (*Continued.*)

Appendix A

2-dimensional Taylor Series:

$$\left.\begin{aligned}
\psi_{I+1,J} &= \psi_{I,J} + \Delta\left.\frac{\partial\psi}{\partial x}\right)_{I,J} + \frac{\Delta^2}{2}\left.\frac{\partial^2\psi}{\partial x^2}\right)_{I,J} + \frac{\Delta^3}{6}\left.\frac{\partial^3\psi}{\partial x^3}\right)_{I,J} + O(\Delta^4)\\
\psi_{I-1,J} &= \psi_{I,J} - \Delta\left.\frac{\partial\psi}{\partial x}\right)_{I,J} + \frac{\Delta^2}{2}\left.\frac{\partial^2\psi}{\partial x^2}\right)_{I,J} - \frac{\Delta^3}{6}\left.\frac{\partial^3\psi}{\partial x^3}\right)_{I,J} + O(\Delta^4)\\
\psi_{I,J+1} &= \psi_{I,J} + \Delta\left.\frac{\partial\psi}{\partial y}\right)_{I,J} + \frac{\Delta^2}{2}\left.\frac{\partial^2\psi}{\partial y^2}\right)_{I,J} + \frac{\Delta^3}{6}\left.\frac{\partial^3\psi}{\partial y^3}\right)_{I,J} + O(\Delta^4)\\
\psi_{I,J-1} &= \psi_{I,J} - \Delta\left.\frac{\partial\psi}{\partial y}\right)_{I,J} + \frac{\Delta^2}{2}\left.\frac{\partial^2\psi}{\partial y^2}\right)_{I,J} - \frac{\Delta^3}{6}\left.\frac{\partial^3\psi}{\partial y^3}\right)_{I,J} + O(\Delta^4)
\end{aligned}\right\}\text{diamond stencil}$$

$$\left.\begin{aligned}
\psi_{I+1,J+1} &= \psi_{I,J} + \Delta\left(\frac{\partial\psi}{\partial x} + \frac{\partial\psi}{\partial y}\right)_{I,J} + \frac{\Delta^2}{2}\left(\frac{\partial^2\psi}{\partial x^2} + 2\frac{\partial^2\psi}{\partial x\partial y} + \frac{\partial^2\psi}{\partial y^2}\right)_{I,J}\\
&\quad + \frac{\Delta^3}{6}\left(\frac{\partial^3\psi}{\partial x^3} + 3\frac{\partial^3\psi}{\partial x^2\partial y} + 3\frac{\partial^3\psi}{\partial x\partial y^2} + \frac{\partial^3\psi}{\partial y^3}\right)_{I,J} + O(\Delta^4)\\
\psi_{I+1,J-1} &= \psi_{I,J} + \Delta\left(\frac{\partial\psi}{\partial x} - \frac{\partial\psi}{\partial y}\right)_{I,J} + \frac{\Delta^2}{2}\left(\frac{\partial^2\psi}{\partial x^2} - 2\frac{\partial^2\psi}{\partial x\partial y} + \frac{\partial^2\psi}{\partial y^2}\right)_{I,J}\\
&\quad + \frac{\Delta^3}{6}\left(\frac{\partial^3\psi}{\partial x^3} - 3\frac{\partial^3\psi}{\partial x^2\partial y} + 3\frac{\partial^3\psi}{\partial x\partial y^2} - \frac{\partial^3\psi}{\partial y^3}\right)_{I,J} + O(\Delta^4)\\
\psi_{I-1,J+1} &= \psi_{I,J} - \Delta\left(\frac{\partial\psi}{\partial x} - \frac{\partial\psi}{\partial y}\right)_{I,J} + \frac{\Delta^2}{2}\left(\frac{\partial^2\psi}{\partial x^2} - 2\frac{\partial^2\psi}{\partial x\partial y} + \frac{\partial^2\psi}{\partial y^2}\right)_{I,J}\\
&\quad - \frac{\Delta^3}{6}\left(\frac{\partial^3\psi}{\partial x^3} - 3\frac{\partial^3\psi}{\partial x^2\partial y} + 3\frac{\partial^3\psi}{\partial x\partial y^2} - \frac{\partial^3\psi}{\partial y^3}\right)_{I,J} + O(\Delta^4)\\
\psi_{I-1,J-1} &= \psi_{I,J} - \Delta\left(\frac{\partial\psi}{\partial x} + \frac{\partial\psi}{\partial y}\right)_{I,J} + \frac{\Delta^2}{2}\left(\frac{\partial^2\psi}{\partial x^2} + 2\frac{\partial^2\psi}{\partial x\partial y} + \frac{\partial^2\psi}{\partial y^2}\right)_{I,J}\\
&\quad - \frac{\Delta^3}{6}\left(\frac{\partial^3\psi}{\partial x^3} + 3\frac{\partial^3\psi}{\partial x^2\partial y} + 3\frac{\partial^3\psi}{\partial x\partial y^2} - \frac{\partial^3\psi}{\partial y^3}\right)_{I,J} + O(\Delta^4)
\end{aligned}\right\}\text{square stencil}$$

Appendix B

Basic Equations	Multi- Level Spectral Model Equations	Spectral (Working) Equations
Vorticity Equation (Eq. 7.64, p. 121) $\frac{\partial \zeta}{\partial t} = -\vec{\nabla}\cdot(\zeta+f)\vec{V} - \hat{k}\cdot\vec{\nabla}\times\left(RT\vec{\nabla}q + \dot{\sigma}\frac{\partial \vec{V}}{\partial \sigma} - \vec{F}\right)$	(Eq. 7. 91, p. 124) $\frac{\partial \zeta}{\partial t} = -\alpha(A, B)$	(Eq. 7.113, p. 128) $\frac{\partial \zeta_l^m}{\partial t} = -\{\alpha(A, B)\}_l^m$
Divergence Equation (Eq. 7.65, p. 121) $\frac{\partial D}{\partial t} = \hat{k}\cdot\vec{\nabla}\times(\zeta+f)\vec{V} - \vec{\nabla}\cdot\left(RT\vec{\nabla}q + \dot{\sigma}\frac{\partial \vec{V}}{\partial \sigma} - \vec{F}\right) - \vec{\nabla}^2\left(\phi + \frac{\vec{V}\cdot\vec{V}}{2}\right)$	(Eq. 7.92, p. 124) $\frac{\partial D}{\partial t} + \vec{\nabla}^2 P = \alpha(B, -A) - a^2\vec{\nabla}^2 E$	(Eq. 7.114, p. 128) $\frac{\partial D_l^m}{\partial t} - a^{-2}l(l+1)P_l^m = \{\alpha(B,-A)\}_l^m + l(l+1)E_l^m$
Hydrostatic Law (Eq. 7.68, p. 121) $\sigma\frac{\partial \phi}{\partial \sigma} = -RT$	(Eq. 7.98, p. 126) $T = q\sigma\frac{\partial T^*}{\partial \sigma} - \frac{\sigma}{R}\frac{\partial P}{\partial \sigma}$	$T_l^m = \sigma\left(\frac{\partial T^*}{\partial \sigma}q_l^m - \frac{1}{R}\frac{\partial P_l^m}{\partial \sigma}\right)$
First Law of Thermodynamics (Eq. 7.66, p. 121) $\frac{\partial T}{\partial t} = -\vec{\nabla}\cdot\vec{V}T + TD + \dot{\sigma}\gamma - \frac{RT}{C_p}\left(D + \frac{\partial \dot{\sigma}}{\partial \sigma}\right) + h$	(Eq. 7.103, p. 126) $\sigma\frac{\partial^2 P}{\partial t\partial \sigma} + R\gamma^* W = R\alpha(UT', VT') - RB_T$	(Eq. 7.115, p. 128) $\sigma\frac{\partial^2 P_l^m}{\partial t\partial \sigma} + R\gamma^* W_l^m = R[\alpha(UT', VT') - B_T]_l^m$
Mass Continuity (Eq. 7.67, p. 121) $\frac{\partial q}{\partial t} = -\vec{V}\cdot\vec{\nabla}q - D - \frac{\partial \dot{\sigma}}{\partial \sigma}$	*Mass Continuity tendency* (Eq. 7.105, p. 126) $\frac{\partial q}{\partial t} = W_S$	(Eq. 7.118, p. 128) $\frac{\partial q_l^m}{\partial t} = (W_S)_l^m$
	Pseudo-vertical velocity (W) (Eq. 7.108, p. 127) $\frac{\partial W}{\partial \sigma} + D = B_W$	(Eq. 7.117, p. 128) $\frac{\partial W_l^m}{\partial \sigma} + D_l^m = (B_W)_l^m$
Moisture Conservation (Eq. 7.79, p. 122) $\frac{\partial S}{\partial t} = -\vec{\nabla}\cdot\vec{V}S + SD - \dot{\sigma}\frac{\partial S}{\partial \sigma} + h - h_m - \left(\frac{RT}{C_p} - \frac{RT_d^2}{\varepsilon L}\right)\left(D + \frac{\partial \dot{\sigma}}{\partial \sigma} - \frac{\dot{\sigma}}{\sigma}\right)$	(Eq. 7.96, p. 125) $\frac{\partial S}{\partial t} = -\alpha(US, VS) + B_S$	(Eq. 7.119, p. 128) $\frac{\partial S_l^m}{\partial t} = \{-\alpha(US, VS) + B_S\}_l^m$
$\dot{\sigma}$ - Equation (Eq. 7.83, p. 123) $\dot{\sigma} = (\sigma - 1)\left(\hat{D} + \hat{\vec{V}}\cdot\vec{\nabla}q\right) + \left(\hat{D}^\sigma + \hat{\vec{V}}^\sigma\cdot\vec{\nabla}q\right)$	(p. 125) $\dot{\sigma} = (\sigma - 1)\left(\hat{G} + \hat{D}\right) + \hat{G}^\sigma + \hat{D}^\sigma$	$\dot{\sigma}_l^m = (\sigma - 1)\left(\hat{G}_l^m + \hat{D}_l^m\right) + \hat{G}_l^{\sigma m} + \hat{D}_l^{\sigma m}$

$q = \ln p_s$, (p_s = surface pressure), h represents diabatic heat sources and sinks, S = dew point depression, $\varepsilon = 0.622$, L = latent heat of phase change, h_m is moisture source or sinks, pseudo pressure function $P = \phi + qRT^*$, A, B, G, E, B_T, B_S on p.125, $B_W = -G$, γ = static stability, T^* = reference atmospheric temp., pseudo vertical velocity (W) is $W = \dot{\sigma} + \sigma\frac{\partial q}{\partial t}$, $\alpha(A,B) = \frac{1}{\cos^2\theta}\left(\frac{\partial A}{\partial \lambda} + \cos\theta\frac{\partial B}{\partial \theta}\right)$, $\hat{M} = \int_0^1 M d\sigma$ and $\hat{M}^\sigma = \frac{1}{\sigma}\int_0^\sigma M d\sigma$, other symbols have their usual meaning.

References

Aberson, S. D. and Franklin, J. L. 1999, Impact on hurricane track and intensity forecasts of GPS dropwindsonde observations from the first season flights of the NOAA Gulfstream IV jet aircraft. *Bull. Amer. Meteorol. Soc.*, **80,** 421–427.

Aberson, S. D., 2002, Two years of operational hurricane synoptic surveillance. *Wea. Forecasting*, **17**, 1101–1110.

Anthes, R.A. 1977, A cumulus parameterization scheme utilizing a one dimensional cloud model. *Mon. Weather Rev.*, **105**, 270-286.

Arakawa, A. 1966, Computational design for long-term numerical integrations of the equations of atmospheric motion. *J. Comput. Phys.*, **1**, 119-143.

Arakawa, A. and Schubert, W. H. 1974, Interaction of cumulus cloud ensemble with the large-scale environment. *J. Atmos. Sci.*, **31**, 674-701.

Baer, F. 1964, Integration of the spectral vorticity equation. *J. Atmos. Sci.*, **21**, 260-276.

Baer, F. 1977, Adjustment of initial conditions required to suppress gravity oscillations in nonlinear flows. *Contrib. Atmos. Phys.*, **50**, 350-366.

Bishop, C. H. and Toth, Z. 1999, Ensemble transformation and adaptive observations. *J. Atmos. Sci.*, **56**, 1748–1765.

Bishop, C.H., Etherton, B. J., and Majumdar, S. J. 2001, Adaptive sampling with the ensemble transform Kalman filter. Part I: Theoretical aspects. *Mon. Wea. Rev.*, **129,** 420–436.

Blackadar, A. K. 1962, The vertical distribution of wind and turbulent exchange in a neutral atmosphere. *J. Geophys. Res.*, **67**, 3095-3102.

Borges, M. D. and Hartmann, D. L. 1992, Barotropic instability and optimal perturbations of observed nonzonal flow. *J. Atmos. Sci.*, **49**, 335–354.

Bourke, W. 1972, An efficient, one-level, primitive-equation spectral model. *Mon. Weather Rev.*, **100**, 683-689.

Bourke, W. 1974, A multi-level spectral model. I. Formulation and hemispheric integrations. *Mon. Weather Rev.*, **102**, 687-701.

Bowie, E. H., 1922, Formation and movement of West Indian hurricanes. *Mon. Wea. Rev.*, **50**, 173–190.

Bristor, C. L., 1958, Effect of data coverage on the accuracy of 500 mb forecasts. *Mon. Wea. Rev.*, **86**, 299–308.

Buizza, R. and Palmer, T. N. 1995, A singular vector structure of the atmospheric global circulation. *J. Atmos. Sci.*, **52**, 1434-1456.

Burpee, R. W., Franklin, J. L., Lord, S. J., Tuleya, R. E., and Aberson, S. D. 1996, The impact of omega dropwindsondes on operational hurricane track forecast models. *Bull. Amer. Meteorol. Soc.*, **77**, 925 – 933.

Businger, J. A., Wyngaard, J. C., Izumi, Y., and Bradley, E.F. 1971, Flux profile relationships in the atmospheric surface layer. *J. Atmos. Sci.*, **28**, 181-189.

Chang, L. W. 1978, Determination of surface flux of sensible heat, latent heat, and momentum utilizing the Bulk Richardson number. *Pap. Meteorol. Res.*, **1**, 16-24.

Charney, J. G. 1955, The use of the primitive equations of motion in numerical prediction. *Tellus*, **7**, 22-26.

Charnock, H. 1955, Wind stress on the water surface. *Q. J. R. Meteorol. Soc.*, **81**, 639-640.

Chou, M. D. 1984, Broadband water vapor transmission functions for atmospheric IR flux computations. *J. Atmos. Sci.*, **41**, 1775-1778.

Chou, M. D. and Arking, A. 1980, Computation of infrared cooling rates in the water vapor bands. *J. Atmos. Sci.*, **37**, 855-867.

Chou, M. D. and Peng, L. 1983, A parameterization of the absorption in the 15μm CO_2 spectral region with application to climate sensitivity studies. *J. Atmos. Sci.*, **40**, 2183-2192.

Cocke, S. 1998, Case study of Erin using FSU nested regional spectral model. *Mon. Weather Rev.*, **122**, 3-26.

Cooley, J. W. and Tukey, J. W. 1965, An algorithm for the machine computation of complex Fourier series. *Math. Comp.*, **19**, 297-301.

Courant, R., Friedrichs, K. O., and Levy, H. 1928, Uber die partiellen differenzengleichungen der mathematischen physik. *Math. Annalen*, **100**, 32-74.

Craig, R. 1945, A solution of the nonlinear vorticity equations for atmospheric motion. *J. Meteorol.*, **2**, 175-178.

Daley, R., Girard, C., Henderson, J., and Simmonds, I. 1976, Short-term forecasting with a multi-level spectral primitive equation model. *Atmosphere*, **14**, 98-116.

Davies, R. 1982. Documentation of the solar radiation parameterization in the GLAS climate model. Technical memo 83961, NASA, Goddard Space Flight Center, Greenbelt, Maryland.

Derber, J. and Bouttier, F. 1999, A reformulation of the background error covariance in the ECMWF global data assimilation system. *Tellus*, **51A,** 195–221.

Dickinson, A. and Temperton, C. 1984. The operational numerical weather prediction model. Technical Note No. 183, Meteorological Office, Bracknell, England.

Donner, L. J. 1988, An initialization for cumulus convection in numerical weather prediction models. *Mon. Weather Rev.*, **116**, 377-385.

Ebisuzaki, W. and Kalnay, E. 1991. Ensemble experiments with a new lagged analysis forecasting scheme. Research Activities in Atmospheric and Oceanic Modeling. Report No. 15, WMO. [Available from WMO, C. P. No. 2300, CH1211, Geneva, Switzerland.]

Eliasen, E., Machenhauer, B., and Rasmussen, E. 1970. On a numerical method for integration of the hydrodynamical equations with a spectral representation of the horizontal fields. Report No. 2, Institut für Teoretisk Meteorologi, Kobenhavns Universitet, Denmark.

Epstein, E. S. 1969, Stochastic dynamic prediction. *Tellus*, **21**, 739-759.

Fjørtoft, R. 1953, On the changes in the spectral distribution of kinetic energy for two-dimensional, nondivergent flow. *Tellus*, **5**, 225-230.

Gregg, W. R., 1920, Aerological observations in the West Indies. *Mon. Wea. Rev.*, **48**, 264.

Haltiner, G. J. and Williams, R. T. 1980. *Numerical Prediction and Dynamic Meteorology,* second edition, John Wiley & Sons, New York.

Harshvardhan and Corsetti, T. G. 1984. Longwave parameterization for the UCLA/GLAS GCM. Technical Memo 86072, NASA, Goddard Space Flight Center, Greenbelt, Maryland.

Haurwitz, B. 1940, The motion of atmospheric disturbances on the spherical earth. *J. Marine Res.*, **3**, 254-267.

Hoffman, R. N. and Kalnay, E. 1983, Lagged average forecasting, an alternative to Monte Carlo forecasting. *Tellus*, **35A**, 100-118.

Hoke, J. E. and Anthes, R. A. 1976, The initialization of numerical models by dynamic initialization technique. *Mon. Weather Rev.*, **104**, 1551-1556.

Holton, J. R. 1992. *An Introduction to Dynamic Meteorology*, Volume 48 of *International Geophysics Series*, third edition, Academic Press, San Diego.

Hoyer, J. M. 1987. The ECMWF spectral limited area model. Proc. ECMWF Workshop on Techniques for Horizontal Discretization in Numerical Weather Prediction Models, Shinfield Park, Reading, U.K., 343-359.

Joseph, J. H. 1966. Calculation of radiative heating in numerical general circulation models. *Numerical Simulation of Weather and Climate.* Technical Report No.1, Department of Meteorology, UCLA.

Juang, H. and Kanamitsu, M. 1994, The NMC Nested Regional Spectral Model. *Mon. Weather Rev.*, **122**, 3-26.

Kalnay, E. and Toth, Z. 1991. Estimating the growing modes of the atmosphere: The breeding method. Research Highlights of NMC Development Division: 1990-1991, 160-165. [Available from the National Center for Environmental Prediction, Washington, D. C.]

Kasahara, A. and Puri, K. 1981, Spectral representation of three-dimensional global data by expansion in normal mode functions. *Mon. Weather Rev.*, **109**, 37-51.

Katayama, A. 1966, On the radiation budget of the troposphere over the northern hemisphere (I). *J. Meteorol. Soc. Japan*, **44**, 381-401.

Kitade, T. 1983, Nonlinear normal mode initialization with physics. *Mon. Weather Rev.*, **111**, 2194-2213.

Krishnamurti, T.N., Ramanathan, Y., Pan, H., Pasch, R. J., and Molinari, J. 1980, Cumulus parameterization and rainfall rates I. *Mon. Weather Rev.*, **108**, 465-472.

Krishnamurti, T. N., Low-Nam, S., and Pasch, R. 1983, Cumulus parameterization and rainfall rates II. *Mon. Weather Rev.*, **111**, 816-828.

Krishnamurti, T. N., Ingles, K., Cocke, S., Pasch, R., and Kitade, T. 1984, Details of low latitude medium range numerical weather prediction using a global spectral model II. Effect of orography and physical initialization. *J. Meteorol. Soc. Japan*, **62**, 613-649.

Krishnamurti, T.N. and Bedi, H.S. 1988, Cumulus parameterization and rainfall rates: Part III. *Mon. Weather Rev.*, **116**, 583-599.

Krishnamurti, T. N., Bedi, H. S., Heckley, W., and Ingles, K. 1988, Reduction of the spinup time for evaporation and precipitation in a spectral model. *Mon. Weather Rev.*, **116**, 907-920.

Krishnamurti, T. N., Xue, J., Bedi, H. S., Ingles, K., and Oosterhof, D. 1991a, Physical initialization for numerical weather prediction over the tropics. *Tellus*, **43A-B**, 53-81.

Krishnamurti, T. N., Yap, K. S. and Oosterhof, D. K. 1991b, Sensitivity of tropical storm forecast to radiative destabilization. *Mon. Weather Rev.*, **119**, 2176-2205.

Krishnamurti, T. N., Xue, J., Bedi, H.S., Ingles, K., and Oosterhof, D. 1991c, Physical initialization for numerical weather prediction over the tropics. *Tellus*, **43AB**, 53-81.

Krishnamurti, T. N., Bedi, H. S., and Ingles, K. 1993, Physical initialization using SSM/I rain rates. *Tellus*, **45A**, 247-269.

Krishnamurti, T. N. Bedi, H. S. Oosterhof, D., and Hardiker, V. 1994a, The formation of hurricane Frederic of 1979. *Mon. Weather Rev.*, **122**, 1050-1074.

Krishnamurti, T. N., Rohaly, G.D., and Bedi, H.S. 1994b, On the improvement of precipitation forecast skill from physical initialization. *Tellus*, **46A**, 598-614.

Krishnamurti, T. N., Kishtawal, C.M., LaRow, T., Bachiochi, D., Zhang, Z., Williford, C.E., Gadgil, S., and Surendran, S. 1999, Improved skills for weather and seasonal climate forecasts from multimodel superensemble. *Science*, **285** (5433), 1548-1550.

Krishnamurti, T. N., Kishtawal, C.M., LaRow, T., Bachiochi, D., Zhang, Z., Williford, C.E., Gadgil, S., and Surendran, S. 2000a, Multimodel superensemble forecasts for weather and seasonal climate. *J. Climate*, **13**, 4196-4216.

Krishnamurti, T. N., Kishtawal, C.M., Shin, D.W., and Williford, C.E. 2000b, Improving tropical precipitation forecasts from a multianalysis superensemble. *J. Climate*, **13**, 4217-4227.

Krishnamurti, T. N., Surendran, S., Shin, D.W., Correa-Torres, R., Vijaya Kumar, T. S. V., Williford, C.E., Kummerow, C., Adler, R.F., Simpson, J., Kakar, R., Olson, W., and Turk, F.J. 2001, Real time multianalysis/multimodel superensemble forecasts of precipitation using TRMM and SSM/I products. *Mon. Wea. Rev.*, **129**, 2861-2883.

Krishnamurti, T. N., Stefanova, L., Chakraborty, A., Vijaya Kumar, T. S. V., Cocke, S., Bachiochi, D., and Mackey, B. 2002, Seasonal forecasts of precipitation anomalies for North American and Asian monsoons. Submitted to *J. Meteorol. Soc. Japan*.

Kuhn, P. M. 1963, Radiometersonde observations of infrared flux emissivity of water vapor. *J. Appl. Meteorol.*, **2**, 368-378.

Kuo, H. L. 1951, Dynamical aspects of the general circulation and the stability of zonal folw. *Tellus*, **3**, 268-284.

Kuo, H. L. 1965, On formation and intensification of tropical cyclones through latent heat release by cumulus convection. *J. Atmos. Sci.*, **22**, 40-63.

Kuo, H. L. 1974, Further studies of the parameterization of the influence of cumulus convection on large-scale flow. *J. Atmos. Sci.*, **31**, 1232-1240.

Kuo, H. L. and Low-Nam, S. 1990, Prediction of explosive cyclones over the western Atlantic with a regional model. *Mon. Weather Rev.*, 118, 3-25.

Lacarra, J. F. and Talagrand, O. 1988, Short-range evolution of small perturbations in a barotropic model. *Tellus*, **40A**, 81-95.

Lacis, A. A. and Hansen, J. E. 1974, A parameterization for the absorption of solar radiation in the earth's atmosphere. *J. Atmos. Sci.*, **31**, 118-133.

Leith, C. E. 1974, Theoretical skill of Monte Carlo forecasts. *Mon. Weather Rev.*, **102**, 409-418.

Lord, S. J. 1978. *Development and Observational Verification of a Cumulus Cloud Parameterization.* Ph.D. thesis, UCLA.

Lorenz, E. N. 1955, Available potential energy and the maintenance of the general circulation. *Tellus*, **7**, 157-167.

Lorenz, E. N. 1960a, Energy and numerical weather prediction. *Tellus*, **12**, 364-373.

Lorenz, E. N. 1960b, Maximum simplification of the dynamic equations. *Tellus*, **12**, 243-254.

Lorenz, E. N. 1965, A study of the predictability of a 28-variable atmospheric model. *Tellus*, **17**, 321-333.

Lorenz, E. N. 1969, The predictability of a flow which possesses many scales of motion. *Tellus*, **21**, 289-307.

Louis, J. F. 1979, A parametric model of the vertical eddy fluxes in the atmosphere. *Bound. Layer Meteorol.*, **17**, 187-202.

Machenhauer, B. 1974. On the use of the spectral method in numerical integrations of atmospheric models. In *Difference and Spectral methods for Atmosphere and Ocean Dynamics Problems*. USSR Academy of Sciences, Siberian Branch, Novosibirsk. September 1973.

Machenhauer, B. 1977, On the dynamics of gravity oscillations in a shallow water model, with application to normal mode initialization. *Contrib. Atmos. Phys.*, **50**, 253-271.

Machenhauer, B. and Daley, R. 1972. A baroclinic primitive equation model with a spectral representation in three dimensions. Report No. 4, Institut für Teoretisk Meteorologi, Kobenhavns Universitet, Denmark.

Manobianco, J. 1989, Explosive east coast cyclogenesis: Numerical experimentation and model based diagnostics. *Mon. Weather Rev.*, **117**, 2384-2405.

Merilees, P. E. 1968, The equations of motion in spherical form. *J. Atmos. Sci.*, **25**, 736-743.

Miyakoda, K. and Moyer, R. W. 1968, A method of initialization for dynamic weather forecasting. *Tellus*, **20**, 115-128.

Molteni, F., Buizza, R., Palmer, T. N., and Petroliagis, T. 1996, The ECMWF ensemble prediction system: Methodology and validation. *Q. J. R. Meteorol. Soc.*, **122**, 73-120.

Mullen, S. L. and Baumhelfner, D. P. 1994, Monte Carlo simulations of explosive cyclogenesis. *Mon. Weather Rev.*, **122**, 1548-1567.

Murreau, R., Molteni, F., and Palmer, T. N. 1993, Ensemble prediction using dynamically conditioned perturbations. *Q. J. R. Meteorol. Soc.*, **119**, 229-323.

Neamtan, S. M. 1946, The motion of harmonic waves in the atmosphere. *J. Meteorol.*, **3**, 53-56.

Nitta, T. and Hovermale, J. B. 1969, A technique of objective analysis and initialization for the primitive forecast equations. *Mon. Weather Rev.*, **97**, 652-658.

Orszag, S. A. 1970, Transform method for the calculation of vector-coupled sums: Application to the spectral form of the vorticity equation. *J. Atmos. Sci.*, **27**, 890-895.

Palmer, T. N., Gelaro, R., Barkmeijer, J., and Buizza, R. 1998, Singular vectors, metrics, and adaptive observations. *J. Atmos. Sci.*, **55**, 633–653.

Phillips, N. A. 1957, A coordinate system having some special advantages for numerical forecasting. *J. Meteorol.*, **14**, 184-185.

Phillips, N. A. 1960, On the problem of initial data for the primitive equations. *Tellus*, **12**, 121-126.

Platzman, G. W. 1960, The spectral form of the vorticity equation. *J. Meteorol.*, **17**, 635-644.

Puri, K. and Miller, M. J. 1990, The use of satellite data in the specification of convective heating for diabatic initialization and moisture adjustment in numerical weather prediction models. *Mon. Weather Rev.*, **118**, 67-93.

Rabier, F., Courtier, P., Pailleux, J., Talagrand, O., and Vasiljevic, D. 1993, A comparison between four-dimensional variational assimilation and simplified sequential assimilation relying on three-dimensional variational analysis. *Q. J. R. Meteorol. Soc.*, **119**, 845-880.

Reynolds, C. A., Webster, P. J., and Kalnay, E. 1994, Random error growth in NMC's global forecasts. *Mon. Wea. Rev.*, **122**,1281–1305.

Riehl, H. and Shafer, R. J. 1944, The recurvature of tropical storms. *J. Meteorol.*, **1**, 42–54.

Robert, A. J. 1966, The integration of a low order spectral form of the primitive meteorological equations. *J. Meteorol. Soc. Japan*, **44**(Series 2), 237-245.

Robert, A. J. 1969. The integration of a spectral model of the atmosphere by the implicit method. In *Proceedings, WMO/IUGG Symposium on Numerical Weather Prediction Tokyo*, Meteorological Society of Japan, VII-19-VII-24.

Roberts, R. E., Selby, J. E. A. and Biberman, L. M. 1976, Infrared continuum absorption by atmospheric water vapor in the 8-12 μm window. *Appl. Opt.*, **15**, 2085-2090.

Rodgers, C. D. 1967. The radiative heat budget of the troposphere and lower stratosphere. *Planetary Circulation Project*, Report No. 2, MIT, Department of Meteorology.

Rodgers, C. D. 1968, Some extensions and applications of the new random model for molecular band transmission. *Q. J. R. Meteorol. Soc.*, **94**, 99-102.

Saltzman, B. 1957, Equations governing the energetics of the larger scales of atmospheric turbulence in the domain of wave number. *J. Meteorol.*, **14**, 513-523.

Saltzman, B. 1959, On the maintenance of the large-scale quasi-permanent disturbances in the atmosphere. *Tellus*, **11**, 425-431.

Sasaki, Y. 1958, An objective analysis based on the variational method. *Soc. Japan*, **36**, 77-78.

Silberman, I. S. 1954, Planetary waves in the atmosphere. *J. Meteorol.*, 11, 27-34.

Slingo, J. 1985, A new cloud cover scheme. *ECMWF* (*European Centre For Medium-Range Weather Forecasts*) *Newsletter*, No. 29, 14-18.

Starr, V. P. 1953, Note concerning the nature of the large-scale eddies in the atmosphere. *Tellus*, **5**, 494-498.

Starr, V. P. and White, R. M. 1954, Balance requirements of the general circulation. *Geophys. Res. Pap.*, No. 35, 57 pp.

Stefanova, L. and Krishnamurti, T.N. 2001, Interpretation of seasonal climate forecasts using Brier skill score, FSU superensemble and the AMIP-1 dataset. Submitted to *J. Climate*, 19 February 2001.

Stephens, G. L. 1984, The parameterization of radiation for numerical weather prediction and climate models. *Mon. Weather Rev.*, **112**, 826-867.

Tatsumi, Y. 1986, A spectral limited area model with time-dependent lateral boundary conditions and its application to multilevel primitive equation model. *J. Meteorol. Soc. Japan*, **64**, No. 3, 637-663.

Temperton, C. 1976, Dynamic initialization for barotropic and multilevel models. *Q. J. R. Meteorol. Soc.*, **102**, 297-311.

Thompson, P. D. 1961. *Numerical Weather Analysis and Prediction*, Macmillan, New York.

Toth, Z. and Kalnay, E. 1991. Ensemble forecasting: Are two members enough? Research Highlights of the NMC Development Division: 1990-1991, 439-443. [Available from the National Center for Environmental Prediction, Washington, D. C.]

Toth, Z. and Kalnay, E. 1993, Ensemble forecasting at NMC: The generation of perturbations. *Bull. Amer. Meteorol. Soc.*, **74**, 2317-2330.

Treadon, R.E., 1996, Physical initialization in the NMC global data assimilation system. *Meteorol. Atmos. Phys.*, **60** (1-3), 57-86.

Tribbia, J. J. and Baumhelfner, D. P., Estimates of predictability of low-frequency variability with a spectral general circulation model. *J. Atmos. Sci.*, **45**, 2306-2317.

Vijaya Kumar, T. S. V., Krishnamurti, T. N., Fiorino, M., and Nagata, M. 2002, Multimodel superensemble forecasting of tropical cyclones in the Pacific. Submitted to *Mon. Weather Rev.*

Williamson, D. L. 1976, Normal model initialization procedure applied to forecasts with the global shallow water equations. *Mon. Weather Rev.*, **104**, 195-206.

Williford, C.E., Krishnamurti, T. N., Correa-Torres, R. J., Cocke, S., Christidis, Z., and Vijaya Kumar, T. S. V. 2002, Real-time multimodel superensemble forecasts of Atlantic tropical systems of 1999. Accepted for publication, *Mon. Weather Rev.*

Yanai, M., Esbensen, S., and Chu, J. H. 1973, Determination of bulk properties of tropical cloud clusters from large-scale heat and moisture budgets. *Mon. Weather Rev.*, **30**, 611-627.

Yun, W.T and Krishnamurti, T. N. 2002, Improvement of seasonal and long-term forecasts using multimodel superensemble technique. Submitted to *J. Climate.*

Zhang, A. 1997. Hurricane ensemble prediction using EOF-based perturbations. Ph.D. Thesis, 1997, 174 pp. [Available from the Department of Meteorology, Florida State University, Tallahassee, FL 32306.]

Zhang, Z. and Krishnamurti, T. N. 1997, Ensemble forecasting of hurricane tracks. *Bull. Amer. Meteor. Soc.*, **78**, 2785-2795.

Zhang, Z. and Krishnamurti, T.N. 2000, Adaptive observations for hurricane prediction. *Meteorol. Atmos. Phys.*, **74**, 19-35.

Zhu, Y., Iyengar, G., Toth, Z., Tracton, M. S., and Marchok, T. 1996. Objective evaluation of the NCEP global ensemble forecasting system. Preprints, *15th Conf. on Weather Analysis and Forecasting,* Norfolk, VA, Amer. Meteor. Soc., J79–J82.

Index

ATMOSPHERIC AND OCEANOGRAPHIC SCIENCES LIBRARY

1. F.T.M. Nieuwstadt and H. van Dop (eds.): *Atmospheric Turbulence and Air Pollution Modelling.* 1982; rev. ed. 1984 ISBN 90-277-1365-6; Pb (1984) 90-277-1807-5
2. L.T. Matveev: *Cloud Dynamics.* Translated from Russian. 1984 ISBN 90-277-1737-0
3. H. Flohn and R. Fantechi (eds.): *The Climate of Europe: Past, Present and Future.* Natural and Man-Induced Climate Changes: A European Perspective. 1984 ISBN 90-277-1745-1
4. V.E. Zuev, A.A. Zemlyanov, Yu.D. Kopytin, and A.V. Kuzikovskii: *High-Power Laser Radiation in Atmospheric Aerosols.* Nonlinear Optics of Aerodispersed Media. Translated from Russian. 1985 ISBN 90-277-1736-2
5. G. Brasseur and S. Solomon: *Aeronomy of the Middle Atmosphere.* Chemistry and Physics of the Stratosphere and Mesosphere. 1984; rev. ed. 1986 ISBN (1986) 90-277-2343-5; Pb 90-277-2344-3
6. E.M. Feigelson (ed.): *Radiation in a Cloudy Atmosphere.* Translated from Russian. 1984 ISBN 90-277-1803-2
7. A.S. Monin: *An Introduction to the Theory of Climate.* Translated from Russian. 1986 ISBN 90-277-1935-7
8. S. Hastenrath: *Climate Dynamics of the Tropics,* Updated Edition from *Climate and Circulation of the Tropics.* 1985; rev. ed. 1991 ISBN 0-7923-1213-9; Pb 0-7923-1346-1
9. M.I. Budyko: *The Evolution of the Biosphere.* Translated from Russian. 1986 ISBN 90-277-2140-8
10. R.S. Bortkovskii: *Air-Sea Exchange of Heat and Moisture During Storms.* Translated from Russian, rev. ed. 1987 ISBN 90-277-2346-X
11. V.E. Zuev and V.S. Komarov: *Statistical Models of the Temperature and Gaseous Components of the Atmosphere.* Translated from Russian. 1987 ISBN 90-277-2466-0
12. H. Volland: *Atmospheric Tidal and Planetary Waves.* 1988 ISBN 90-277-2630-2
13. R.B. Stull: *An Introduction to Boundary Layer Meteorology.* 1988 ISBN 90-277-2768-6; Pb 90-277-2769-4
14. M.E. Berlyand: *Prediction and Regulation of Air Pollution.* Translated from Russian, rev. ed. 1991 ISBN 0-7923-1000-4
15. F. Baer, N.L. Canfield and J.M. Mitchell (eds.): *Climate in Human Perspective.* A tribute to Helmut E. Landsberg (1906–1985). 1991 ISBN 0-7923-1072-1
16. Ding Yihui: *Monsoons over China.* 1994 ISBN 0-7923-1757-2
17. A. Henderson-Sellers and A.-M. Hansen: *Climate Change Atlas.* Greenhouse Simulations from the Model Evaluation Consortium for Climate Assessment. 1995 ISBN 0-7923-3465-5
18. H.R. Pruppacher and J.D. Klett: *Microphysics of Clouds and Precipitation*, 2nd rev. ed. 1997 ISBN 0-7923-4211-9; Pb 0-7923-4409-X
19. R.L. Kagan: *Averaging of Meteorological Fields.* 1997 ISBN 0-7923-4801-X
20. G.L. Geernaert (ed.): *Air-Sea Exchange: Physics, Chemistry and Dynamics.* 1999 ISBN 0-7923-5937-2
21. G.L. Hammer, N. Nicholls and C. Mitchell (eds.): *Applications of Seasonal Climate Forecasting in Agricultural and Natural Ecosystems.* 2000 ISBN 0-7923-6270-5

22. H.A. Dijkstra: *Nonlinear Physical Oceanography.* A Dynamical Systems Approach to the Large Scale Ocean Circulation and El Niño. 2000 ISBN 0-7923-6522-4
23. Y. Shao: *Physics and Modelling of Wind Erosion.* 2000 ISBN 0-7923-6657-3
24. Yu.Z. Miropol'sky: *Dynamics of Internal Gravity Waves in the Ocean.* Edited by O.D. Shishkina. 2001 ISBN 0-7923-6935-1
25. R. Przybylak: *Variability of Air Temperature and Atmospheric Precipitation during a Period of Instrumental Observations in the Arctic.* 2002 ISBN 1-4020-0952-6
26. R. Przybylak: *The Climate of the Arctic.* 2003 ISBN 1-4020-1134-2
27. S. Raghavan: *Radar Meteorology.* 2003 ISBN 1-4020-1604-2
28. H.A. Dijkstra: *Nonlinear Physical Oceanography.* A Dynamical Systems Approach to the Large Scale Ocean Circulation and El Niño. 2nd Revised and Enlarged Edition. 2005 ISBN 1-4020-2272-7 Pb; 1-4020-2262-X
29. X. Lee, W. Massman and B. Law (eds.): *Handbook of Micrometeorology.* A Guide for Surface Flux Measurement and Analysis. 2004 ISBN 1-4020-2264-6
30. A. Gelencsér: *Carbonaceous Aerosol.* 2005 ISBN 1-4020-2886-5
31. A. Soloviev and L. Roger: *The Near-Surface Layer of the Ocean. Structure, Dynamics and Applications.* 2006 ISBN 1-4020-4052-0
32. G.P. Brasseur and S. Solomon: *Aeronomy of the Middle Atmosphere. Chemistry and Physics of the Stratosphere and Mesosphere.* 2005 ISBN 1-4020-3284-6; Pb 1-4020-3285-4
33. B. Wozniak and J. Dera: *Light Absorption and Absorbants in Sea Water.* 2006 ISBN 0-387-30753-2
34. A. Kokhanovsky: *Cloud Optics.* 2005 ISBN 1-4020-3955-7
35. T.N. Krishnamurti, H.S. Bedi, V.M. Hardiker, and L. Ramaswamy: *An Introduction to Global Spectral Modeling, 2nd Revised and Enlarged Edition.* 2006 ISBN 0-387-30254-9